Meaning in Life

What makes a person's life meaningful? Thaddeus Metz offers a new answer to an ancient question which has recently returned to the philosophical agenda. He proceeds by examining what, if anything, all the conditions that make a life meaningful have in common. The outcome of this process is a philosophical theory of meaning in life. He starts by evaluating existing theories in terms of the classic triad of the good, the true, and the beautiful. He considers whether meaning in life might be about such principles as fulfilling God's purpose, obtaining reward in an afterlife for having been virtuous, being attracted to what merits attraction, leaving the world a better place, connecting to organic unity, or transcending oneself by connecting to what is extensive. He argues that no extant principle satisfactorily accounts for the three-fold significance of morality, enquiry, and creativity, and that the most promising theory is a fresh one according to which meaning in life is a matter of intelligence contoured towards fundamental conditions of human existence.

Meaning in Life

An Analytic Study

Thaddeus Metz

UNIVERSITY PRESS

OXFORD
UNIVERSITY PRESS

Great Clarendon Street, Oxford, OX2 6DP,
United Kingdom

Oxford University Press is a department of the University of Oxford.
It furthers the University's objective of excellence in research, scholarship,
and education by publishing worldwide. Oxford is a registered trade mark of
Oxford University Press in the UK and in certain other countries

© Thaddeus Metz 2013

The moral rights of the author have been asserted

First published 2013
First published in paperback 2015

All rights reserved. No part of this publication may be reproduced, stored in
a retrieval system, or transmitted, in any form or by any means, without the
prior permission in writing of Oxford University Press, or as expressly permitted
by law, by licence or under terms agreed with the appropriate reprographics
rights organization. Enquiries concerning reproduction outside the scope of the
above should be sent to the Rights Department, Oxford University Press, at the
address above

You must not circulate this work in any other form
and you must impose this same condition on any acquirer

Published in the United States of America by Oxford University Press
198 Madison Avenue, New York, NY 10016, United States of America

British Library Cataloguing in Publication Data
Data available

Library of Congress Cataloging in Publication Data
Data available

ISBN 978-0-19-959931-8 (Hbk.)
ISBN 978-0-19-874801-4 (Pbk.)

Links to third party websites are provided by Oxford in good faith and
for information only. Oxford disclaims any responsibility for the materials
contained in any third party website referenced in this work.

Preface

I began writing what would eventually form chapters of this book over ten years ago, and while I owe all those who have helped me in this endeavour a debt of gratitude, I am sure to neglect to mention some of them, including, I fear, some who have made a major difference. Nonetheless, it seems better to err on the side of recognizing those whom I recall having helped me, rather than to recognize no one at all so as to avoid the non-recognition of some.

It first occurred to me that I could write seriously on the topic of the meaning of life upon hearing a talk by Susan Wolf in the mid-1990s. I was then a graduate student in philosophy at Cornell University, and she visited that department to give a colloquium on the role of judgements of life's meaning in the work of Bernard Williams and the way a different, more plausible conception of it would affect his conclusions.[1] She approached the topic of life's meaning in the same analytic manner as I had been taught to deal with issues of well-being and morality, something I had not seen done before. Although I was in the middle of a doctorate on a different topic, her presentation stayed with me, and eventually prompted me to write my first essay on meaning in life in 1997, which ended up winning a Jacobsen Prize, an essay competition for best paper on the nature and value of life. That was some inspiration to continue.

After that, I decided to write a string of articles and chapters that I envisioned would eventually become part of this book, and I thank the many people who provided input on them, acknowledging here those (besides anonymous referees) who took the time to provide written comments or to speak with me one-on-one: Alexander Alexis, Lucy Allais, Nafsika Athanassoulis, David Benatar, Lisa Bortolotti, Johan Brännmark, David Brink, Alex Broadbent, Eric Brown, David Copp, John Cottingham, Roger Crisp, Garrett Cullity, Christopher Eberle, John Martin Fischer, Larry James, Ward Jones, John Kekes, Stephen Kershnar, Sigurdur Kristinsson, Laurance Kuper, Alexandros Kyriakou, Michael Lacewing, Iddo Landau, James Lenman, Mark Leon, Jerrold Levinson, David Martens, Darrel Moellendorf, Yujin Nagasawa, Mark Oppenheimer, Michael Pendlebury, David Phillips, Duncan Pritchard, Philip Quinn, Piers Rawling, David Schmidtz, Joshua Seachris, Frank Snyckers, Johan Snyman, Jim Stone, Eleonore Stump, Frans Svensson, Pedro Tabensky, Torbjörn Tännsjö, Brooke Alan Trisel, Neil Van Leeuwen, Samantha Vice, Christopher Wellman, Eric Wiland, and Wai-hung Wong.

I also thank my editor at Oxford University Press, Peter Momtchiloff, who provided helpful guidance on the book proposal, as well as the Stellenbosch Institute for

---

[1] For the eventual publication, see Wolf (1997c).

Advanced Study (STIAS), which gave me a fellowship to support the writing of it. With regard to the book *qua* book, I am pleased to acknowledge the input of two careful and thoughtful reviewers for Oxford University Press, whose comments have led to significant changes in the manuscript, as well as the advice of Hennie Lötter, who also read the entire draft (and without financial inducement). In addition, I have benefited from the help of Adrian Erasmus, who collated the references and secured permissions to use materials from the following essays:

- Metz, Thaddeus (2012) 'Understanding the question of life's meaning', in *Exploring the Meaning of Life: An Anthology and Guide*, Seachris, Joshua (ed.), Oxford: John Wiley & Sons, pp. 23–7.
- Metz, Thaddeus (2011) 'The good, the true and the beautiful: towards a unified account of great meaning in life', *Religious Studies*, 47: 389–409.
- Metz, Thaddeus (2010) 'The meaning of life', in *Oxford Bibliographies Online: Philosophy*, Pritchard, Duncan (ed.), Oxford: Oxford University Press, <http://oxfordbibliographiesonline.com/display/id/obo-9780195396577-0070>
- Metz, Thaddeus (2009) 'Imperfection as sufficient for a meaningful life: how much is enough?', in *New Waves in Philosophy of Religion*, Nagasawa, Yujin and Wielenberg, Erik (eds), New York: Palgrave Macmillan, pp. 192–214.
- Metz, Thaddeus (2009) 'Happiness and meaningfulness: some key differences', in *Philosophy and Happiness*, Bortolotti, Lisa (ed.), New York: Palgrave Macmillan, pp. 3–20.
- Metz, Thaddeus (2008) 'God, morality and the meaning of life', in *The Moral Life: Essays in Honour of John Cottingham*, Athanassoulis, Nafsika and Vice, Samantha (eds), New York: Palgrave Macmillan, pp. 201–27.
- Metz, Thaddeus (2007) 'God's purpose as irrelevant to life's meaning: reply to Affolter', *Religious Studies*, 43: 457–64.
- Metz, Thaddeus (2007) 'New developments in the meaning of life', *Philosophy Compass*, 2: 196–217.
- Metz, Thaddeus (2007) 'The meaning of life', in *Stanford Encyclopedia of Philosophy*, Zalta, Edward (ed.), <http://plato.stanford.edu/entries/life-meaning/>
- Metz, Thaddeus (2005) 'Introduction' (to a special issue on 'Meaning in Life'), *Philosophical Papers*, 34: 311–29.
- Metz, Thaddeus (2005) 'Critical notice: Baier and Cottingham on the meaning of life', *Disputatio*, 19: 215–28.
- Metz, Thaddeus (2003) 'The immortality requirement for life's meaning', *Ratio*, 16: 161–77.
- Metz, Thaddeus (2003) 'Utilitarianism and the meaning of life', *Utilitas*, 15: 50–70.
- Metz, Thaddeus (2002) 'Recent work on the meaning of life', *Ethics*, 112: 781–814.
- Metz, Thaddeus (2001) 'The concept of a meaningful life', *American Philosophical Quarterly*, 38: 137–53.

- Metz, Thaddeus (2000) 'Could God's purpose be the source of life's meaning?', *Religious Studies*, 36: 293–313.
- Metz, Thaddeus (1998) 'Contributions toward a naturalist theory of life's meaning', *Dialogue and Universalism*, 8: 25–32.

I do not refer specifically to these previously published works when I use material from them, except where it is important to indicate when certain ideas first appeared in print. For instance, sometimes I respond to critics of essays of mine, while at other times I need to clarify that I am not cribbing from others who have made points similar to ones I am making.

I would also like to thank the several sources of funding awarded to assist with various chapters, beginning with the most recent: Incentive Funding for 'A' Rated Researchers from the South African National Research Foundation (2008-2012); a Research Promotion Grant from the University of the Witwatersrand Faculty of Humanities Research Committee (2006); Research Awards from the University of Missouri Research Board (2004, 1999); a Summer Stipend to a Senior Researcher from the United States National Endowment for the Humanities (2004); a Research Fellowship from the University of the Witwatersrand Research Committee (2002-2003); and Research Awards from the University of Missouri-St Louis Senate Research Panel (2002, 1997). This book would not have appeared, or at least not nearly as 'soon' as it has, without the generous amount of time free from teaching and the other forms of support that these bodies have given me. Similar remarks go for the Research Professorship offered by Dean Rory Ryan at the University of Johannesburg in 2009. I am truly grateful.

Although I have still probably failed to convey to my mother, Ellen Metz, my brother, Joshua Metz, and my parents in-law, Aziz and Zohra Hassim, precisely what it is that I do as a philosopher, they have always been encouraging and expressed pride in me for the work that I do. My companion, Adila Hassim, entered my life at the time that I embarked on this intellectual journey, and has for over a decade, with little visible annoyance, put up with random solicitations from me for her reactions to a variety of thought experiments, hypotheticals, intuitions, arguments, and principles. Her intellect and love—both powerful—have enriched this book and my life; I am lucky that she chose to be with me, and that she has chosen to stay. Finally, there are my two young boys, destroyers of men and worlds, occasions for redemption and virtue, bodies with beautiful eyes and insightful minds, and sites of hand-holding and giggles. I behold all of you, but dedicate this, my first book, to my first-born son, Mika'il Metz.

Contents

1 Introduction ... 1
  1.1 Finding meaning through the search for it ... 1
  1.2 Clarifying the question ... 3
  1.3 Answering the question ... 6
  1.4 Overview of the answer ... 9

Part I Meaning as One Part of a Good Life

2 The Concept of Meaning ... 17
  2.1 The meaning of 'meaning' ... 17
  2.2 Concept and conceptions of meaning ... 18
  2.3 Unpromising analyses ... 21
  2.4 Monist analyses ... 24
  2.5 A pluralist analysis ... 34
  2.6 From concept to conception ... 35

3 The Bearer of Meaning ... 37
  3.1 'Life' as what is meaningful ... 37
  3.2 Mereology and meaning: whole, part, extent, balance ... 38
  3.3 Rejecting the pure whole-life view ... 40
  3.4 Rejecting the pure part-life view ... 49
  3.5 Developing the mixed view ... 51

4 The Value of Meaning ... 59
  4.1 Values in life ... 59
  4.2 That pleasure and meaning differ ... 60
  4.3 How pleasure and meaning are similar ... 62
  4.4 How pleasure and meaning differ ... 65
  4.5 Happiness and meaning as basic goods ... 73

Part II Supernaturalist Theories of Meaning in Life

5 Purpose Theory I: Questioning Motivations ... 77
  5.1 Overview ... 77
  5.2 Supernaturalism: God- and soul-centred theories ... 79
  5.3 Weaker arguments for purpose theory ... 82
  5.4 The strongest argument for purpose theory ... 84
  5.5 How to account for an invariant morality ... 87

6 Purpose Theory II: Advancing Objections — 98
- 6.1 Aiming to undermine purpose theory — 98
- 6.2 God's purpose v. God's morality — 99
- 6.3 God's purpose v. God's omnipotence — 104
- 6.4 God's purpose v. God's eternality — 105
- 6.5 A new argument against purpose theory — 106
- 6.6 Reconsidering the qualitative properties and purposiveness — 114
- 6.7 The end of purpose theory — 117

7 Non-purposive Supernaturalism — 119
- 7.1 How to be a supernaturalist — 119
- 7.2 God-centred theory grounded on the qualitative properties — 120
- 7.3 An analysis of soul-centred theory — 122
- 7.4 Arguments for soul-centred theory — 124
- 7.5 Arguments against an immortality requirement — 133
- 7.6 The perfection thesis as the fundamental motivation for supernaturalism — 137

8 Rejecting Supernaturalism — 139
- 8.1 How to be a naturalist — 139
- 8.2 Weak arguments against supernaturalism — 140
- 8.3 Meaning because there is nothing supernatural — 142
- 8.4 Meaning even if there were nothing supernatural — 144
- 8.5 Meaning without knowledge of the supernatural — 145
- 8.6 Developing the imperfection thesis — 146
- 8.7 Implications of the imperfection thesis — 158

Part III Naturalist Theories of Meaning in Life

9 Subjectivism — 163
- 9.1 Overview — 163
- 9.2 Naturalism: subjectivism and objectivism — 164
- 9.3 Types of subjectivism — 165
- 9.4 Arguments for subjectivism — 169
- 9.5 The argument against subjectivism — 175
- 9.6 The 'whac-a-mole' problem — 179

10 Objectivism I: Being Attracted, Meriting Attraction, and Promoting Consequences — 180
- 10.1 Objective naturalism — 180
- 10.2 Subjective attraction to objective attractiveness — 182
- 10.3 Making people in the world better off — 184
- 10.4 Making better people in the world — 192
- 10.5 Agent-relativity as intrinsic to meaning — 197

11 Objectivism II: Non-consequentialism — 199
- 11.1 Honing in on that sublime nature and lofty goal — 199
- 11.2 Non-consequentialist objective naturalism — 200

	11.3	Existing non-consequentialist theories	200
	11.4	Towards a new non-consequentialism	218
12	Objectivism III: The Fundamentality Theory	219
	12.1	Time to get deep	219
	12.2	Desiderata for an attractive theory	220
	12.3	The fundamentality theory	222
	12.4	Refining the fundamentality theory	233
	12.5	Reconsidering the fundamentality theory	236
13	Conclusion: The Fine Game of Nil	240
	13.1	Remaining concerns	240
	13.2	Supernaturalism redux	242
	13.3	The point of view of the universe	244
	13.4	Concerns about remaining	247

Epilogue   249
Bibliography   251
Index   263

1

Introduction

'You lay a heavy question on me: *What is the meaning or purpose of life?* It seems to me to be the only question worth asking and one that probably a person ought not spend too much time answering.'

David Small[1]

'The meaning of life, you ask? Being aware of it, I should say.'

Edward Albee[2]

1.1 Finding meaning through the search for it

I have not taken David Small's counsel not to spend too much time trying to answer the question of what would make a life meaningful. Indeed, after having spent about ten years doing so, I realize that I am sympathetic to the old saying that the meaning of life lies in the search for it, which Edward Albee expresses with grace. Now, no one really buys that, at least as a complete account of meaning in life; I cannot recall a single philosophical work I have read on the topic that even mentions the idea that life's meaning is comprised of the quest for it, let alone takes the idea seriously. However, as I begin to write a book that systematically searches for the most justified answer to the question of what constitutes meaning in life—and as you begin to read it—I want to consider whether there is a kernel of truth to be had here.

I confess that what has largely motivated me to devote a substantial portion of my research time over the past decade to issues of meaningfulness has been an unarticulated sense that doing so would itself be a meaningful enterprise, the sort of thing comparable to, say, rearing children with love or creating a work of art. I have not thought that I have had a moral obligation to write the material that would eventually appear in this book. Nor have I thought that I have been doing something supererogatory, a morally praiseworthy project above and beyond the call of duty. Furthermore, it has not occurred to me that I am becoming a more excellent person, perfecting valuable facets of my human nature, by conducting research into the particular topic

---

[1] Quote solicited by Moorhead (1988: 179).   [2] Quote solicited by Moorhead (1988: 10).

of life's meaning. In addition, it is doubtful that I am a person who has become all that much happier or better off by virtue of undertaking this research. As most scholars will tell you, it feels great to make an apparent discovery and to receive positive news from publishers, but those feelings are butterflies, beautiful creatures that quickly fly away, abandoning one to the unattractive, heavier feelings of guilt, shame, anxiety, and worry that one is not writing enough, or that one is not writing well enough, or that editors will not accept for publication what one has written well enough, or that people will not read what one has written well enough and published, or that they will read what one has written well enough and published, but not recognize that it is written well enough to critically discuss. That spiral continues for a while.

Instead of being motivated by duty, supererogation, excellence, or happiness, I have been driven to write about what makes a life meaningful because, at some level, I have believed that doing so would itself be meaningful, even if far from the level of the works of Nelson Mandela, Mother Teresa, Albert Einstein, Charles Darwin, Pablo Picasso, Fyodor Dostoyevsky, and the like, in our stereotypical apprehension of them. I am not saying that I thought that conducting research into the meaning of life would be an instrument by which I could eventually obtain meaning from some other, more practical endeavours. The idea has not been, for instance, that my research would eventually lead me to discover with some degree of assurance what constitutes meaning in life, thereby enabling me to close my laptop, get up from my desk and go and make my life meaningful. Although I do believe there are better and worse answers to the question of life's meaning, and I like to think I am often able to distinguish between them with justification, it is not as though I have felt that knowledge of what makes a life meaningful is a necessary tool without which I could not acquire meaning in light of it. Instead, my view has been that finding full-blown *knowledge* of what makes a life meaningful would be meaningful *for its own sake*, and, furthermore, that *searching* for knowledge of meaning would also be meaningful for its own sake, regardless of whether it successfully lands me with knowledge (at least, if I had a reasonable chance of success).

I suspect most readers' orientation to this book is similar. Perhaps some are looking for definite answers, champing at the bit to get to the end of the book, at which point their doubts will have been settled and they will have found a north star by which to guide their travels. I do argue against a wide array of accounts of what makes a life meaningful and proffer my own, which I find most promising. However, the field of systematic enquiry into meaningfulness is too young to expect very firm results at this point in time. It is only in the past three or four decades that a distinct field has arisen with thoughtful and intricate argumentation, and it is rare that more than 10 books, chapters, or articles squarely on the meaning of life in English are published in a given year. I conclude that the theory of meaning in life that I favour is the most defensible, given the current state of the academic literature, but I make no stronger claim than that.

Perhaps an offer of the theory of meaning that is most justified relative to the body of English-speaking philosophical writings that exist as of early 2012 is enough to motivate some who feel lost to read this book. My point is that they, along with what I suspect are most readers, are probably also reading this book out of the hunch that thinking carefully about what makes a life meaningful would *itself* make their life at least *somewhat* more meaningful. A philosophical theory of life's meaning should account for that intuition.

In the rest of this chapter, I spell out what I aim to accomplish in this book. I begin by providing some clarity to the nature of the main question I seek to answer (1.2), after which I indicate the methods I use to answer it (1.3). Then, I provide a brief overview of the chapters, which includes a sketch of the answers to the question that I reject and of the answer that I defend (1.4).

1.2 Clarifying the question

In this section, I bring out what I mean in posing the question of what would make a life meaningful. Some of this is a matter of indicating what the concept of life's meaning inherently includes (or what talk of it essentially connotes), but I do not here aim to provide a thorough, positive analysis of the concept (or definition of terms), something I undertake in Chapter 2. Instead, my focus here is more negative, in the sense of primarily indicating what the concept does *not* essentially include.

Most people, or at least philosophers, interested in topics readily placed under the rubric of 'the meaning of life' ultimately want to know what, if anything, would confer meaning on their own lives and the lives of those people for whom they care.[3] Of course, some, perhaps even a substantial minority, might also or instead be interested in considerations of whether the universe has a meaning or of whether the human species does. However, I do not address these 'holist' or 'cosmic' questions in this book. For one, although there is clearly a body of literature addressing these topics,[4] it is small compared to the number of 'individualist' writings, of which there are well enough to warrant treatment on their own. For another, often asking 'What is the point of it all?' or 'How did we get here?' is a function of a deeper concern to know how, if at all, the existence of individual human beings can be significant.

In any event, in this book I am strictly concerned to address this individualist orientation, so that another title for it could have been *Meaning in a Life*. When I speak of 'meaningful' and synonyms such as 'significant', 'important', 'matters' and the like, the only bearer that I have in mind is a human person's life. This includes the phrase 'the meaning of life', which several in the literature, unlike me, use to connote ideas about human life as such, not a given human's life. I set aside such holist understandings of the question of life's meaning, so as to make progress on the individualist construal.

---

[3] I disagree with many theorists of meaning in life who take the 'first-person' perspective, i.e., the question of whether one's *own* life is meaningful, to be basic in some respect (e.g., Wong 2008). Most of us are concerned about whether, say, the lives of our spouses and children are meaningful, and not merely because the meaning of our own life might depend on the meaningfulness of theirs.

[4] Wisdom (1965); Britton (1969); Edwards (1972); Munitz (1986); Cooper (2003: 126–42); Seachris (2009).

Note that talk of an individual's 'life' being meaningful is vague, and admits of two different understandings, both of which I address. One sense I label 'whole-life', which concerns the respects in which a person's life as an entirety can be meaningful, perhaps from a deathbed perspective, looking back on the story one's life makes up. Contrast this with a 'part-life' sense, where one considers how a segment of a life can be meaningful or not. There are those in the literature who restrict talk of 'meaning' to lives as a whole; they believe that the only bearer of meaning is a person's entire existence as a pattern or narrative, and not any subset of it. That belief could be true, but I submit that, if it were true, it would not be so *by definition* of phrases such as 'meaning in life'. A time of engaging in prostitution to feed a drug addiction can be characterized without logical contradiction as a 'meaningless period in one's life', as something distinct from a 'meaningless life as a whole'. Or at least that is the way I elect to use the relevant terms. When the distinction is not important to the issue at hand, I gloss it by speaking merely of 'life' being meaningful. At other times, though, particularly when I address whole-lifers who provide substantive arguments for the view that only a complete life is something that can have meaning or not, I make the distinction salient.

Nearly all those writing on meaning in life believe that it comes in degrees, so that, say, some lives as a whole are more meaningful than others (perhaps *contra* Sartre 1946; Britton 1969: 189, 192). Note that calling someone's life 'more important' or 'more significant' than another's might be thought to imply some kind of assessment from a moral perspective, but in the present context it does not. One can coherently hold the view that some people's lives are less meaningful, important, etc. than others, or even downright meaningless, and still maintain that people have an equal moral status grounding obligations to help and not to harm. Consider a consequentialist moral theory according to which each individual counts for one in virtue of having a capacity for a meaningful life (cf. Railton 1984), or a Kantian perspective that people have a dignity in virtue of their capacity for autonomous choices and meaning is a function of exercising this capacity (Nozick 1974: 48-51). On both views, ethical norms could counsel an agent to help people with comparatively meaningless or unimportant lives, precisely because they are peers with regard to moral standing.

Another element of the question of life's meaning that I take for granted in my enquiry is that meaningfulness, whether of the whole of one's life or a part, is desirable. More strongly, when I speak of 'meaning' and cognate terms, by definition I mean something that is finally, and not merely instrumentally, valuable. Such talk connotes something that is good for its own sake, and not merely as a means to something else distinct from it. There are some in the literature who deem talk of 'meaning' to indicate something that could be evaluatively neutral, a property that, say, an evolutionary biologist would be comfortable invoking. However, I stipulate here that by 'meaning' I am in part essentially conveying something that is worthy for its own sake, something that provides a person with at least some (*pro tanto*) basic reason to prize it.

Furthermore, I, like most in the field, take specific exemplary instances of great meaning to have been realized by the likes of Mandela, Mother Teresa, Einstein,

Darwin, Picasso, and Dostoyevsky. I often use the phrase 'the good, the true, and the beautiful' to capture their accomplishments, and the reader should read that figuratively, as a rough way of referring to certain kinds of moral achievement, intellectual reflection, and aesthetic creation. Hence, the reader is cautioned, for instance, not to use a plain sense of 'the good' that would be overly broad for connoting something more than do-gooding, such as pleasure. Insofar as 'the good' least controversially confers meaning on life, it picks out ethical accomplishments such as maintaining integrity in the face of temptation, going beyond the call of duty to help others, helping to pull off a just political revolution, and, at a more everyday level, sustaining friendly and loving relationships. Similarly, the reader should note that 'the true' refers to thinking that need not be literally the truth, and 'the beautiful' picks out artwork that could be original, revealing, and emotionally stirring without necessarily being alluring.

Note that it does not follow from this definition of 'meaningful' that I have analytically ruled out the possibility that Adolf Hitler's life was meaningful. To maintain that talk of 'meaning' essentially connotes something desirable of the sort widely taken to be instantiated by helping others does not logically imply that immoral projects cannot be desirable in the relevant way. For all I have said, Hitler's life *conceptually* could have been meaningful and even in virtue of having achieved his end of wiping out Europe's Jewish population, a view that some philosophers have found palatable.[5] To say that Hitler's existence was 'significant', I am contending, would indicate that it was valuable merely *in a certain respect*, which would leave open the judgements that what he did was seriously immoral and that he had the most reason not to do it.

Just as it is not a logical contradiction to speak of an 'immoral but meaningful life', so it is not to speak of an 'unhappy but meaningful life'. There are many conditions that appear suitably described as 'meaningful misery', with some examples being people who: take care of a sick, elderly parent when doing so would prevent them from pastimes they would find enjoyable; struggle against injustice at the cost of their own peace and harmony; and sacrifice life or limb so that others will survive. If one believes that these lives are not in fact meaningful, that would take argument to show and could not be established merely by definition upon accurately describing them as 'unhappy'.

Similarly, it does not do violence to ordinary linguistic usage, at least among philosophers, to speak of a 'happy but meaningless life', with prominent examples from the literature including: being subject to manipulation and passivity, but feeling cheery because of consumerism and spectacular culture (Marcuse 1964; Debord 1967); spending life in an experience machine that gives the occupant the vivid impression he is doing sophisticated and interesting things that he is not (Nozick 1974: 42-5); rolling a rock up a hill for eternity and enjoying it because of the way the gods have structured one's brain (Taylor 1987: 679-81); and being taken in by charlatans who make one feel special, say, falsely believing in the fidelity of one's beloved or in the divine status of a charismatic leader (Wolf 1997a: 218).

[5] Ellin (1995: 326); Kekes (2000: 30); Belliotti (2001: 7); Frankfurt (2002: 248-9); Wong (2008: 141).

So far, I have suggested that talk of 'meaningfulness' (and cognate terms) does not by definition connote anything about rightness or happiness. Conversely, I do not think that 'absurd' is best understood to be a synonym of 'meaningless'. The concept of absurdity, if one is usefully narrow when analysing it, essentially involves the idea of incongruity (Feinberg 1980; Martin 2002: 219-24), whereas the concept of meaninglessness does not. For example, Albert Camus (1942) famously maintains that a world without God and a soul fails to fit our expectations for order and justice in the universe, and Thomas Nagel (1971, 1986: 208-32) argues that certain unavoidable standpoints that we take on our lives render them absurd, since they posit contradictory judgements about whether they matter. Even if life were absurd in these ways, it would not follow, by definition, that no individual lives can be meaningful.

Finally, consider the concept of futility (analysed with care by Trisel 2002). Futility is more or less the idea of a repeated failure to obtain one's ends. Given that analysis, a human being's life logically could be meaningless but not futile, for she might have done a good job of realizing ends such as urinating in snow and chewing gum. Conversely, it seems that a life could conceivably be futile but meaningful, for instance, if one believes that meaning in one's life can come from states that one cannot bring about, perhaps being one of God's chosen people.

To sum up, the concept of a meaningful life includes the idea of a human person's existence (and not human life as such) exhibiting a kind of non-instrumental value to some degree, where this value: is exemplified by the good (morality), the true (enquiry), and the beautiful (creativity); is not by definition the same as happiness or rightness; and is logically compatible with absurdity and futility. The related concept, of a meaningless life, indicates the comparative lack of this value in a person's existence, does not necessarily include the presence of futility or absurdity (but might), and is exemplified by a life spent in an experience machine or forever rolling a stone.

This rough account of what I mean when asking what would make a life meaningful is substantially negative in the sense of mainly indicating what the question does not essentially include. As such, it is merely a beginning, with the more positive project of articulating an explicit and comprehensive account of it being undertaken in the following chapter. For now, my goal is merely to put enough flesh onto the skeleton of the question of how (if at all) one's life can be meaningful in order to clarify my project in this book.

1.3 Answering the question

One could seek to answer the question of what constitutes meaning in life by presenting a list of specific ways to do so. Few in the literature have sought to do that, and even one of those who has admits, 'Lists are boring. They fail to make us stop and think. They fail to illuminate underlying structure' (Schmidtz 2001: 177). The philosophical mind, or at least one major sort of it, seeks more than a list because it seeks order, roughly explanatory unity, amongst diversity. It naturally asks this of a list of meaningful conditions: is there something that all the elements on the list have in common? An answer to this question is what I often call a 'theory' or 'principle' of meaning in life. Prominent theories of life's meaning, stated in unrefined ways, include the views that it is constituted by conforming to God's will, putting one's immortal soul into a state of perfection, fulfilling one's strongest desires, or obtaining objective goods in this earthly life. An acceptable theory would be a general principle that entails, and provides a convincing explanation of, the many particular ways in which life can be meaningful.

There might not exist an acceptable theory of life's meaning, in the final analysis, viz., there might be nothing other than a list to be had. However, in order to know that no principle does an acceptable job of accounting for all the elements on the list, one must search for one that does, which is my foremost aim in this book.[6] And a thorough search by professional philosophers has not been undertaken long at all, having begun in earnest only in the 1980s or so.

Finding an acceptable theory would be intellectually satisfying, akin to discovering that the rain, the ocean, and the liquid that runs from the tap and one's eyes are all H_2O. It is a matter of dispute how strongly to take this analogy between scientific claims of identity and value-theoretic claims of it. I favour 'realism' (of the naturalist variety), which is roughly the view that the analogy is strong; there exists a feature in the world to be identified with meaningfulness in the way that H_2O is identified with water. A bit more carefully, realism is the view that talk about 'life's meaning' refers to a property that obtains independent of our beliefs about it that we can learn about over time largely through probabilistic evidence. Just as we can learn through empirical and fallible methods that the various things we call 'water' are all constituted by the chemical composition H_2O, so, according to realism about value, we can learn in similar ways that the conditions of life we are most firmly inclined to call 'meaningful' are, at bottom, a certain basic composition of being and doing.

I articulate a kind of realist metaphysic later in the book (5.5),[7] but do not defend it here against the myriad challenges to it; my present aim is to motivate the project of ascertaining what, if anything, all the various meaningful conditions of life have in common, and that would be philosophically important not only if realism were true, but also if it were not. The main competitor to realism is constructivism, roughly the view that a better analogy with a general account of meaning in life would be with the basic rules of grammar. Similar to the way that human beings unconsciously create languages and can find fundamental rules entailing and explaining a wide array of meaningful sentences, we as a society, without much awareness, invent meaningful lives that might well admit of some explanatory unity. This kind of relativism would still make good sense of the theoretical project I undertake in this book, for would it

---

[6] Some would also suggest, plausibly, that one requires a theory in order to come up with a definite list.
[7] For other applications of realism to issues of meaning in life, see Post (1987: 317–26) and Smith (1997: 211–21).

not be fascinating to discover a general principle that underlies the particular judgements that people in a large and long-standing tradition are inclined to make, upon reflection? If the basic rules governing English would be revealing to apprehend, then so would a fundamental principle capturing judgements of meaning in life of those who have thought most about them, even if they were limited to Western culture. Relativism is consistent with a society's pursuit of deep self-understanding.

I have suggested that the goal of seeking an acceptable theory of meaning in life should appeal to readers with a wide array of metaphysical commitments. In addition, the method I use to achieve this goal should be attractive to those with various epistemological views. In a nutshell, I *argue* for one theory to the exclusion of others. An argument is a collection of claims in which some claims putatively provide reason to believe another claim, part of which reason is constituted by the fact that the supporting propositions are *less* controversial than the proposition being supported. A good argument is one in which, for all we can tell, the premises are true, lend evidentiary support to the conclusion, and are weaker than the conclusion.

Often my premises include what I, with the field, call an 'intuition', that is, a judgement of a particular instance of what does or does not confer meaning on life, which judgement is purportedly less controversial than the general principle that is being evaluated in light of it. For instance, it would be a strike against a theory if it entailed that a life of torturing babies for fun would be superlatively meaningful, and it would, in contrast, be a mark in favour of a theory if it could plausibly explain why having composed *The Brothers Karamazov* conferred meaning on Dostoyevsky's life.

Appealing to such intuitions might suggest a commitment to either foundationalism or coherentism; but in fact it entails neither. I make no suggestion that intuitions are self-justifying, beyond doubt or anything 'foundational'. Conversely, I make no suggestion that intuitions are ultimately justified by virtue of their logical and explanatory fit, or 'reflective equilibrium', with a wide array of other claims. I set aside the debate between foundationalism and coherentism and even their common rival, reliabilism, noting that any plausible philosophical theory of justification or knowledge will entail that argumentation, particularly of the sort that appeals to intuition, is a source of justification.[8] I rest content by evaluating theories of life's meaning with claims that are plausible, viz., are themselves supported by good arguments, and are initially more compelling than the theories they are invoked to appraise. This method, which prizes logical virtues such as clarity, evidence, principle, counterexample, and inference to the best justification, is the one used as standard by contemporary analytic philosophers, regardless of their epistemological views, such that some would aptly characterize my project as 'analytic existentialism' (Benatar 2004: 1-3).

In sum, this book is addressed in the first instance to the professional scholar, and is principally devoted to organizing, clarifying, evaluating, and surpassing the theories

---

[8] Argumentation of the kind that appeals to intuition is, after all, how epistemologists characteristically aim to justify their favoured theories of justification!

of life's meaning prominent in the philosophical literature. To obtain focus and to make my task manageable, I focus on English-speaking journals and books written by academic philosophers. I do discuss works originally written in, say, French and German, but they are mostly classic sources. To the best of my knowledge, the most systematic attempt to develop an acceptable *theory* of meaning in life has been undertaken by contemporary Anglo-American analytic philosophers.[9] Furthermore, the amount of literature they have produced on the topic is large enough to work through and evaluate on its own. There is, of course, also literature on meaning in life in non-philosophical fields such as psychology and religion, which I draw upon here and there, but not in any thorough way. Although I have worked to make this book accessible to a wide audience, by defining my terms and minimizing the use of jargon, it is by a philosopher about philosophy for philosophers.

1.4 Overview of the answer

The book proceeds developmentally, with later chapters assuming and building on claims established in earlier ones; that is, I have sought to provide more than a mere patchwork of essays. It would, nonetheless, be feasible for the reader already well acquainted with contemporary normative philosophy to jump around as befits her interests, for I routinely indicate the numbered sections where certain claims are more fully articulated and defended.

The book consists of three major parts, with the first part fleshing out meaningfulness as an evaluative category distinct from others, discussing features of it that are compatible with the overwhelming majority of competing theories of what constitutes it. In the second chapter, I continue the project begun here of indicating what talk of 'life's meaning' connotes. I first critically examine the major definitions in the literature, which attempt to provide a single necessary and sufficient condition for a theory to be about meaning as opposed to some other value, and I argue that they are all vulnerable to counterexamples. I then present my own, complex analysis that is a 'family resemblance' view, holding roughly that theories of meaning in life are united by virtue of being answers to a variety of related and substantially overlapping questions that cannot be reduced to anything simpler. Such questions include: 'Which ends intrinsically merit striving for, beyond one's own pleasure? How should a human person transcend her animal nature? What are the features of a life that warrant great esteem or admiration?'

In the third chapter, I take up the issue of which facets of a life are capable of exhibiting meaning or the lack of it: only its parts, only the life as a whole, or both? I defend the latter, mixed answer, providing reason to reject both a pure whole-life

---

[9] Philosophers in, say, the French, sub-Saharan, and Chinese traditions have tended to address the topic of life's meaning using more particularist, phenomenological, or hermeneutic approaches.

view, according to which only the narrative relationships among the parts of a life are what can be meaningful (or meaningless), as well as a pure part-life view, according to which only segments of a life in themselves are what can be meaningful. I maintain that both a slice of one's life, as well as a life's overall pattern, can exhibit meaningfulness or meaninglessness. I develop a model to capture the whole-life dimensions of meaning, laying out, in a way that exhibits a progressive logic, several distinct respects in which the pattern of a life can plausibly affect its meaning.

I will have argued in the second chapter that a large part of what we mean by 'meaning in life' is a kind of final good distinct from pleasure as such. In the fourth chapter, the last of part one, I highlight some of the substantive value-theoretic implications of the contrast. After first arguing that there is a largely unrecognized disvaluable dimension of meaning, parallel to the way that pain is the opposite of pleasure, I focus on the respects in which meaningfulness and pleasure differ with regard to issues such as: which attitudes it is appropriate to have towards them; how much luck can influence them; and when we should prefer them in a life. Regarding the last issue, I demonstrate that the category of meaning in life ultimately explains much of the literature addressing Derek Parfit's fascinating discussion of 'bias towards the future', providing unity to sundry intuitions about when pleasure, relationships, creativity, and other goods, as well as bads, are preferable in a life.

The second and third parts of the book critically discuss theories of life's meaning, which, at the broadest level, I differentiate metaphysically, by the kind of property taken to constitute it. The second part, comprising Chapters 5 through 8, addresses supernaturalist theories according to which meaning in life consists of engagement with a spiritual realm. Such a general view implies that if neither God nor a soul nor any other spiritual being or force existed, then all our lives would be meaningless. My aims in this part include specifying the most defensible versions of supernaturalism, bringing out what fundamentally motivates them, and ultimately concluding that life can be meaningful in the absence of anything supernatural.

I begin by addressing the most influential form of supernaturalism, the God-centred theory that meaning in life is constituted by fulfilling His purpose (or carrying out His commands). In Chapter 5, I spell out this 'purpose theory' in detail and critically discuss the major rationales for it from the literature. I use the most space to address the most powerful and interesting argument for purpose theory, namely, that God's will alone could ground an objective or universal morality that is necessary for life to make any sense. I grant that—*contra* standard '*Euthyphro*' objections in the literature—facts about God probably could entail an ethic that applies to all human persons, but I argue that they would not best explain it, and hence that a divine command morality is unlikely to justify purpose theory.

In Chapter 6, I consider arguments against purpose theory, first rebutting some influential objections to it in the literature, many of which include the claim that it would be disrespectful and hence immoral for God to give us a purpose. I then present a new, and what I take to be more powerful, objection to the idea that God's purpose grounds meaning. In a nutshell, I argue that the best explanation of why God alone might constitute meaning in life is that God has perfections that we cannot conceivably exhibit, features such as simplicity and infinity, which are widely taken to be incompatible with purposiveness. With this argument, I maintain that the reason to believe God-centred theory in general undercuts the particular, purposive version of it that has dominated the field.

In the seventh chapter, I articulate the conceptions of meaning in life that a God-centred theorist ought to find attractive, supposing that my argument against purpose theory is sound. I also aim in this chapter to specify the most defensible versions of soul-centred theory, views that do not take God, but rather an immortal, non-physical substance constitutive of one's identity, to be the key to meaning in life. I address major rationales in the literature for thinking that having a soul is necessary for meaning in life, pointing out that they are all deductively invalid and inductively weak, reconstructing them, and then showing this: any promising argument for a soul-centred theory is equally an argument for a God-centred theory, because all such arguments ultimately rely on the 'perfection thesis', the claim that meaning in life is possible only if one engages with some maximally conceivable value.

Before questioning the perfection thesis that underlies supernaturalism in Chapter 8, I conclude Chapter 7 by arguing that extant objections to soul-centred theory are weak. In particular, I take up some influential arguments for thinking that immortality would be sufficient for meaninglessness, providing new reasons to reject them. For example, against the claim that immortality would get boring, I do not argue, as most do, that it need not, but rather that, even if it did, it could still be meaningful. And against those who argue that immortality would make us unappreciative and unmotivated, I point out that it is belief in immortality, not immortality itself, that would be the culprit.

Chapters 5 through 7 will have demonstrated that the most promising motivation for holding a God-centred or soul-centred theory of meaning in life is the 'perfection thesis', the idea that meaning requires engagement with some kind of maximally conceivable or ideal value. In Chapter 8, I provide reasons to favour some version of the 'imperfection thesis' over the perfection thesis, defending the idea that there can be meaning in life in the absence of relating to anything perfect or spiritual. After rejecting the adequacy of extant arguments against the perfection thesis and supernaturalism, I present a new one, to the effect that most readers cannot coherently hold these views, given plausible beliefs to which they are already committed. Basically, if we had conclusive evidence for supernaturalism and its perfection thesis, then we would know that both perfection and meaningfulness exist, but while we do know that the latter exists, we do not know that the former does. My overall conclusion is that while neither God nor a soul is necessary for meaning in life, they would probably enhance it, something those who reject supernaturalism have tended not to explain, or even to acknowledge.

At the end of Chapter 8, I also begin to develop the imperfection thesis in detail, with particular focus on how much less than perfect value one must engage with in order for one's life to count as meaningful *on balance*, as opposed to meaningful merely *to some extent*. I construct a new account according to which a life is meaningful all things considered if it is not far from the maximum goodness constitutive of meaning available to humanity in the physical universe, given the laws of nature. I argue that this principle avoids counterexamples to several existing rivals, such as the views that a life is on balance meaningful if it realizes a certain percentage of meaning available to the individual, or if it has (much) more than the average amount of meaning for members of the species.

Part three, composed of Chapters 9 through 13, addresses the other major perspective on meaning in life, naturalism, the view that certain, imperfect ways of living in a purely physical world would be sufficient to make life meaningful. Overall, I argue against the most plausible extant versions of naturalism, and then develop a new naturalist theory that both avoids and explains the problems facing others and that is, I contend, the most justified theory of meaning in life at present.

The ninth chapter takes up subjective naturalism, the theory that a person's life is more meaningful, the more it obtains the objects of propositional attitudes such as desires or goals. I first reject arguments for subjectivism, after which I reject subjectivism itself for the usual reason, that it has counterintuitive implications regarding which lives are meaningful. I address underexplored versions of subjectivism that are more likely to avoid this objection, e.g., intersubjectivism, but ultimately argue that they fail to do so.

The tenth chapter begins my attempt to develop the best version of the broad objective naturalist view that one's existence is significant insofar as one lives in a physical world in ways that are valuable not merely because they are the object of anyone's propositional attitudes. Here, I critically examine two of the three major forms of objectivism that are in the literature. I start by refuting the view of life's meaning that currently dominates the field, the theory that the combination of subjective attraction to objective attractiveness is necessary and sufficient for meaning. I argue that however 'subjective attractiveness' is construed, it is not necessary for meaning, and then point out that everyone's talk of 'objective attractiveness' is much too vague, requiring another theory altogether to fill it out.

The second major form of objectivism that I address is consequentialism, the main version of which is the utilitarian view that meaning comes from making those in the world better off. I reject this theory not only on the ground that there are other plausible ways to obtain meaning besides promoting *well-being*, but also because of its consequentialist structure. Regarding the latter, I argue that there are three major respects in which *promoting* well-being (or some other final good such as virtue) insufficiently captures meaning in life, and that certain 'agent-relative' ways of responding to final goodness are essential for an adequate theory. I do acknowledge, though, that under some circumstances the long-term consequences can confer meaning on life, and that a satisfactory theory must account for this judgement.

In Chapter 11, I turn to the third major variant of objectivism, according to which meaning in life is constituted primarily by some kind of non-consequentialist relationship to certain values. According to this broad perspective, meaning comes from positive engagement with mind-independent goods in the physical world, where this is not merely a matter of seeking to promote them however one can, wherever one can and as much as one can. I argue that all the major extant non-consequentialist theories are counterintuitive, unable adequately to capture the exemplars of meaning, viz., the true, the good, and the beautiful.

Next, in Chapter 12, I develop a novel, non-consequentialist theory which, I argue, avoids all the problems facing its rivals. According to my favoured principle, one's life is more meaningful, the more one contours one's rational self (in a way that does not violate certain moral constraints) towards fundamental objects, i.e., conditions of human life that are largely responsible for many of its other conditions. I demonstrate how this 'fundamentality theory' plausibly accounts for the meaningfulness of the good, the true, and beautiful, and avoids the objections to other theories while incorporating their kernels of truth. For examples of the latter, I explain how the fundamentality theory accommodates the ideas that meaning in life could be enhanced by, but does not require: relating to God, exhibiting subjective attraction to what one is doing, and improving others' lives.

Parts two and three will have discussed the theoretical conditions that could make life meaningful, and in them I will have provisionally accepted the claim that some lives have meaning in them. Confirming that some people's lives are in fact meaningful is my aim in the concluding chapter of the book. Nihilists or pessimists are those who contend that our lives are all utterly meaningless, i.e., that some necessary conditions for a meaningful life fail to obtain for any human beings. In Chapter 13 I critically discuss two influential arguments for nihilism, aiming to provide reason to doubt them. In both cases, I show that nihilists implicitly appeal to claims that are in tension with each other, so that nihilist arguments fail to give one sufficient reason to reject the pre-theoretic judgement, shared by most readers, that some lives are indeed meaningful.

I close the book with an epilogue that returns to the issue that opened the present, introductory chapter. It addresses the respect in which the fundamentality theory of what constitutes meaning in life that I articulate and defend explains the intuition that the search for meaning is itself a source of meaning. I conclude that the book's central thesis justifies its own composition.

I

Meaning as One Part of a Good Life

2
The Concept of Meaning

> 'Well, my goodness, "What is the meaning of life?" you ask. What is the meaning of "meaning" in your question? And whose life? A worm's?'
>
> John Updike[1]

2.1 The meaning of 'meaning'

In the previous chapter I began to clarify what it means to ask the question of how (if at all) one's life can be meaningful. I indicated that this question, for the purposes of my project, is solely about an individual person's life insofar as it manifests, to a certain degree, a kind of value that is not merely instrumental and that is not by definition the same as rightness or happiness. Such a characterization, as I pointed out, mainly indicates what is not essential to talk of 'meaning in life', and in this chapter I aim to say something more positive and systematic about this phrase. I did note that part of its sense is that certain kinds of moral achievement, intellectual reflection, and artistic creation are widely taken to be exemplars of meaning. However, it would be still more useful if, without begging the question in favour of any particular theory of life's meaning, I indicated additional facets of what talk of 'life's meaning' essentially means. What exactly are we asking when enquiring into life's meaning, or, equivalently, just what are theories of life's meaning *about*?

If you are reading this book, then you no doubt have some sense of what is involved in the question of life's meaning, having elected to read this book instead of one about some other value such as justice or virtue. However, there are two reasons to be like John Updike in wanting to become extremely clear and precise about what is being asked when posing the question of life's meaning. First, doing so would enable one to differentiate enquiry into the meaning of life from related sorts of intellectual investigation and so should help to prevent conflation by a given enquirer and to ensure that disputants are debating the same thing, rather than speaking past one another. Second, many of those with a developed philosophical sensibility might want to know for its own sake what talk of 'meaning in life' means. They would find it intrinsically

---

[1] Quote solicited by Moorhead (1988: 200).

interesting to specify what the difference is between asking about what makes life meaningful, on the one hand, and about what makes life sacred (for example), on the other.

To achieve the goal of articulating what we are asking when enquiring into the meaning of life, it will not do, of course, merely to say that a meaningful life is one that is 'important' or 'significant' or that it is an existence that 'matters' or 'has a point'. These terms are synonyms of 'meaningful'. I am pursuing something more revealing than simply enumerating terms that have a sense identical (or very similar) to that of the word 'meaningful'; I am instead seeking to expound this sense. What do these words essentially connote?

Most attempts to answer this question have proposed a single necessary and sufficient condition, accounts of the form that a theory is about the meaning of life if and only if it addresses one property. I argue against salient monist views to be found in the literature, before offering in their place a 'family resemblance' model, according to which there is a cluster of related and overlapping ideas associated with talk of 'meaning in life'.

I begin by indicating the method I use to clarify what we have in mind when theorizing about meaning in life (2.2), after which I raise some proposed clarifications from the literature that are easily dispatched (2.3). Next, I examine more promising, monist proposals, which appeal to ideas about purposiveness, self-transcendence, and the aptness of emotions such as awe and esteem, respectively (2.4). Each of these three ideas captures a large array of theoretical work in the literature, but I maintain that none on its own captures everything. I then explore a way to invoke multiple properties to capture the sense of talk about 'life's meaning' (2.5). I conclude by indicating some implications of the discussion for the central aim of this book, finding the theory of meaning in life that is most defensible relative to the English-speaking philosophical literature (2.6).

2.2 Concept and conceptions of meaning

One way of putting this chapter's undertaking that will be familiar to analytic philosophers is to invoke the distinction between the *concept* of life's meaning and *conceptions* of it. A 'conception' of life's meaning is just another word for a theory of it. It is a general and fundamental account of the conditions that constitute a significant existence, an attempt to describe the underlying structure of meaningful lives in as few principles as possible. Given the aim of contributing to debates among professional philosophers who speak of 'life's meaning', I must bring out what they mean by this phrase, or what they implicitly have in mind when using it, leaving aside other senses that, say, some religious scholars or lay-people might have in mind. So, in this chapter what I do is ascertain whether there is something common to, and unique to, the conceptions of life's meaning to be found in at least the Anglo-American philosophical literature from the past 100 or so years. That core element would be what is called the 'concept' of a meaningful life. The concept of life's meaning is what all and only the competing conceptions of a meaningful life are about; it is that which makes a given theory one of meaningfulness as opposed to one of rightness or happiness. Here, I analyse the concept that unites competing conceptions of life's meaning, i.e., that makes them about the same thing.

The first order of business, then, is to sketch briefly the major conceptions of life's meaning that are out there. If one wanted to devote an entire book solely to finding the concept unifying competing conceptions of life's meaning, then one could adumbrate literally all the theories of life's meaning in a painstaking attempt to ascertain what their common denominator is. However, I shall not be so systematic as that. Instead, for the purposes of this chapter, I focus on three broad perspectives, and ask what it might be that makes them all accounts of life's meaning in contrast to something else. What follows is an overview of the three most influential and widely held self-described stances on life's meaning that one finds in at least English-speaking philosophy from the twentieth and twenty-first centuries.

Supernaturalism is one such theoretical orientation. On this overarching view, one's existence is significant just insofar as one has a certain relation with some spiritual realm. If neither a god nor a soul existed, or if they existed but one failed to relate to them in the right way, then one's life would be meaningless. There are several familiar types of supernaturalism. One variant is the theory that meaning is constituted by actions that conform to God's will, which is the source of justice in this world, while another theory is that one's life matters insofar as one does what it takes to commune with God upon physical death. Theists in the existentialist tradition, including Søren Kierkegaard (1850), Leo Tolstoy (1884), Martin Buber (1923) and Emil Fackenheim (1965) are the ones who most prominently hold supernaturalism. Several contemporary philosophers of religion, including Robert Adams (1999) and John Cottingham (2003, 2005), have espoused the view, and laypeople frequently express such ideas. One often hears something like, 'What is the point of living (or in a particular way) if I will not survive the death of my body?' or 'Nothing matters if nothing is really right or wrong, and only God could make that the case.'

Naturalism denies supernaturalism in the sense of rejecting its implication that life would be meaningless without a god or a soul. Instead, the naturalist maintains that life could be meaningful in a purely physical world, roughly, the spatio-temporal universe as known by scientific means. Naturalists can in principle (but rarely do in practice) grant that relating to a god or a soul could confer some meaning on a life; they simply dispute that such a relation is a necessary condition of a life's having any meaning.

The two most influential broad kinds of naturalism are subjectivism and objectivism. Subjectivism is the abstract idea that meaningful conditions vary, depending on the subject. According to the standard form of this theoretical stance, what is meaningful for a given person is entirely a function of that towards which she has a certain 'pro-attitude', e.g., wanting something and getting it, or setting something as an end and achieving it. William James (1899), Jean-Paul Sartre (1946), Richard Taylor (1970), and Harry Frankfurt (1982) are influential subjectivists about meaning in life. Of these thinkers, Taylor has been addressed the most, at least in the English-speaking literature, with his discussion of the mythic figure of Sisyphus being particularly engaging. Sisyphus, recall, was imagined to have been condemned by the Greek gods to roll a stone up a hill for eternity. Taylor and other subjectivists often defend their view by appealing to the intuitive claim that a life filled with boredom or frustration could not avoid being meaningless. The problem with Sisyphus, according to Taylor, is that he is dissatisfied pushing a rock.

Those who hold objectivism include some classic German thinkers such as Karl Marx (1844) and Friedrich Nietzsche (1886), on my reading of them, and, more recently, philosophers such as Robert Nozick (1981: 594-619, 1989: 162-9), Peter Singer (1993: 314-35, 1995: 171-235) and Susan Wolf (1997a, 1997b). They say that certain features of our natural lives can make them meaningful, but not merely by virtue of a positive attitude towards them. Their idea is that not just any condition could confer meaning on a person's life, no matter what her mental orientation towards it; to appeal to earlier examples, a life cannot matter simply by virtue of urinating in snow and chewing gum, however much those activities might be wanted or sought out. Instead, there are particular ways of being and behaving in the physical world that constitute a significant existence apart from being the object of anyone's pro-attitude. There are forms of life that individuals sometimes fail to want or to pursue, but that they should if they want their lives to matter.

Supernaturalism, subjective naturalism, and objective naturalism, then, are the three broad theoretical perspectives on meaning that one most often encounters in the English-speaking literature. They are the general views that those who describe themselves as investigating 'life's meaning' most often critically discuss. Of course, not all theories of meaning fit under one of these three headings; non-naturalism (8.7) and intersubjectivism (9.3) are just two that are left out. However, working with the most dominant perspectives will be enough to make headway on the goal of analysing the concept of life's meaning.

Supposing, then, that supernaturalism, subjectivism, and objectivism are the major conceptions of life's meaning, there are two clear criteria for an analysis of their concept. First, an analysis of the concept of meaning should allow for the logical possibility of supernaturalism, subjectivism, and objectivism, and influential instances of them, being conceptions of meaning. If an analysis of the concept implied that one of these three perspectives, or a clear version of them, were not an account of meaning at all, we would have serious grounds for questioning the analysis. In addition to giving us conditions that fit all theories of meaning, an analysis ideally ought, second, to indicate conditions that fit only theories of meaning. An analysis should single out meaning in life from other final goods. For example, I have noted that the sense of the question of what would make a life significant differs from that of other evaluative questions, such as what would make a life moral or pleasing (1.2). Therefore, failing to indicate the respect in which supernaturalism, subjectivism, and objectivism, are all accounts of meaningfulness rather than of rightness or happiness would also be a strike against the analysis.

2.3 Unpromising analyses

I start with the non-starters. They are worth addressing because first-rate philosophers have advocated some of them or they were held for long periods of the twentieth century. Part of making progress, or knowing that one is doing so, involves explaining where previous accounts have gone wrong. So, I briefly recount the errors, in order to bring out what the field has learned. One error is to claim that there is something non-sensical about asking whether life has a meaning (2.3.1), while another is to grant that asking about life's meaning is sensical but to fail to capture its sense with much plausibility (2.3.2).

2.3.1 Non-sense

It used to be common for philosophers to think that statements about life's meaning are not well-formed propositions; that is, many held the view that there is something non-sensical about assertions about 'the meaning of life'. Such talk was viewed as expressing something that could neither be literally true or false, nor count as some other sort of 'validity-claim' to be rationally appraised (cf. Habermas 1983). At best, such talk was seen as a matter of a speaker exhorting others to do something or of expressing the speaker's emotions, not as stating a belief held by the speaker. There were two influential arguments for the view that asking about 'the meaning of life' lacks cognitive sense, both of which are typically rejected these days.

First, some suggested that to ask whether life is meaningful is to make a category mistake somewhat akin to asking whether the number two is green or whether a telephone conversation is salty. The latter questions might be answered 'no', but they are probably best considered an abuse of language. Just as the number two is not the sort of thing that *could be* green, so some say that life is not the sort of thing that *could be* meaningful. Since to be meaningful is just to be a symbol, and since a human person's life cannot be a symbol, a human person's life is not the sort of thing that can be meaningful, so the argument goes. As the writer William Gass has remarked, 'There is no meaning to life. It is not a sign.'[2]

Although there have been those who have responded by trying to show that life can be a kind of symbol,[3] most reject this argument, since there is no reason to believe that all senses associated with the term 'meaningful' include the idea of being symbolic (e.g., Smart 1999). After all, synonyms of 'meaningful life' include phrases such as 'significant existence' and 'way of being that matters', which do not inherently connote

---

[2] Quoted in Moorhead (1988: 70); see also Ellin (1995: 323–4).
[3] See, e.g., Gewirth (1998: 184–5) and Thomson (2003: 12), and note that Mahatma Gandhi is reported to have once said, 'My life is my message.'

something symbolic. 'Meaningful' in the context of language can be defined as one thing, but the same word in a different context, of evaluating a life, is properly defined as something else. Note that when speaking of 'meaning' with regard to life it is natural to think of a gradient property, something that comes in degrees (1.2), whereas talk of 'meaning' in terms of a sentence suggests something that is all or nothing; either it is meaningful or it is not.[4]

A second argument for finding talk of 'life's meaning' non-sensical in some way is grounded on the philosophical movement known as 'positivism', which includes the view that a statement expresses something that can be true or false only if it can be conclusively proven to all rational enquirers on the basis of sense experience (Ayer 1947; cf. Wittgenstein 1965). The claim that there is a glass on the table is capable of truth just insofar as it is something that could be demonstrated (verificationism), that could be demonstrated in principle to any inquisitor (logicism), and that could be demonstrated in principle to any inquisitor on the basis of sensory information, e.g., touch or sight (empiricism). Given these criteria, statements using the phrase 'life's meaning' probably do not express something that can be true or false; for it is unlikely that such statements can in principle be conclusively proven to anyone who would care to investigate them, let alone on the basis of the five external senses and one's internal sense of oneself. And if theoretical claims about life's meaning are 'non-cognitive', are not the kinds of things that are even capable of truth or falsity, then a question about life's meaning is similarly not going to be the kind of thing with a content admitting of a 'yes' or a 'no' in response.

Rather than seek to meet positivist criteria for statements to express something capable of truth, most philosophers now reject them. Few these days believe that a statement is capable of being true or false only if one can understand how it (or its rejection) could in principle be established with (near) certainty, let alone to everyone in terms of sensory information. After all, as has often been pointed out, this very theory of language is self-refuting; it does not satisfy its own criteria, in that the theory itself cannot be proven to all rational enquirers in light of sense-data. The bulk of contemporary philosophers believe that one can make judgements that admit of truth or falsity about conditions that probably cannot be proven to all enquirers, and certainly not merely on the basis of immediate sensory experience. Most accept that one can say something capable of truth or falsity about entities such as God and atoms, relationships such as gravity and causation, and states such as having dignity or a meaningful life.[5] Whether we can be confident of having acquired the truth about such things is another matter, of course.

---

[4] I am not sure this is his point, but it came to me upon reading Kauppinen (2012: 353–4).

[5] There are additional arguments for non-cognitivism about value judgements that continue to be taken seriously. For influential post-positivist work, see Blackburn (1984) and Gibbard (1990). However, I focus on texts and issues that are fairly unique to issues of meaning, and seek to avoid more broadly meta-ethical debates. As far as the contemporary literature on life's meaning goes, the dominant view is that people hold and express beliefs about it.

2.3.2 Inaccurate sense

Recent enquiry has focused on determining which proposition philosophers are expressing when they say that a life has meaning (or lacks it), or how to understand the question that is being answered with such a proposition. Here is a quick discussion of analyses that I submit are clearly too narrow.

First off, a handful of philosophers have sought to interpret statements about life's meaning so that they in fact do satisfy the logicist, verificationist, and empiricist criteria for expressing a statement with truth-conditions. For example, explicitly responding to positivism, some have suggested that to speak of a person's life being 'meaningful' is just a matter of saying that a person feels satisfied upon achieving her aims (Wohlgennant 1981; see also Hepburn 1966).

Although this analysis probably satisfies the positivist criteria, the trouble is that it is much too cramped. It counterintuitively implies, for one, that it would be logically contradictory to say that a person's life could be meaningful despite feeling dissatisfied with her achievements. For example, this analysis implies that Vincent van Gogh, at least in our stereotypical understanding of him, by definition could not have had a meaningful life, which seems counterintuitive; after all, many believe that van Gogh actually did have a meaningful life, in virtue of the great works of art he created and their posthumous reception. It might in fact be true that for one's life to be meaningful, one must feel satisfied with one's achievements (e.g., Wolf 1997a),[6] but my point is that this would not be true *by definition* of the word 'meaningful'.

Second, two of the most powerful philosophical minds in the pre-war era have taken the question about the meaning of life essentially to connote something about God. For example, according to Bertrand Russell, 'Unless you assume a God, the question (of life's meaning) is meaningless, and, like Laplace, "*je n'ai pas besoin de cette hypothese*"' (quoted in Moorhead 1988: 165). Similarly, Ludwig Wittgenstein remarks, 'To believe in a God means to understand the question about the meaning of life' (1914-1916: 74). These quotations suggest the view that enquiring about one's meaning in life is identical to asking how one is or should be related to God, the source of oneself, and the universe. This view gains plausibility from the fact that the question of life's meaning is often associated with the question, 'Why am I here?', and this question, in turn, is naturally understood as asking for the reason for which one was created.

Some might have exclusively theistic ideas in mind when seeking to learn about the meaning of life, but they will not do as an analysis of the essential elements of the concept that underlies contemporary philosophical dispute about the issue. The obvious problem with the present understanding of the concept of life's meaning is that it *a priori* excludes the possibility of naturalist conceptions of meaning.[7] If asking about life's meaning logically involved asking about one's relation to God, then no naturalist

---

[6] I take up this substantive issue in 10.2.
[7] As well as soul-centred supernaturalist conceptions that do not appeal to God (7.3).

theory of significance would be conceptually possible and naturalists and supernaturalists would be talking past each other. However, there are many naturalist accounts in the literature, accounts that supernaturalists themselves have argued provide substantively inadequate answers to the question of life's significance. This question therefore cannot itself be understood in theistic, or even more broadly supernaturalist, terms. For naturalists and supernaturalists to disagree about a common subject matter, it must be captured in a way independent of the controversial elements of these competing theories.[8]

2.4 Monist analyses

Suppose, then, that the question of life's meaning is capable of truth and is not merely about whether one is doing God's bidding or satisfied with one's achievements. In the literature there are three particularly attractive candidates for the way to understand the question or, equivalently, to specify what makes a conception one of life's meaning as opposed to something else. I argue here that none on its own can capture the range of theories that intuitively are about life's meaning. In the next section, I consider how one might plausibly combine them.

2.4.1 Purposiveness

To many in the field, there is a close connection between meaningfulness and purposiveness. Kai Nielsen has most clearly articulated the position that to enquire about the meaning of life is merely to ask which purposes a person should adopt and realize:

> When we ask: 'What is the meaning of life?' or 'What is the purpose of human existence?' we are normally asking, as I have already said, questions of the following types: 'What should we seek?' 'What ends—if any—are worthy of attainment?'....(T)his question is in reality a question concerning human conduct (Nielsen 1981: 240, 246).[9]

By this analysis, the concept of a meaningful life is that of a person who has done well at achieving goals that she should strive to achieve. This analysis of the concept implies that different conceptions of a meaningful life are to be understood as competing theories of the ends that humans should pursue. Roughly, typical supernaturalists say that humans should fulfil the purpose God has assigned them, standard subjectivists maintain that people should adopt whatever goals they are inclined to upon reflection, and typical objectivists hold that one should undertake activities such as taking up a caring profession and educating oneself. It appears, then, that the present analysis provides a nice account of what the three major theories of life's meaning have in common.

---

[8] A third, more idiosyncratic analysis of the concept of life's meaning that I maintain is clearly weak comes from Baier (1997: 45–70). For discussion, see Metz (2005a: 216–19).
[9] See also Ayer (1947: 226–7); Morris (1992: 56–7); Hartshorne (1996: 10–11); Gewirth (1998: 186–9).

However, this analysis is deeply flawed, in two respects. First, the analysis is too broad for failing to demarcate the domain of meaning. Consider these goals: scratching an itch, rearing a child with love and insight, waiting in a queue to deposit money in the bank, eating ice cream, acquiring an education, staying alive. All these goals are worth pursuing, but not all of them are *prima facie* candidates for conferring meaning on one's life.

Second, the analysis is too narrow in not admitting the logical possibility of a person's life being meaningful in virtue of conditions that she cannot control. For example, consider an 'aristocratic' theory of meaning according to which one's life is significant by virtue of having been born into a certain clan. Being part of a particular bloodline might be thought to make one's life matter, but this is not an end that an individual can pursue. This essentialist view, although largely bizarre to a modern sensibility,[10] does not seem logically contradictory, as the present analysis entails. For another example, consider people who think their lives are meaningful because they are God's chosen people. Being deemed special by God is not a state of affairs that an individual can bring about (let us suppose), and hence the conceptual possibility of its conferring meaning on a life cannot be accommodated by analysing meaning in terms of ends that should be adopted and realized.[11]

Consider, now, an improved variant of the purposive analysis. Sometimes purposiveness is understood in terms of serving a function. So, one might suggest that the concept of a meaningful life is that of an existence that plays a role in the realization of valuable ends, even if that role is not to will them into being. G. E. Moore suggests such a view when he says,

I have been very much puzzled as to the meaning of the question 'What is the meaning or purpose of life?'.... But at last it occurred to me that perhaps the vague words of this question are often used to mean no more than 'What is the use of a man's life?'.... A man's life is of some use, if and only if the *intrinsic* value of the Universe as a whole (including past, present, and future) is greater, owing to the existence of his actions and experiences, than it would have been if, other things being equal, those actions and experiences had never existed.[12]

It is not clear whether Moore's statement about a person's being useful insofar as she promotes intrinsic value is intended to be part of the thin concept of meaning or a thick conception of it. Deeming it to be inherent to the concept is reasonable, since if a meaningful life is one that is useful, it presumably will be useful in a relevant sense only insofar as it produces intrinsic value, by which Moore roughly means something that is valuable for its own sake (finally valuable) in virtue of its inherent nature. To think of

---

[10] And, yet, consider the pride people take in what their long dead relatives have accomplished.

[11] One might suggest that even these conditions may be said to be 'chosen', insofar as a person elects to stay alive. However, such a response merely buys a little time, since the examples may be effectively reformulated. For instance, this manoeuvre would not work for the view that meaning comes from God's loving one's immortal soul.

[12] Quoted in Moorhead (1988: 128–9); see also Moore (1903: 35).

a meaningful life as one that is useful for ends, regardless of whether they are desirable or not, would fail to account for the normative element of meaning. Therefore, I read Moore's remarks as suggesting that we analyse the concept of a significant life in terms of an existence that promotes intrinsically valuable states of affairs.

The present analysis plausibly avoids the two problems facing the previous attempt to cash out meaning in terms of purpose. First, it can weed out at least many of the intuitively non-meaningful goals adumbrated above by including the idea that the meaningful ones are those worth pursuing for their own sake, something that waiting in line clearly is not and that the mere fact of being alive probably is not as well. Second, since the Moorean view has a wide notion of valuable end in terms of intrinsic goodness that is promoted, it does not logically tie the meaningful conditions of an individual's life to her choices and so can account for the logical possibility of meaning being something that is endowed instead of pursued. It allows for the conceptual possibility of a person's life becoming more meaningful merely by *constituting* a desirable state of affairs, such as being the object of God's love, where this state of affairs need not have been the product of a voluntary decision. And, like the previous analysis, this one *prima facie* captures the subject matter that is common to supernaturalism, subjectivism, and objectivism. Roughly, supernaturalists would think of the relevant intrinsic value in terms of being the object of God's will, subjectivists would construe it in terms of preference fulfilment, and objectivists would have a mind-independent account of the good.

Despite these advantages, the present analysis is questionable. Analysing the concept of meaning in terms of the promotion of intrinsic value is to think of meaning as an essentially 'teleological' relation, in which meaning can be conferred on a life just insofar as it promotes a state of affairs that is good for its own sake. However, this *by definition* rules out non-teleological conceptions of meaning. For a first example, those inclined to think that Hitler's life was meaningful tend to think that a person's existence can be more significant by virtue of having made a large impact on human history or because of a compelling or interesting life-story (cf. 1.2).

In addition, there are conceptions of meaning that do invoke intrinsic value, but cash out the meaningful relation to it in terms of honouring rather than promoting. For example, some objectivist views, held by Kant (1788, 1790) and some neo-Kantians (e.g., Pogge 1997), are that life has importance insofar as it treats rational nature with respect. Implicit in these views is the idea that rational nature is superlatively good for its own sake, viz., has a dignity, but the meaning is thought to be constituted by avoiding degradation of this good, not fundamentally by promoting it by, say, creating lots of children. If it is logically possible to conceive of respecting people's intelligence as making one's life meaningful, then the concept of meaning is not merely the idea of promoting intrinsic value.

There is an additional problem that faces both Nielsen's and Moore's analyses of the concept of meaning in terms of purposiveness. So far, I have provided reason to think that both are too narrow, for failing to capture conceptions that are intuitively about meaning in life. However, another objection to them both is that they are, as they stand, too broad, failing to differentiate meaning from other desirable facets of a life such as pleasure, or at least certain kinds of it. In Chapter 1, I noted that to enquire into meaningfulness is *not by definition* to enquire into happiness (1.2), but, here, I make a stronger point, namely, that to enquire into meaningfulness is *by definition not* to enquire into a certain subjective conception of happiness,[13] and that a satisfactory analysis must capture this facet.

Here is what I have in mind. Spending time in an 'orgasmatron' (from Woody Allen's movie *Sleeper*) or some other experience machine would be enormously pleasurable and might even be worth doing for its own sake on occasion, but doing so would not make one's life more meaningful. Eating chocolate and ice cream would taste delicious, and would be desirable in itself, but would not enhance the significance of one's existence (Wolf 1997a: 210, 216). One could feel secure and content consequent to a sense of being special, while unknowingly being deceived and manipulated by a cheating spouse or a cult leader, but these feelings would not make one's life important (Wolf 1997a: 218). And being subjugated and manipulated while feeling upbeat because of psychotropic drugs (Huxley 1932) would not be a way for one's life to matter.

I submit that none of these experiences conceptually *could* make one's life matter. In light of such cases, I draw the conclusion that one's own pleasure *qua* pleasure is not a *prima facie* candidate for meaning-conferral. Note that such a view allows for the possibility that pleasure *under certain conditions* could enhance meaning in a life, perhaps when it is consequent to morally right actions or to consensual sex with a beloved. Holding that it is logically contradictory to think that one's pleasure *in itself*, the mere experience, is meaningful is consistent with believing that pleasure *in relation* to other things could sometimes be so.

One might object to the *a priori* exclusion of pleasure as such from the concept of meaning by pointing out that many people would not find life worthwhile in the absence of pleasure. If life would not be worth continuing without pleasures (even in the absence of pains), then it is reasonable to suggest that they might be part of its meaningfulness.

However, a good explanation of the relevance of pleasure is that it is necessary for obtaining meaning in life, as opposed to being something that constitutes it. To carry on and find meaning in life, humans may well need to experience pleasure to some degree, but it does not follow that pleasure *qua* pleasure is what would make their existence significant.[14]

This reflection on pleasure raises a problem for the purposive analyses of meaning, in that they do not exclude one's pleasure from the logical possibility of meaningfulness.

---

[13] For suggestions of such a perspective, see Nozick (1974: 42–5); Bond (1983: 119–22); Landau (1997: 266–7); Wolf (1997a); Thomson (2003: 68–79); Baggini (2004: 89–105, 124–39).

[14] In addition, I argue elsewhere that worthwhileness, in the sense of what makes life worth starting or continuing, is distinct from meaningfulness, so that the former does not entail the latter (see Metz 2012).

Tasting chocolate is an end that is worth pursuing to some degree, as per Nielsen, or is a state of affairs that is intrinsically valuable in some respect, à la Moore. If some element of a purposive analysis is going to work, it must specify a *kind* of final purpose or goodness, one that excludes the individual's pleasure. The concept of meaning in a human person's life essentially includes, in part, the idea that we are *not* discussing her pleasure *qua* pleasure, or at least pleasures such as eating sweets or having sex with a machine. That requires indicating, at a minimum, that the relevant ends to be pursued, or the relevant intrinsically valuable states of affairs to be promoted, are not those constituted by such experiences.

Although such a characterization would solve the problem of the purpose analysis of the concept of life's meaning being too broad, it remains, for all that has been said, too narrow, leaving out several conceptions in which meaning is not a function of promoting the good. It is worth considering whether another analysis can do any better.

2.4.2 Transcendence

Another major analysis of the concept of a meaningful life can be gleaned from some of Robert Nozick's remarks (1981: 574–9). It is not clear that Nozick is intending to provide an analysis of the concept of meaning. However, it does not matter what Nozick's aims are, for his comments provide a *prima facie* attractive analysis of the concept, in terms of transcending limits. The initial motivation for construing theories of meaning as providing substantive accounts of how properly to transcend limits comes from reflection on the use of the word 'meaning' in other contexts. If we ask for the meaning of a word, we are told about its relationship with other words or with objects in the world. If we ask what inflation means for the economy, we are told about its effects on something else such as unemployment or interest rates. Nozick proposes that we likewise think of asking for the meaning of a life as a matter of asking how it 'connects up to what is outside it' (1981: 601).[15]

In order to evaluate this proposal, I first need to sharpen it. Exactly which limits are relevant to the issue of life's being significant and how must one cross them? Breaking the speed limit and pinching a stranger are ways of 'crossing boundaries', but these actions do not conceptually count as candidates for meaning. Conceiving of meaning as merely a function of connection with something external is not sufficient to capture the evaluative dimension of meaning, and, more generally, does not express anything fairly exclusive to meaning in life.

Therefore, consider this proposal: the concept of what makes a life meaningful is the idea of connecting with something valuable for its own sake beyond one's person. As Nozick also says, 'meaning is a transcending of the limits of your own value, a transcending of your own limited value' (1981: 610). People often think of meaning in terms of an intense relationship with something greater than oneself, where 'greatness'

---

[15] For similar ideas, see Wolf (1997a: 222); Runzo (2000: 188); Cooper (2005); Ford (2007: 254).

has an evaluative sense, not merely one of magnitude or of something external to oneself.[16]

One way of connecting with final value beyond one's person would be to promote it, but this need not be the only way. Supposing that the words 'connecting' and 'transcending' mean something like responding positively, one could also connect with external value by honouring it, meaning that the present analysis avoids part of the narrowness objection to the previous, purposive one.[17] In addition, since one's pleasure is a final value internal to one's body, the present analysis is able to exclude it *a priori* from meaningfulness, and thereby avoid the broadness objection.

However, there are counterexamples that tell against the transcendence analysis. Consider supernaturalist theories according to which a person's life is meaningful insofar as she honours her soul or realizes what she essentially is *qua* spiritual substance endowed with a freedom independent of the laws of nature (e.g., Swenson 1949; Rao 1965). As neither conception is a matter of connecting with a value external to oneself, the present analysis counterintuitively implies that these activities are conceptually incapable of making a life meaningful.

There are objective naturalist accounts, too, that the present analysis wrongly entails are not theories of meaning at all. Consider that the following could in principle make one's life somewhat more meaningful: publicly standing by what one reflectively believes to be right, exhibiting courage and performing a difficult act because it is right, being true to oneself, overcoming addiction, not letting oneself be bossed around, discovering new particles and confirming the existence of certain laws of nature. Since integrity, virtue, authenticity, autonomy, self-respect, and knowledge are internal to a person, or at least do not essentially involve a relationship to an external final good, and since they are *prima facie* candidates for a meaningful life, the concept of a meaningful life cannot just be that of an existence that has 'a connection with an external value' (Nozick 1981: 610).

If a transcendence analysis is going to work, it must allow for the logical possibility of a person's life being meaningful for connecting to both internal and external goods. And such an analysis must carefully specify which internal goods are relevant, to be able to continue to differentiate meaningfulness from pleasure. In light of these concerns, I go beyond Nozick's remarks and address the following transcendence analysis:[18] the concept of meaning is the idea of connecting with final goods beyond one's animal self. The animal self is constituted by those capacities that we share with (lower) animals, roughly those not exercising reason. These include the fact of being alive, the instantiation of a healthy body, and the experience of pleasures. These internal conditions may well be good for their own sake, but they do not seem to be the sorts of values with which one must connect to make one's existence significant. To say that

---

[16] For interesting, systematic reflection on greatness as a distinct value, see Rees (2011).
[17] Though the case of Hitler remains at large.
[18] Which I first articulated in Metz (2001a: 146–7) and which Neil Levy (2005: 177–80) has also advanced.

the concept of meaning is the idea of relating positively to non-instrumental goods beyond one's animal self is to say that while merely staying alive or feeling sensations logically cannot make one's life meaningful, connecting with other internal goods, say, those involving rationality or spirituality, and with all sorts of external goods, can do so.

By virtue of distinguishing between the animal self and the rational self, the present transcendence analysis allows conceptual space for certain internal goods to confer meaning on a life. Honouring one's soul, realizing one's higher nature, developing excellences, exhibiting integrity, acting autonomously and even acquiring knowledge logically could confer meaning on a life by virtue of being a relevant way to transcend one's animal nature. Yet these internal goods exclude pleasure as such, in the way I have argued the concept of meaning should. Finally, the present analysis articulates a reasonable understanding of what supernaturalism, subjectivism, and objectivism have in common. On this view, supernaturalists who prescribe communing with God or honouring one's soul, subjectivists who advocate striving to achieve whatever ideals one adopts upon reflection, and objectivists who recommend creating artworks or promoting justice, are all indicating ways to connect with value beyond one's animal self. This is the most promising analysis proposed so far.

The most important objection to the present transcendence analysis is that it cannot account for certain influential forms of subjectivism. For example, Taylor (1970) once defended a kind of subjective theory that the present analysis must deem not to be one about meaning at all. Recall that, according to Taylor, one's existence is more significant, the more one gets whatever one passionately desires. It does not matter for Taylor how one's desires have been formed, in particular, whether one has reflected on them; he imagines that Sisyphus' life would be as meaningful as it could be if the gods were to implant in him the intense desire to roll his stone. Assuming desire satisfaction is a good at all, it is an internal one that is a part of the animal self. Hence, the present analysis of the concept of meaning in terms of connecting with value beyond one's animal self cannot count Taylor's view as a theory of meaning.

Some might be tempted to bite the bullet by denying that Taylor's subjectivism counts as a *prima facie* candidate for meaning in life. Levy, for one, is willing to accept the idea that it is about something other than meaning in life, perhaps happiness or rational choice instead (2005: 180n11). And to buttress his view, Levy could note that there are several subjectivist views for which this transcendence analysis can account. For example, consider subjectivist theories that cash out meaning in terms of the realization of goals that agents have adopted upon careful appraisal. These views include a cognitive element, differing from Taylor's purely conative view. For another example, consider the difference between holding meaning to consist of desire satisfaction and deeming it to consist of an agent satisfying her desires. The former does not essentially involve any exercise of rational choice; a genie in a bottle could fulfil a person's desires, as could a 'result machine' (Nozick 1974: 42-45). However, the latter inherently includes a volitional element, the notion that someone must strive to acquire meaning, which might put it within the ambit of the present analysis. Perhaps it is not a great strike against it, then, that it cannot accommodate Taylor's theory, at least if it can accommodate other subjectivist views.

However, the problem with this reply is the existence of so many in the field who have thought that desire satisfaction is indeed the kind of thing that logically could be the source of meaning in life. Taylor's discussion of Sisyphus is probably the most widely reprinted and read discussion of the meaning of life in the past 40 years.[19] His theory lies at the heart of recent debates about what can make someone's existence significant. It is difficult to rest content with an analysis that entails that the many who consider Taylor's theory to be about meaning are conceptually confused. It is worth seeing whether there is some other analysis that both has the advantages of the best version of the transcendence analysis and avoids the disadvantage of not accommodating desire satisfaction.

2.4.3 Apt emotions

A third promising analysis of the concept of life's meaning is one that, so far as I can tell, Charles Taylor (1989: 3-90) first suggested and that has been articulated by myself (2001a: 147-50) and more recently by Antti Kauppinen (2012: 353-8). It is the idea that talk of 'meaning' with regard to a life essentially connotes those facets of it that warrant certain emotional responses. Taylor speaks of a life that engages with goods that are 'worthy of love and devotion' or that 'command our awe'; I have discussed those parts of a life 'in which people would be justified in taking great pride'; and Kauppinen appeals to 'feelings of fulfilment and admiration being appropriate'. Notice that the present analysis does not construe talk of 'meaning' in terms of when people *in fact* have emotions of love, awe, pride, admiration, or fulfilment; it rather indicates that talk of 'meaning' essentially indicates a life in response to which such emotions would be *suitable*, regardless of whether they are in fact manifested.

In order to make the discussion manageable, in the following I focus on the emotions of pride and admiration. Note that we often associate enquiry into life's meaning with deathbed reflection and eulogies. To access whether one's life is meaningful or not, a person tends to imagine that she is at the end of her days on earth and considers how she would appraise her life from that perspective. And a characteristic time to consider whether another's life has been meaningful is after she has died and one is reflecting with others about what her life was like. The most salient emotions in these contexts are, respectively, pride and admiration. From a first-person perspective, of thinking about one's own life, to judge it to have been meaningful apparently is to think that it is sensible for one to feel great pride or esteem about it. And from a third-person perspective, to judge another's life to have been meaningful appears to be a matter of thinking that it would be apt to admire it. In the following, therefore, I consider this analysis of the concept of life's meaning: the idea of those aspects of a

---

[19] Perhaps after Tolstoy's *My Confession* (1884).

life for which the person whose life it is may sensibly have great esteem and for which others may sensibly have great admiration.[20] For additional sake of economy, I tend to focus on the first-person perspective, concentrating on the idea of conditions meriting great esteem.

Let me say more about what 'esteem' and 'pride' mean. I use the terms interchangeably to connote a certain perceptual-affective response to an object that one judges to be related to oneself. One has both a sense and a feeling of esteem. In taking pride in something, one both judges that it is worthy[21] and feels satisfaction about it. For example, if a person takes pride in having reared her children, then she judges her behaviour to have been all things considered well done, and feels pleased for having done it. It would be odd to say that a person has esteem for an action when she deems it on the whole bad or does not feel good for having done it. In short, esteem is a matter of high regard and high spirits about facets of one's life.

Esteem is not the same as self-esteem. Self-esteem involves having not only a sense that one is worthy, but also a sense that one can do something worthy. Esteem differs in that it does not fundamentally involve self-confidence; one can take pride in having done something and still feel incapable of doing much else. One should also not reduce esteem to a mere sense of satisfaction. Esteem includes this, but also includes an element of judgement. A person may be pleased that he finally finished washing the dishes, but he presumably will not have esteem for doing so, since he does not regard that action to be particularly choice-worthy.

Analysing the concept of meaning in terms of greatly estimable (or admirable) conditions appears to avoid the major problems facing the purpose and transcendence analyses. Since something other than achieving ends and promoting value could in principle merit great esteem, the esteem analysis allows for non-purposive conceptions of meaning. Since pleasurable experiences as such are not the kind of thing that in principle could merit great esteem, the esteem analysis does a good job of accounting for the judgement that talk of 'meaning' inherently connotes facets of a person's life other than them. And since it seems logically possible for desire satisfaction to merit great esteem (or admiration), the esteem analysis accommodates theories of meaning that the transcendence analysis could not.

In addition, the esteem analysis provides a promising way to understand what the three major theories of life's meaning have in common. It construes theories of meaning as providing rival accounts of those aspects of a life in which a person could reasonably take great pride, or for which an outsider could reasonably have great admiration.

---

[20] Another reason for focusing on these emotions and not so much on the ones Taylor invokes is that his do not easily account for subjectivist views; he, like Levy, is committed to the view that desire satisfaction is not a logically possible contender for what makes a life meaningful, maintaining that only 'qualitative' goods that are desire-independent in principle could merit awe and love (Taylor 1989: 3–90).

[21] Cf. James (2005), who, while not purporting to give an analysis of the concept of meaning in life, argues that a necessary condition for a meaningful achievement is that it would give one reason to think more of oneself (where a meaningless project gives one reason to think worse of oneself).

By this analysis, supernaturalists are contending that if there were no purely spiritual realm, then there would be nothing about one's life that could be worthy of substantial esteem; subjectivists are maintaining that substantially estimable conditions vary, depending on a person's stronger wants or higher-order ends; and objectivists hold that certain features of our physical lives warrant great pride, but not merely in virtue of the subject's pro-attitude towards them.

Attractive as it is, the esteem analysis has a difficult time accommodating all major theories that are intuitively ones about meaning in life. If the concept of what makes a life meaningful were merely the idea of those conditions of a life warranting great esteem or admiration, then it obviously would be logically contradictory to maintain that a condition is meaningful and yet does not warrant great esteem (admiration). But there are at least two sorts of cases where it does not seem logically contradictory to think that a condition is, or at least could be, meaningful but does not warrant such emotions.

One kind of counterexample involves cases in which meaning is thought to be a partial function of happiness, understood as pleasure, consequent to some kind of project. Consider the theory that enjoying worthwhile activities is what makes a life meaningful, which is probably the most commonly held approach to meaning these days (10.2). According to this view, doing something worthwhile or good for its own sake is not enough for one's life to be meaningful; one must also be satisfied doing it. Think, too, about the theory that a person's life is meaningful insofar as she is justly rewarded, perhaps in an afterlife, for having performed right actions and exhibited an upright character. To some, life would make sense only if happiness were ultimately proportioned to morality; it would be meaningless for the wicked to prosper and the upright to suffer. Now, it seems logically consistent to suppose that life could be more meaningful in virtue of these two happiness-related conditions and to deny that there is anything about the enjoyment or reward (as opposed to the worthwhile activities, moral deeds, and good character) in which to take great pride.

I am not sure how damning the counterexample is. The friend of the esteem analysis might fairly urge us not to focus on the pleasure in itself, but rather the whole of pleasure-in-response-to-activity-deserving-of-it. Perhaps the latter is a good candidate for esteem and admiration.

A second, more worrisome kind of counterexample concerns cases in which living in a certain environment might sensibly be thought to confer meaning on an individual's life. Consider the view that living among natural objects would be somewhat more meaningful than living among plastic replicas of them. Or think about the idea that a person's life would be more meaningful for living among old, hand-worked crafts and once-off architectural constructions than for living among new, mass-produced works. Or note the perspective that one's existence would be more significant for being part of a long-standing lineage in which ancestors are remembered and progeny are forthcoming.

A person could plausibly think that being part of a natural ecosystem, participating in a rich culture, or having a continuity with human history confers meaning on a life and deny that there is anything worthy of substantial pride about such conditions, particularly since they do not (in these examples) concern a person's decision-making or self more broadly. But the esteem analysis implies that one would be conceptually confused to maintain that something could be meaningful and yet not be worthy of great pride.

2.5 A pluralist analysis

In the previous section I examined the three most promising candidates for unifying theories of life's meaning with a single property, and I argued that all such monist analyses fail to capture all and only theories of life's meaning. Most importantly, each of the ideas, of purposiveness, transcendence, and esteem, faces the problem of entailing that certain conceptions that are intuitively about life's meaning are not about it. It is natural at this point to begin to doubt that there is any analysis available that can provide a single common denominator among all the diverse theories of meaning. The next step, then, is to consider an analysis of the concept of life's meaning that is complex for appealing to more than one property.

Specifically, I advance a family resemblance approach, according to which enquiry into life's meaning is, roughly, about a cluster of ideas that overlap with one another. To ask about meaning, I submit, is to pose questions such as: which ends, beside one's own pleasure as such, are most worth pursuing for their own sake; how to transcend one's animal nature; and what in life merits great esteem or admiration. If a theory is a competent answer to one of these questions, then one should deem it to count as being about meaning in life.[22]

One readily sees the overlap between the three properties of purposiveness, self-transcendence, and esteem/admiration. To pursue higher-order aims beyond experiencing pleasure as such is a large part of what would be involved in transcending one's animal nature, given that sensation is something that both animals and we

---

[22] A rival pluralist analysis in the literature comes from Garrett Thomson (2003: 8–13), who believes that talk of 'meaning' in life connotes the degree to which it achieves some purpose, contains value, and is intelligible, and from Arjan Markus (2003), who maintains that such talk indicates the extent to which a life fulfils a purpose, is worth living, and has parts that cohere in some way. I take the three recurrent themes here to be: purpose, value, pattern.
There are two problems with this perspective. First, if a theory is about meaning in life just insofar as it addresses *all three* ideas of purposiveness, value, *and* pattern, as appears to be Markus' intention (2003: 127), then the analysis is much too narrow; there are many theories that are intuitively about meaning in life that fail to address all three elements. For example, I argued above that there are *prima facie* candidates for meaning that do not address purposiveness in any straightforward sense (2.4.1). Second, if instead a theory counts as being about meaning in life if it addresses *at least one* of these three ideas, then concerns about broadness arise: a theory of happiness is about the respect in which a life is 'valuable' to no less a degree than a theory of meaningfulness.

experience. In addition, transcending one's animal nature would be a *prima facie* good candidate for conditions worthy of great esteem, given that we tend to think that there is something about human nature that is higher than the animal. Furthermore, pursuing aims beyond experiencing pleasure as such would on the face of it be apt for conditions worthy of admiration, given that we do not think it suitable to admire someone simply by virtue of feeling it. My suggestion, here, is not that the three properties are ultimately identical, but rather that there is substantial coincidence between them.

Recall that the esteem analysis seemed to avoid the weightiest counterexamples to the other two views, but that it faced its own counterexamples. Grouping the properties of purpose, transcendence, and esteem together does an elegant job of dealing with the latter. Even if living among natural objects, participating in an aesthetic environment imbued with character, and being part of a human lineage are not reasonable grounds for esteem, they could be considered final goods beyond one's animal self. And if these were ends that one could bring about, they would count as purposes beyond one's pleasure that merit pursuit.

The family resemblance proposal is attractive, and I do not have definitive grounds to reject it. It might be that there are plural necessary and sufficient conditions for the concept of meaning such that a theory is about meaning if and only if it addresses either X (purposiveness) or Y (self-transcendence) or Z (admiration) and nothing else. A weaker view, more at home in the family resemblance analogy, would be that the three elements specified here make up a *very large portion* of the ideas that theories of life's meaning are about, but that there are probably additional ideas, as yet unspecified (and perhaps unspecifiable in a conclusive manner). I intend to leave space open for more properties that might be found to capture salient facets of self-described theoretical debate about meaning.[23]

2.6 From concept to conception

The important point to draw away from this chapter is that, regardless of whether necessary and sufficient conditions for the concept of life's meaning have been established, substantial progress has been made on obtaining clarity about what we mean when posing the question of whether life is meaningful. To use an analogy, even if a cartographer has not detailed all of a certain territory, if his map has covered much of it, progress has been made. I submit that there exists a much richer understanding of talk of 'life's meaning' to be had now, upon canvassing various suggestions about its possible necessary and sufficient conditions.

I began the chapter by noting that the previous one had established that enquiry into life's meaning is at least about a gradient final good in a human person's life that is not by definition identical to rightness or happiness. What else has been learned now? I like to think that the field should: not worry that such enquiry is non-sensical; not

---

[23] A promising fourth candidate would be about issues of pattern or narrative in a life.

reduce talk of 'life's meaning' to talk of conditions that can be experienced through the senses, on the one hand, or that are about God, on the other; be hesitant to think that all theories of life's meaning can be unified by virtue of being about a single property such as purposiveness beyond pleasure *qua* pleasure, transcendence of one's animal nature, or greatly estimable or admirable conditions. That said, more positively, the field may be said to have learned that much self-described philosophical discussion about 'meaning in life' is about the cluster of these three properties (even if not all of it is). The appeal to family resemblance is clear and precise enough to have answered Updike's questions, and to have indicated what is involved in the project of critically discussing theories of what makes a life meaningful.

I have indicated that it is worth searching for an acceptable theory of life's meaning because it would be intellectually important if it turned out to exist and because the search has only just begun (1.3). One might think that a pluralist account of the concept of life's meaning, advanced in this chapter, gives us strong reason to doubt the existence of a conception of life's meaning that is monist, sought for in Chapters 5 to 12. Admittedly, it does provide some ground for doubt. As one scholar has recently remarked, 'If there are indeed so many questions of the meaning of life, it is no wonder that no pithy formula can contain the answer to all of them' (Mawson 2010: 21).

However, if a cluster analysis or family resemblance model in particular of what talk of life's meaning connotes were true, it would not necessarily follow that pluralism is true of what this talk denotes. It could be that a single property grounds the best answer to all the variegated questions associated with enquiry into life's meaning, particularly if there are major overlaps between the questions, as I have maintained there are. At the very least, it is worth acting as though there is such a property to be discovered, for, even if there turned out not to be, we would have learned much about the substantive nature of meaning in life by searching for it.

3

The Bearer of Meaning

'Every moment is a dot in the painting of our lives.'

Kathleen Arnason ('Thank God for Selfish People')

3.1 'Life' as what is meaningful

The previous two chapters largely aimed to clarify the sense of the question of what, if anything, makes life meaningful. I maintained that this question concerns the nature of a gradient final good that can be exhibited by a human person, which final good is at least to a large degree exhausted by properties such as achieving purposes beyond one's pleasure as such, transcending one's animal self, and realizing conditions worthy of great esteem or admiration. I differentiated this evaluative and individualist issue, the exclusive focus of the book, from descriptive and holist ones regarding why there is a universe rather than nothing at all or what the point of the human race is. Focusing squarely on a certain desirable feature in our lives, in this chapter I address precisely which facets of them are capable of being meaningful or of lacking meaning.

Recall that when we say that a 'life' is meaningful in some respect we might be speaking of the life as a whole or merely a part of it (1.2). What I call 'pure whole-lifers' maintain that the only bearer of meaning is an entire life composed of certain relationships between its parts. Typically, they maintain that what can make a life meaningful is solely a function of the narrative structure among the parts, viz., a story or biography characterizing one's existence that admits of aesthetic properties of the sort Kathleen Arnason is alluding to in the quote at the beginning of the chapter. In contrast, 'pure part-lifers' maintain that the only bearer of meaning is a part of a life 'in itself', usually a spatio-temporal segment such as the fulfilment of a desire or the performance of an activity. In this chapter, I argue against both of these extreme views, defending instead an impure or mixed one; I conclude that both parts of a life and life as a whole are capable of exhibiting meaning of different kinds.

Most of those working in the field would probably, upon reflection, share this mixed view. However, as will become clear there are some contemporary pure whole-lifers whose arguments deserve attention. In addition, philosophers have overwhelmingly reflected on the part-life dimension of meaning; the whole-life dimension is hungry for attention from them, and I aim to feed it (though admittedly not to the point of satiation).

I begin by clarifying the nature of the debate between whole- and part-lifers (3.2), after which I articulate recent arguments in favour of a pure whole-life view (3.3).[1] I consider and ultimately reject two powerful and interesting lines of argument from the past decade or two for a pure whole-life view, arguments that have so far not been critically discussed. However, I also contend, in the following section, that there are strong reasons to reject a pure part-life view (3.4). I distinguish and organize the respects in which a life *qua* whole can plausibly exhibit meaning, providing a framework in which to place the scattered discussions in the literature. I close the chapter by intending to advance discussion with regard to some unanswered queries about a whole-life perspective, namely, how it relates to the part-life dimension when appraising a life's meaning *in toto*, how to understand what grounds narrative structure in a life, and how consciously and intentionally one ought to seek out such structure (3.5).

3.2 Mereology and meaning: whole, part, extent, balance

I have noted that the debate between pure whole-lifers and pure part-lifers is internal to an individualist approach to meaning in life; it is not to be confused with the difference between holism and individualism (1.2). There is an additional distinction that needs to be drawn up front, in order to avoid confusion between the present issue and a related one that I discuss later (8.6).

The present distinction concerns the *bearer* of life's meaning; that is, what it is about a life that is capable of being either meaningful or meaningless. Pure whole-lifers claim that only life as an entire period can be something that counts as 'meaningful' or not. For example, some (pure) whole-lifers claim that a life is meaningful just insofar as it is a matter of personal development over the course of one's existence (Blumenfeld 2009), and many others invoke the concept of a life-story (Velleman 1991; Fischer 2005).[2] A pure whole-life view is not merely the claim that the same part of a life can have a different significance, or different degree of it, depending on where it fits in the whole or on which whole it fits into. It is the stronger view that, strictly speaking, parts are not capable of being significant or trivial, which concepts apply only to life

---

[1] Pre-war authors articulating a whole-life standpoint, if not a pure one, include G. E. Moore (1903) and C. I. Lewis (1946).

[2] The reader might like more detail, here, but whole-lifers have not provided a very specific account of the shape that a life must take in order to be meaningful. Most simply invoke abstract ideas regarding 'narrative structure' and the like, though a notable exception is Kauppinen (2012). For (critical) discussion of the concept of narrative as applied to life, see Brännmark (2003); Vice (2003); Fischer (2005); Williams (2007); Velleman (2009: 185–206).

as a narrative or some other all-inclusive pattern; any meaning ascribed to a segment of life is sensibly done just insofar as its relationship to other parts is being considered.

Pure part-lifers, in contrast, maintain that only some subset of a person's existence can be what is meaningful or meaningless. What counts as a relevant subset has not been analysed in the literature, but that project would be worth undertaking on some occasion. Sometimes the subset is deemed to be a mere sliver of space-time, such as the satisfaction of a single desire or the promotion of a momentary experience of pleasure in another's life. However, at least for the sake of this chapter, I also consider periods of life, such as one's adolescence, to count as a part of life that is to be contrasted with life as a whole.

A different distinction concerns the *amount* of meaning in a life, regardless of *in what* the meaning inheres, a part or the whole. Sometimes we want to know whether a condition is something that makes a life *somewhat* more meaningful, more meaningful *to a certain degree*. Often I refer to such a condition as meaningful '*pro tanto*' (to an extent), the Latin phrase that is now often invoked among moral philosophers. A pure whole-lifer would maintain that only facts about life as a patterned whole can make a life to any degree more meaningful, while a pure part-lifer maintains the opposite.

The contrast of considering whether a life is only somewhat more meaningful is asking whether a life is *on balance* meaningful. The latter is to enquire as to whether a life has *enough* meaning in it to count as meaningful *period*, paralleling what one typically means in saying that a relative who has died had a 'happy life'. It might appear that such an enquiry implies a commitment to a whole-life perspective of where meaning is located, but it does not. It is open to a pure part-lifer to maintain that a person's life is on balance meaningful in virtue of the large sum of higher-order desires of hers that have been satisfied or of pleasures she has promoted in the world. Furthermore, holding a pure whole-life view does not commit one to thinking that a given life is necessarily meaningful on balance, for it could be that this life has little or none of the relevant whole-life patterns such as narrative structure.

In sum, the whole- and part-life distinction about *what bears meaning* crosscuts the *pro tanto* and on balance distinction about *how much meaning* is to be found in a life, so that the following four possibilities are logically available. First, one might be enquiring into what would make a life somewhat more meaningful, and hold the view that only parts of a life in themselves can do so. Second, one might be enquiring into what would make a life somewhat more meaningful, and hold the view that only the relationships between the parts of an entire life can do so. Third, one might be judging whether a life is meaningful on balance, while maintaining that this would be strictly a function of the aggregate of the parts of a life. Fourth, one might be judging whether a life is meaningful on balance, while maintaining that this would be entirely a function of the degree of relevant relationships to be found between the parts of an entire life.

In the following, I reject all four of these perspectives, holding that life *both* with respect to its parts and as a whole can confer meaning (to some extent and on balance). For most of the book, I set aside discussion of how much meaning a life must have in order for it to count as meaningful on balance.[3] With the rest of the field, I am primarily interested in the respects in which a life can obtain somewhat more meaning, in the present chapter ascertaining whether a life can be *pro tanto* meaningful in virtue of its parts, the relationship between its parts, or both. As stated above, I defend the latter, mixed view.

3.3 Rejecting the pure whole-life view

I pointed out earlier that if a pure whole-life view were true, it would not be so by definition of the word 'meaningful' (1.2). One is not logically contradicting oneself to suggest that a life could obtain more meaning in it not merely in virtue of the relationships between its parts, but in virtue of the parts considered in themselves. There are thoughtful and powerful arguments in the literature, though, for thinking that the pure whole-life view might be substantively correct. I believe the various considerations can be reduced to two particularly interesting rationales for the pure whole-life view. One purports to establish it in light of reflection on the differences between our rational nature, on the one hand, and the nature of animals and other non-personal entities, on the other (3.3.1). The other appeals to the way we tend to appraise lives, concluding in favour of a pure whole-life view given the putative fact that only whole-life judgements of meaning in a life can be made with confidence (3.3.2).

In constructing and evaluating these two arguments, the reader should keep in mind that I approach the literature with a certain degree of liberality, using texts merely as a means to the end of constructing a philosophically revealing defence of the pure whole-life view of meaning in life. So, I am not concerned to represent authors, or complete discussions by them, with accuracy, but instead extract ideas as I see fit to present coherent and promising arguments for certain views. Hence, even if an author rejects a *pure* whole-life view, if his text suggests reasons for taking that extreme position seriously, I draw on them. Furthermore, if an author is discussing a pure whole-life perspective outside the context of life's meaning, and rather in that of, say, well-being, morality, or intrinsic value, I apply his ideas to my subject matter.

3.3.1 Rationality is a function of unity

One major rationale for a pure whole-life account of the bearer of meaning in life appeals to salient differences between animals, which purportedly cannot have a significant existence, and persons, beings who are self-aware and can deliberate prior to acting, who can. I encounter two versions of this argument, neither of which I find convincing.

First off, some in the literature can be read as seeking to explain what feature is missing from certain beings that cannot exhibit meaning. The suggestion is that animals, as

---
[3] Addressing it only in 8.6 and at the end of 12.4.

well as other entities in the vegetable and mineral kingdoms, are not the kinds of things that can have a meaningful—or even a meaningless—existence. 'We may be inclined to see, for example, a spider's life as meaningless because it cannot be meaningful, but strictly speaking it is neither meaningful nor meaningless; it is simply not subject to this kind of evaluation' (Wong 2008: 131). Supposing that is true, what is it that the spider lacks such that it cannot bear meaning? According to some whole-lifers, it is the capacity for viewing one's life as a whole (Velleman 1991), for 'narrative structure' in one's life (Brännmark 2003: 63, 76), or for a 'biographical life' (Wong 2008: 131). The pure whole-lifer might be tempted to suggest that it is *only* this sophisticated capacity that best explains why human persons can have meaningful lives and why spiders, for example, cannot.

However, the latter, strong claim is open to question. It is true that human persons are able to adopt a point of view that includes their lives as a whole and that, so far as we know, no other being on the planet can. And it is plausible to think that this capacity, at least in combination with a kind of free will (Fischer 2005: 388-90), is essential for a being's existence to be capable of any kind of meaning. Human beings who lack the concept of their lives as a whole, or even those who have it but who cannot remember more than five minutes of their past at a time (as in the film *Memento*), might well be beyond the pale with regard to the possibility of exhibiting meaning in their lives (or at least a meaningful life on balance). To put my objection tersely, however, a necessary condition need not be a constitutive condition. Even if the ability to live a certain life-story were necessary for meaning of any sort (which I grant, here, for the sake of argument), it would not necessarily follow that meaning inheres only in the use of that ability. I now suggest that this ability is necessary for some *other* ability that is also (or instead) the bearer of meaning in life.

I ultimately contend that the sophisticated contouring of one's rational faculties towards fundamental conditions of human life is the major property by virtue of which a person's life is meaningful (Chapter 12). Focus, for now, simply on the idea that one's existence might be more significant, the more it exercises intelligence in certain ways, including by creating works of art, advancing intellectual reflection, and going beyond the call of duty in the moral sphere. Now, there is nothing inherently whole-life oriented about these actions; in principle, one could do them for a limited period, and then go and eat doughnuts in a jacuzzi for the rest of one's life. Nonetheless, in order for an individual to engage in these kinds of activities during a part of her life to a substantial degree, it is reasonable to think that she must be 'a being who refers to a life as "my life"' (Wong 2008: 131) and who can act from that perspective. One reason for thinking so is that, without being able to act in light of a comprehensive standpoint on one's existence, one would fail to make sensible strategic calculations about how to use one's limited time and other resources, and hence would not be able to effectively direct one's rational faculties in the relevant sort of part-life ways.

I conclude, then, that even supposing for the sake of argument that the ability to exhibit a life-story is necessary for any sort of meaning in a being's life, we need not be committed to the view that the only bearer of meaning in life is the exercise of that ability. Indeed, there is some reason to think that the ability to exhibit a life-story is necessary for meaning in life (at least in part), because it makes possible the effective realization of some other, part-life feature, such as using one's intelligence in certain ways.

Consider, now, the second version of the present rationale for a pure whole-life view, appealing to features of human beings not found in animal life. Pedro Tabensky has most clearly expressed it in a thoughtful article titled 'Parallels between Living and Painting' (2003). Adopting an Aristotelian (and Kantian) angle, Tabensky maintains that what crucially differentiates us from animals is our rational nature. We are capable of meaning in our lives, for him, insofar as we are rational, and animals are incapable of it insofar as they are not. So far, there is nothing to differentiate Tabensky's view from the part-life view I proposed in the previous two paragraphs. However, Tabensky maintains that once we agree that meaning is a function of our rationality, we are then committed to a pure whole-life view as to what it is about our lives that is capable of exhibiting meaning.

Tabensky's argument starts from the key premise that what it is to be rational is to seek coherence, system, or organization. '(W)hat ultimately defines our rational existence is our directedness toward the ideal of unity' (Tabensky 2003: 62). This appears plausible, when considering salient features of theoretical reason. For one, the most patent form of irrationality is logical contradiction, and, for another, the highest form of rationality arguably is, in Immanuel Kant's terminology, a matter of conducting a 'regress on the conditioned in search of the unconditioned' (1787). For instance, we are thoroughly engaging our rationality when we seek a cause$_1$ for an event, and a cause$_2$ for that cause$_1$, and a cause$_3$ for cause$_2$, and so on in search of an ultimate, initial cause that is itself uncaused. If theoretical reason seeks unity, then it is reasonable to think that practical reason does, too.

Supposing, then, that a (practically) rational life is a unified life, and supposing further that a unified life is one oriented towards a basic goal (parallel to a first cause), and supposing still further that a meaningful life is a rational life, then a meaningful life is one oriented towards a basic goal. From this reasoning, Tabensky draws a pure whole-life conclusion, and does so using moral language, which could be sensibly replaced with meaning-talk:

If life has a goal, then individual ethical acts cannot properly be understood in isolation from the specific niche they occupy within the teleological system that is constitutive of the life of a person. An action could not properly be understood as morally praiseworthy or blameworthy in isolation from an understanding of the overall shape of a human life. Ethical practices cannot be understood in isolation from the totality they ideally constitute (2003: 67).

Illustrating this claim in the context of putative examples of part-life meaningfulness, Tabensky would deny that making an artwork, discovering a law of nature or performing a supererogatory action are meaningful in themselves; if they are meaningful, they are so only insofar as they contribute towards a life that is thoroughly directed towards exhibiting a certain state of affairs that is to be valued for its own sake.

So far as I understand Tabensky's argument in standard argumentative form, it is this: (1) A life is meaningful only to the extent that it is practically rational (or, there is strong reason for a person to pursue meaning in her life). (2) To be practically rational means nothing but to seek unity in a life (or, for there to be strong reason for a person to do something is for the action to be expected to contribute to her life forming a unity). (3) A unified life is nothing more than one that aims to realize a 'single unifying purpose' (Tabensky 2003: 62). (4) Aiming to realize a single purpose is just for one's actions and other parts of one's life to be maximally coordinated so as to realize that end. (5) Therefore, a life is meaningful only to the extent that its actions and other parts are coordinated towards the realization of a single purpose.

The conclusion, (5), is an instance of a pure whole-life view in that what bears meaning is merely a life insofar as its parts jibe together so as to maximize the expected realization of a certain end. If a certain part of a life failed to cohere with the other parts, in the sense of not contributing towards the realization of this ultimate goal, then, Tabensky suggests, 'the overall unity of the composition of life' would be disrupted 'in a manner that is analogous to disruptions in a musical piece by a strident sound, or disruptions in the pictorial harmony of a given piece of art' (2003: 62).

Premises (1) and (4) seem unassailable, and the inference to (5) from the premises as a set seems valid. All that leaves, then, are premises (2) and (3). I first question (3), the kind of unity that is held to constitute rationality, but take my next, and more important, criticism to be of (2), Tabensky's claim that rationality is exhausted by unity.

I doubt that, to count as rational, a person's life must be unified in the sense of being oriented towards a single end, even one that is fairly inclusive. Taking a cue from the parallel of theoretical reason, suppose that one were to 'regress' on causes in search of a single first cause, but that one could never find one, perhaps because one simply does not exist. Suppose either that there is no first cause at all, viz., that there is an infinite chain of causes, or that there are many first causes. If there were no single first cause, there would be no lack of rationality in failing to find one, so long as one regressed on causal conditions in as thorough a way as the subject matter permits. By analogy, in the practical sphere, if there were a single purpose to guide a human person's life, then a rational life might be one that is utterly oriented towards realizing it. However, if there were no single purpose, but rather a variety of different final values, then there would be no lack of rationality, and hence of meaning, in failing to systematize our lives in the way that Tabensky seems to think is essential.

Without going into detail about my favoured theory of meaning in life, I present *prima facie* reason to think that, supposing certain uses of our intelligence are the key to meaning, pluralism about our final ends is plausible. Recall the exemplars of meaning, viz., the true, the good, and the beautiful, that is, the sophisticated exercise of reason in the three realms of the intellectual, the ethical, and the aesthetic (1.2). There is no reason, on the face of it, to think that these three types of intelligence can be unified in a determinate manner such that one's life is meaningful only insofar as a given part of it coheres with other parts so as to maximally realize a single end. Instead, it is just as reasonable to hold that there is a part-life dimension to meaning, where the more one exhibits the true, the good, or the beautiful in one's life, the more meaning there is, *ceteris paribus*.

Tabensky is aware that the weakest part of his argument is the claim that rationality requires unity of the sort he envisions. But his response, as follows, is unconvincing:

> Someone could think of a life as a set of conglomerates, each of which constitutes a system of practices, but which, taken as a set, do not form one unified system. This line of argument is tempting only if we do not consider the fact that what defines us as rational subjects is that, in an important sense, each of us are one. Our engagements are played off each other even if only because our energy reserves are limited, and engaging in one practice has repercussions with regard to the time and energy we can invest in other practices (2003: 62).

This line of argument does provide reason to believe that, to be rational, we must be capable of taking a viewpoint beyond the present moment, and perhaps one that includes our lives as a whole, as discussed in the context of the prior rationale for a pure whole-life view. However, it does not follow that rationality is merely a matter of organizing all the parts of one's life so that they cohere so as to maximally promote a single end.

In reply, Tabensky could accept the idea that meaning in life is not exhausted by coordinating its parts so as to bring about one goal, but retain a pure whole-life view. As an alternative, he might plausibly suggest that a life is meaningful just insofar as its parts are coordinated in order to bring about several ends, ranked in a certain (ideally, cardinal) way. Hence, in order to question the present argument for a pure whole-life view, I must focus on the crucial claim that rationality in a life is nothing more than unity in it.

It is reasonable to think that rationality is to some degree formal in the way Tabensky conceives.[4] I accept that just as one is more theoretically rational, the more one systematizes one's beliefs, so one is more practically rational, the more one's goals form a system. However, there is an additional, substantive dimension to rationality, such that the nature of one's goals admits of being reasonable or unreasonable. It is fair to think that the content of ends is a different dimension by which they admit of rational appraisal, one logically distinct from the way in which ends are organized in relation to one another. For instance, I have suggested that at least some ends are meaningful, and give us good reason to pursue, merely insofar as they involve the exercise of intelligence with regard to the good, the true, and the beautiful. It would follow, then, that insofar as a life is rational, it, at least to some degree, realizes ends with a certain content, independent of how these ends bear on the realization of other ends.

---

[4] See also Hurka (1993) for a thoughtful formal account of rationality.

The outcome of the discussion so far is that, even if we suppose that only human persons are capable of having an existence that matters, we need not hold the view that their existence matters solely in virtue of it as a whole. Although it is plausible to think that exercising our capacity for rationality in some way constitutes meaningfulness, I do not think the present arguments have succeeded in showing that rationality is exhausted by unity of the pure whole-life sort.

3.3.2 Only whole-life judgements are definite

Another recurrent theme in the writings of those who favour a whole-life account, whether pure or not, is that we can know something about the nature of what bears meaning in life from the way we make judgements about it. More specifically, one encounters the idea that we are most confident of the truth of claims about what confers meaning on life when and only when we are making whole-life judgements. The best explanation of why we are most sure of whole-life appraisals of meaning in life, so the argument goes, is that meaning is nothing but the way a whole-life is patterned.

I start with the weakest version of this argument, which includes the claim that, in practice, we simply do not make any appraisals of the parts of a life in themselves. It seems as though one of the most prominent contemporary whole-life theorists, Johan Brännmark, is making this suggestion:

> When, in everyday life, we judge the quality of people's lives we hardly do so by seeing lives as something simply scooping up heaps of units; rather, what we do is to look at a given life as a whole, thus judging the prudential goodness of the life in question directly, not via an independent estimation of the goodness of its parts (2003: 67).

Along these lines, Brännmark draws an analogy between a novel and a life, suggesting that 'when we try to explicate the goodness of a good book, we hardly partition this goodness into discrete parts that we accord to different parts of the book' (2003: 65; see also Brännmark 2001: 222). Instead, as I read Brännmark, just as we judge a novel solely in light of how the parts relate to one another to form a narrative, so we judge lives in the same way.[5]

The glaring problem with this rationale for a pure whole-life view is that it is false to say that laypeople and philosophers never or even rarely make judgements of mere parts, whether they are of lives or of works of art. With respect to novels, it is not unusual to assign merely a part of a book in a course, say, the discussion with the Grand Inquisitor in Fyodor Dostoyevsky's *Brothers Karamazov*. Considering music, it

---

[5] In correspondence, Brännmark disavows the sense I am imputing to these passages; he says that he intends only to say that the *amount* of value in a part cannot be conclusively specified in isolation from other parts. (The latter claim is one that I, a defender of the mixed view, can accept, for I contend that parts have meaning both in themselves and in relation to other parts, requiring an assessment of the latter in order to take all relevant factors into account and thereby be able to specify their degree of value all things considered.) I am not aiming to recount his or anyone's view with accuracy, and am rather culling out what I need in order to construct plausible arguments for a pure whole-life view of what bears meaning.

is common for people to listen merely to a certain part of a symphony, indeed one from its middle; for example, I judge the second movement of Henri Gorecki's *Third Symphony* to be outstanding in itself. And even in the context of painting, one finds oneself attending to a restricted location because of, say, the particular shade of colour or the spatial form of the lines. I naturally do not mean to deny that we also judge works of art *qua* wholes. What I question is the claim that we judge them *only* in those terms, that when we evaluate a part we do so *solely* in relation to all the other parts.

I submit that similar claims are true of lives. It is routine for us to appraise parts of a life without thinking about how they relate to all the other parts. Consider the case of a period of engaging in prostitution to feed a drug habit. Think as well about the following cases that are ubiquitous in the literature: torturing animals for fun; being tortured for the fun it; having been taken in by a charismatic cult leader; spending evenings drinking beer and watching sit-coms; living in an experience machine; being forced to dig a ditch and to fill it up again, only then to be forced to dig a new ditch and to fill it up again; being so depressed as to be unable to get out of bed and to relate to anyone or to do anything. Most of us, without knowing anything further, are readily inclined to deem these episodes to be meaningless periods of a life. Or at the very least we quickly judge them to be parts that, considered outside the context of the pattern of the life as a whole, are much less capable of conferring meaning relative to other parts, also considered outside the context of the pattern of the life as a whole. Parallel remarks go for intuitively meaningful conditions.

The most interesting reply to make to these and related counterexamples is to hold that, even though we might appraise parts of a life considered in themselves, such appraisals are merely *provisional*. If we cannot draw a *firm* conclusion about whether parts of a life have conferred meaning on it without ultimately judging them in relation to its other parts, then, so the argument goes, it is probably the whole-life alone that bears the meaning.

I now critically examine three reasons from the literature for finding this reply attractive. One consideration is that what might initially appear to be meaningless periods in a life can be redeemed and thereby made meaningful. The theme of redemption runs throughout most whole-life discussions, but in his magisterial book, *Sources of the Self*, Charles Taylor appears to explicitly invoke it to support a pure whole-life view.

We want our lives to have meaning, or weight, or substance, or to grow towards some fullness...But this means our *whole* lives. If necessary, we want the future to 'redeem' the past, to make it part of a life story which has sense or purpose, to take it up in a meaningful unity.... To repudiate my childhood as unredeemable in this sense is to accept a kind of mutilation as a person; it is to fail to meet the full challenge involved in making sense of my life (1989: 50-1).

Returning to the putatively meaningless conditions adumbrated above, Taylor would apparently suggest that they could in fact become meaningful if they were to lead to something good. So, if the woman who hired out her sexual organs in order to support a cocaine addiction were eventually to leave that period of her life behind and become a therapist helping people with similar problems, she might look back and be in some sense glad that she underwent what she did, insofar as it was instrumental (and perhaps necessary) for her to become who she is now.

I have two responses to this argument. First, a period of life can have some meaning in virtue of its instrumental value, i.e., insofar as it leads to something desirable later on in the life, and yet still be lacking an important kind of meaning, when considered apart from that relational property. Although the earlier part of a life can be redeemed by a later part, a sensible way of understanding that is to think that the earlier part is not *utterly* meaningless because of that relation, but still seriously lacking a certain amount of meaning when considered in itself.[6]

Second, and more deeply, part of the best explanation of why the therapist is glad that the prior segment in her life led to her being a therapist is that the latter in itself is meaningful and the prior segment in itself was not. What the prostitute-*cum*-therapist would tend to think about her life is that she is glad that she went through a bad (meaningless) period in her life only insofar as it led to something good (meaningful); that is, she would continue to describe the time of being a coked-up hooker as 'bad', or, in present terms, 'meaningless' in some major respect. So, my suggestion is that it is *only because we make some conclusive judgements about whether parts of a life are meaningful or not* that we can make the kind of whole-life judgement Taylor does! To judge a life in terms of redemption implies viewing an earlier part of a life to have been lacking in meaning, considered apart from its relation to the later part; the appeal to redemption, I submit, backfires against the pure whole-life theorist.

Taylor suggests a second reason for thinking that localized judgements of meaning are merely provisional, subject to revision upon consideration of their place in the whole of a life. According to him, in order to grasp the nature of an experience or an event, one cannot remain focused on it alone.

Of course, there are experiences in which...we sense for a minute the incredible fulness and intense meaning of life....But there is always an issue of what to make of these instants, how much illusion or mere 'tripping' is involved in them, how genuinely they reflect real growth or goodness. We can only answer this kind of question by seeing how they fit into our surrounding life, that is, what part they play in a narrative of this life (1989: 48; see also Thomson 2003: 141-3, 150).

If we can confirm that a spatio-temporal period confers meaning on a life only once we have considered it in relation to all its other parts, then it appears that it is ultimately the relationship between all the parts of a life that bears the meaning.

Taylor might be correct that in order to firmly grasp the reality of a part of a life, one must consider, for instance, how it arose and what consequences it has had. However, it does not follow from this point that it is only these kinds of relationships that are capable of being meaningful or meaningless. Consider that in order to know for sure

---

[6] Cf. the interesting discussion of 'Manicheanism' in Adams (2006).

that one is making a scientific discovery, one does need to know a lot of facts 'external' to that specific event. One needs to know that one is aware of all the relevant literature, that the equipment in the lab was built correctly and functioning as intended, that others have replicated the experiment. Having *evidence* that one has made a discovery is, of course, different from having *made the discovery*. My point is that even if, in order to have evidence of having made a discovery one needs to appeal to many factors distinct from the discovery itself, that fact provides little or no reason to deny the claim that the discovery itself was meaningful.

In the literature, one encounters a third version of the rationale that only a whole-life can bear meaning, since judgements of its parts are merely provisional. Jerrold Levinson, who expressly accepts a pure whole-life view of final value, argues that it follows from the purported fact that the value of a given part of life is invariably liable to change, depending on what context it is in. '(T)he problem with anything narrower than a life—such as an object, an experience, a pleasure—serving as a locus of intrinsic value is that the perceived values of all such things are liable to overturning once we connect them to their situational contexts' (2004: 323).[7]

In support of this claim, Levinson presents several cases in which we are initially inclined to think that a spatio-temporal segment is desirable, but then rightly change our minds upon becoming aware of its relationship to other parts of a life. For example, many believe that one's life matters more for taking pleasure in certain pieces of music, but not when doing so immediately upon the death of one's spouse. And many believe the experience of climbing Mt Everest would be worth having, but not if it were merely the product of an 'experience machine'. And it appears reasonable to think that being healthy is something to be valued for its own sake, but not if the only way to avoid an otherwise inescapable, torturous fate would be to have a heart attack. From such cases, Levinson concludes that particulars such as 'the temporal parts of a life have at most *prima facie* intrinsic value' (2004: 327); that is, any part considered in itself can be judged to be valuable merely *on the face of it*. And from the fact that firm judgements of value, ones not subject to revision, are made only in the large context of a life as a whole, Levinson concludes that it is only life as a whole that can be good for its own sake, or, for my purposes, meaningful.

The first way to question Levinson concerns the import of the particular cases he presents. It is correct that one would have most reason to desire that one were not healthy if health prevented one from avoiding a fate worse than death, but it is a leap to infer that health in this case would lose all its final value, as opposed to the more modest claim that the instrumental disvalue of health in this case would outweigh its final value. With respect to the feeling of climbing Mt Everest, it is true that it would

---

[7] Levinson (in correspondence) is reluctant to say that his account of intrinsic (in my terms, 'final') value applies to considerations of meaning in life. However, if I am right that the phrase 'meaning in life' by definition connotes something desirable for its own sake (1.2), then the claim that the only bearer of final value is life as a whole applies *ipso facto* to life's meaning.

be better to feel what it is like to climb Mt Everest while actually doing so than merely to feel what it is like without doing so. But it is a leap to infer that any particular part of a life is meaningful only in virtue of its relations to the other parts; the natural thing to say, after all, is that the part of a life in which one climbs Mt Everest and experiences doing so would be better (have a greater final value) than the part of a life in which one merely has such a feeling from an experience machine. And while some might think that there is nothing remiss in taking pleasure in music right after the death of one's spouse (Reid 2009: 48), even if there were, it would be a leap to infer that a whole-life, something far beyond a 'slice of life', is the only bearer of final value (meaning).

A second way to cast doubt on Levinson's argument is to invoke cases where it appears that, without any further information about a part of a life, one can draw a firm conclusion about its final value, or, for my aims, meaningfulness. Another critic of Levinson points out that awareness of someone's actions such as making great sacrifices to help the sick and poor, as Mother Teresa is understood to have done, is enough to judge firmly that they made her life somewhat more significant, or that knowing that somebody has discovered a cure for cancer would be enough to conclude with confidence that this conferred *pro tanto* meaning on his life (Reid 2009: 49-50). Similarly, regardless of what is antecedent or consequent to the action of torturing a baby for fun, one is able to infer strongly that it has reduced the meaning in one's life.

Furthermore, if there are indeed many cases in which we are able to make firm judgements that a life is *pro tanto* meaningful simply in virtue of a part of it, then we have reason not merely to reject the present argument for a pure whole-life view, but a pure whole-life view itself.[8] The more it appears that actions such as discovering a cure for cancer would make one's existence *pro tanto* more significant, regardless of the overall shape of the life of which it is a part, the more reason there is to think that pure whole-lifers are incorrect. However, it does not follow that a pure part-life view is true, and in the next section I bring out some reasons to reject it as well.

3.4 Rejecting the pure part-life view

In this section, I adumbrate and organize the reasons for thinking that one's existence can be somewhat more significant because of its overall shape. My aims are mainly to distinguish and order the various plausible reasons for thinking that one's existence can be *pro tanto* more significant because of its pattern, beginning with the thinnest idea and ending with the most rich. I do mention, and indicate how to rebut, the general strategy that one would have to invoke in order to cast doubt on the following whole-life claims, but do not aim to do more than that when it comes to defending a mixed view over a pure part-life one.

---

[8] For additional cases, see 4.4.6.

In logical order, beginning with the least controversial and the simplest idea, there is a concern to avoid repetition in a life. Even if the parts of a very repetitive life were quite meaningful in themselves, most would sacrifice some meaning in the parts in order to avoid repetition in the pattern and thereby enhance the importance of the whole (see especially Taylor 1987; Brännmark 2003: 72-5; Blumenfeld 2009). Think of the movie *Groundhog Day*, in which Bill Murray's character relives the same twenty-four-hour period. At a certain point, Murray learns from his mistakes and becomes virtuous. He then rescues the same boy from falling out of a tree, day after day. While each of the rescues confers meaning in itself, viewers are, with Murray's character, relieved once the reliving is over. So, the first step in articulating a mixed theory, with whole-life elements, is the idea of a life that has contrasting elements in it.

Second is the idea that it would be more meaningful for a life, with different kinds of parts, to end on a high note than to have started out good and then declined (Slote 1983: 9-37; Kamm 2003: 221-33). This is one of the most common intuitions motivating the rejection of a pure part-life view. Holding constant the sum of part-life facets, it is, *ceteris paribus*, more meaningful for one's life to get better over time than to get worse.

Third, some maintain that, supposing one's life has better and worse parts and the better parts come later, it would be more meaningful for the comparatively worse parts of a life to have *brought about* the better ones. This is one aspect of the theme of redemption that is so common in whole-life discussions, the thought being that having made mistakes and learned from them, but, say, not having had time to effect a real difference, would be desirable insofar as the bad has *caused* the—albeit limited—good, instead of the good having been caused by something else.[9] Of course, the greater the success that eventually results from failure, the better. On some views, a lasting or substantial good consists of a single purpose (as per a natural reading of Tabensky 2003), but others suggest that a life would be more rich if its bad parts caused a variety of goods that together form an interdependent complexity (Kauppinen 2012). To continue the example, the pattern of one's life would be all the better if one not merely learned from one's mistakes, but also put the learning to good use (Kamm 2003: 222-5; Kauppinen 2012). And, still more, many in the field believe that posthumous influence would confer meaning on one's life. Many of us seek to make ripples from the splash of our lives that would continue once we have gone under. Sundry ripples might be children, books, paintings, tombstones, buildings, or memories. Better that 5000 people benefit from and recognize one's accomplishments now and another 5000 also do so in the next generation than that 10,000 do so now but none does so posthumously. Or so I presume the reader will agree.

---

[9] Harman (1976: 461-3); Velleman (1991); Hurka (1993: 110-11, 119, 121); Kamm (2003: 222-5). See, too, discussions of the respects in which an instrumental relationship can have final value in Nozick (1993: 22, 136) and Kagan (1998).

Fourth, one finds the view that, supposing the worse parts of one's life have caused better and lasting parts towards its end, it would be more meaningful for the latter parts to have been caused in a specific way, say, either by a process of personal growth (Blumenfeld 2009) or in a way that makes for a good autobiography to read (Velleman 1991; Fischer 2005; Wong 2008). To return to the prostitute-*cum*-therapist, it would be more meaningful if it were the fact that this person had been a prostitute that enabled her to be a good therapist to prostitutes, as opposed to some other bad period of her life that had caused this desirable outcome.

Fifth, and finally, there is the idea that the best life-story would be an original or otherwise creative one, not merely an accidental repeat, let alone an intentional copy, of someone else's (cf. Taylor 1987).

It is worth asking whether there are more gradations to add to this schema. For now, though, note that the framework brings out several distinct respects in which a pure part-life view is probably false. Various kinds of relationships between parts of a life do seem relevant to appraising its meaning. The most salient relationships include the absence of similarity in the content of the parts, the temporal order in which parts come, the fact that some parts cause other parts, and the specific types of causation between them.

Pure part-lifers must reply that our judgements about these relational features are confused, such that when we judge there to be more meaning for these reasons, what is actually motivating us is the implicit supposition that there would be a greater sum of meaningful parts. So, a pure part-lifer would have to maintain that we are drawn to judge a life that ends on a high note to be more meaningful only because we are at some level thinking that its aggregate of meaningful parts would be larger than in a life that does not. However, keeping clearly in mind, for the sake of argument, that the sum of meaningful parts is the same in both cases, I am inclined, presumably with most readers, to find more meaning where the parts are related to each other in some ways rather than others.

3.5 Developing the mixed view

I conclude, then, in favour of an impure or mixed view. What matters in life are both its parts and their overall relationship to one another. There arise the questions of how these two dimensions are to be weighed against each other and of how they affect one another. I find myself at this point unable to suggest any precise principles, but can provide some rough suggestions.

First, it seems clear that there is no 'lexical' priority involved here, such that any amount of one dimension, no matter how small, would outweigh the other dimension, no matter how large. The two dimensions seem comparable in importance. And given that, one piece of reasonable advice is to avoid extreme lows with respect to either dimension. So, for example, if necessary and sufficient to avoid an utterly repetitive whole-life, one should be willing to forgo some of the part-life dimension of meaning, even if the parts were quite meaningful in themselves. To invoke *Groundhog Day* again, even though Bill Murray's character will no longer prevent a boy from falling to the ground once he no longer repeats the same twenty-four-hour period, his life is probably more meaningful all things considered for no longer repeating. Parallel remarks go for a life that is exhibiting a terrific pattern, but includes relatively little meaning with regard to the parts in themselves. Perhaps the character in the movie *Seven*, who inflicts appalling kinds of harm on people to call attention to the seven deadly sins, exhibits an original, fascinating, and gripping life-story, but one that should not have been authored, given how meaningless in themselves the parts were.[10]

Another bit of counsel that I can provide would be to take advantage of opportunities to 'max out' on a certain dimension. So, for example, one should be willing to give up some incline with respect to one's whole-life, if it would involve a meaningful 'peak', considered as an isolated part. If one could discover that cure for cancer in one's late 20s, one has most reason of meaning to do it, rather than to do something less substantial that would afford an overall greater upward slope. Similarly, the closer one gets to the higher end of the whole-life schema, the more one should be willing to forgo some meaning in the parts, in order to take advantage of the opportunity to perfect that dimension.

3.5.1 The ground of narrative structure

The whole-life dimension is under-developed in the field; there are a host of additional issues worth addressing to enrich the analysis of whole-life meaning in life, but I elect to focus on two in the rest of this chapter. In this section, I take up the as yet unthematized and, upon reflection, tricky issue of what it is about a life that constitutes a meaning-conferring pattern. To see the problem, consider the most straightforward proposal about what grounds narrative structure in a human person's life:

(GNS_1) A narrative structure is constituted by every spatio-temporal moment of one's life.

If (GNS_1) were true, no life-story would be a good read, or would otherwise exhibit the kind of coherence that is characteristic of a narrative structure. It would have to include daily mention of eight hours of sleep. A good third of the hours spent every day are not only terribly boring, but also fairly constant over the course of one's life. Imagine a novel a third of which were pages with 'zzzzzzzz' on them, perhaps generously peppered with 'snore'; the whole would be marred.

What this example shows is that '*whole*-lifers' are not to be taken literally. A tempting response to make to the problem would be to focus strictly on the 'internal'

---

[10] This last example raises the thorny issue (which I am not yet in a position to resolve) of whether, for someone who accepts a whole-life dimension to meaning, the relevant patterns can obtain independently of the content of the parts. Here, theorists differ, with some judging Hitler's life to have been meaningful in virtue of the autobiography of which he was capable of writing (Wong 2008: 141–2), and others claiming that narrative structure must have the 'relevant contents' in order to be exhibited (Brännmark 2001: 221).

perspective of the agent's life; that is, perhaps what is relevant to a person's autobiography is as follows:

(GNS$_2$) A narrative structure is constituted only by every spatio-temporal moment of which one is aware.

According to (GNS$_2$), one's life-story is solely a function of one's point of view; i.e., of what has entered one's consciousness during one's four score and ten or so. But this will obviously not suffice, since there is plenty of conscious 'dead time', by which I mean periods akin to sleep, e.g., actions such as bathing, voiding one's bowels, brushing one's teeth, walking to the car, choosing a deodorant, picking up dropped items, eating chocolate, waiting in line, daydreaming, etc. As a percentage of one's waking life, the amount these take up is large.

Another natural option, then, is to think of narrative structure this way:

(GNS$_3$) A narrative structure is constituted by every spatio-temporal moment of one's life of which one is aware except for the dead time in it.

How generally to characterize dead time without begging questions is difficult, but a promising way to do so would be to appeal to the concept of meaning I articulated in the previous chapter (2.5). Suppose that dead time is all those conditions of a life that are not *prima facie* candidates for being described as transcending one's animal nature or being worthy of great esteem or admiration.[11] GNS$_3$ is the view that one's life-story is a function of everything in one's conscious life that is, in contrast, fairly described in those ways.

This kind of proposal avoids the problems with the previous two suggestions, but it is too generous, as it stands. Suppose someone elects to spend 99 per cent of her life engaged in dead time, with the remaining one per cent utterly devoted to activities that count as transcendent or estimable when ordered in the right ways. It would follow that her life could be maximally meaningful with regard to the pattern, but that is surely counterintuitive. A substantial portion of one's life must be patterned, viz., one must strive to minimize the amount of dead time in it, so as to live a superlatively meaningful life with regard to its narrative structure.

What might be suggested, then, is a fourth idea:

(GNS$_4$) A narrative structure is constituted by every spatio-temporal moment of one's life of which one is aware except for the unavoidable dead time in it.

This principle evades the objection to the previous one, in that it would straightforwardly count dead time that could have been avoided against the amount of meaning conferred by the pattern of her life, while not counting periods such as sleep and the like. However, it faces the serious objection that a long-term coma could not

---

[11] Which is not to say that dead time is necessarily what I call 'anti-matter' (4.3); it could have a score of zero, rather than a negative number.

affect the pattern of one's life, if one could do nothing to avoid being in one. Surely, though, being in a coma for 25 years would constitute a grave disruption to the unity of one's life.

Therefore, a fifth idea is not to ground narrative structure in the 'live time' of a given individual's whole-life, but rather that of the average of the species. There is a given amount of sleep and other dead time that is typical of the species, and maybe narrative structure should be understood such that:

> (GNS_5) A narrative structure is constituted only by those spatio-temporal moments of one's life of which one is aware beyond the dead time that is average for human beings.

When determining the extent to which a person's life is patterned in the right way, then, one would subtract from her life the dead time in it up to the average that is experienced by the species, and see what pattern she has made of her life with what remains. If she has more dead time than is typical for a human person, say, in virtue of being comatose, then her life-story would be negatively affected, thereby avoiding the problem facing the previous proposal.

Something like (GNS_5) is getting closer to the mark, and I do not seek to improve on it here. What I do is to present counterexamples to it that must be addressed, if a reader elects to take this issue farther. Most urgently, note that it is widely accepted among whole-lifers that the story of one's life includes 'external' factors, i.e., is larger than the facets of which one is, or even could be, conscious. For stock examples, imagine that you were in *The Matrix* right now, or consider that your spouse has betrayed you for many years without your knowledge. A large majority of those who accept a whole-life dimension to meaning maintain that these elements, utterly beyond the purview of the agent's awareness, would form part of the story of her life.

Second, consider that it is also commonly held that posthumous effects of one's life can be part of its narrative structure. People typically think that their existence would be more significant if, upon their deaths, their contributions continued to be recognized and to have some positive influence, made 'ripples'. However, (GNS_5), along with all the previous accounts of the ground of narrative structure, focus only on facets of a person's life insofar as she is alive.

For a third consideration, which pulls in a direction opposite to the first two, imagine that the present generation of human beings were all of a sudden thrown into a temporal freeze; that is, suppose that everyone in the present generation were frozen in time for 50 years, and then came back to normal, without any awareness of what had happened. By (GNS_5), none of these people could have much meaning in their lives by virtue of narrative structure, given that all of them would have much more dead time in their lives than average. However, my intuition is that, despite the massive amount of dead time, narrative structure would not be impaired, or at least not gravely. There appears to be some important difference between the coma, *Matrix*, affair, and ripples cases, on the one hand, and the present, freeze case, on the other. Given that all these scenarios involve aspects of a life beyond the conscious purview of the agent, how do they differ?

I leave this puzzle, now, to deal with another one, regarding the kinds of cognition, intention, and motivation one ought to have with regard to patterning one's life, whatever the relevant ground of that might turn out to be.

3.5.2 *The agent's orientation towards narrative structure*

One objection to the idea that life can be meaningful in virtue of its narrative structure is to note the kinds of undesirable attitudes towards oneself that would seem to be concomitant with awareness of it. For example, some suggest that seeking to impart to one's life the kind of unity that would make one akin to a character in a novel would require a kind of self-consciousness and artificiality inconsistent with meaningfulness (Becker 1992: 32; Vice 2003). One would have to be too aware of oneself and too ready to weed out large parts of one's life to achieve the relevant unity that a character exhibits. And, as the anti-pure-part-lifer Brännmark himself has recognized, such an overly reflective, contrived life would be undesirable on *aesthetic* grounds, a large part of the motivation for thinking that life includes a whole-life dimension (2003: 76).[12]

Now, there is an obvious reply to make to these worries, which involves invoking a distinction familiar to moral philosophers between a theory of X, on the one hand, and a decision procedure about how to obtain X, on the other. The suggestion would be that there is a difference between asking what is in fact meaningful and asking how one should make decisions so as to accrue meaning, where the answer to the latter need not involve consciously aiming at the former. If what is in fact meaningful is, at least in part, narrative structure, it does not follow that one must have it in mind in order to obtain it. Just as people are typically less happy when they try to be happy, or so the 'paradox of hedonism' maintains, so people would typically have more narrative structure in their lives if it were not salient when deciding how to live.

That is the general form of the reply, and in principle I think it is fine. The devil is in the detail of how to flesh it out in a plausible manner. Whole-lifers have made two suggestions about how to do so, one appealing to a lack of *conscious* aiming and the other to a lack of conscious *aiming*, neither of which is compelling, as it stands. With regard to the latter, Frances Kamm claims that narrative structure or some other whole-life feature ought never to be one's reason for action. She says (2003: 222):

I believe that where in a life story some event occurs can be important because the pattern of one's life can be important. (This pattern, however, is something that should come about because of what one does for reasons other than trying to achieve a pattern.)

Kamm is claiming that in order for a pattern to confer meaning on one's life, one ought not intend to realize it for its own sake. Note that she is not denying that agents should be aware of the patterns their lives are forming or could form.

---

[12] Lawrence Becker makes an objection that is ostensibly distinct from the concern about artificiality, but does so merely by posing this rhetorical question: 'Shall we say, "Go make a life like Iago's"?' (1992: 32).

Her claim, as I read it, is that even if one is aware of the potential for pattern in one's life, the pattern ought not be the ultimate motivation for the action; there should be some other consideration that is at bottom driving one's interest in the pattern.

Two objections apply to Kamm's suggestion about how an agent ought to orient herself towards prospective whole-life features. First, Kamm's proposal, at best, would resolve the problem of artificiality, but not the problem of self-absorption. As I have noted, not being moved by considerations of pattern does not imply anything about whether one has the pattern in mind or not. One could avoid ever acting for the sake of a pattern while nonetheless being constantly aware of whether one's life is realizing it.

In addition to not being sufficient to respond to all the critical considerations, it is not even clear that Kamm's suggestion is necessary. There appear to be times when it is apt for a person to choose one path rather than another for reasons of narrative structure. Returning to the case of *Groundhog Day*, if Bill Murray's character were given the choice of returning to normal life or having his life continue to repeat every 24 hours, it would be sensible for him to elect the normal life, and to do so precisely because he thinks the repetition is undercutting the respect in which his whole-life would be meaningful. For another example, people often sensibly think to themselves that they would like to make something good come of the bad parts of their lives. Of course, they do not, and should not, seek out parts *qua* bad so as to have bad parts that can lead to good parts. But, given bad parts that were not intentionally brought about, there can be good reason to then intentionally try to make the bad cause some good. Redemption can be meaningfully sought out.

It is clear from these examples that, if indeed one's whole-life can confer some meaning on one's life, there are times one can intend for it to do so without pulling the rug out from under it as a source of meaning, as well as that a non-Kamm strategy is required in order to deal with the problem of self-consciousness. Brännmark suggests a more promising, second way for the friend of a whole-life dimension to deal with the present objections. He can be read as saying that while it can be sensible to seek out a pattern in one's life, one ought not be aware of it very much. Brännmark recommends keeping the narrative structure of one's life in the 'background', so that one has no more 'than a rough grasp of the outline of the story of one's life' (2003: 76).

Let me be more specific about what Brännmark could plausibly mean. As a metaphor, think of a computer screen, where one's whole-life up to now and its available trajectories are represented on it. There are two salient respects in which one could be aware of what is on the screen. First, one could adjust the strength of the screen's brightness, with Brännmark in effect suggesting that one keep the light fairly dim, so that one cannot see the details clearly, but can obtain only a 'rough grasp'. Second, one could adjust the frequency with which one studies the screen, with Brännmark prescribing only periodic examination: 'The narrative structure that characterizes our lives is best understood as something that does not lie in the foreground of our deliberations about what we ought to do except perhaps on rare occasions' (2003: 76).

Now, having clarified two respects in which we could refrain from *consciously* aiming for a pattern, I point out that it would be unwise to use both at the same time. If one only rarely looked at the screen, and when one did so it were barely lit, then one could not intervene in crucial moments as the above cases indicate can be sensible. So, a whole-lifer ought to recommend that a person fairly constantly maintains at least a dim screen, but illuminates it very brightly from time to time and has a close look, to ensure that the pattern is generally on track.

Such an approach would *largely* resolve the two problems of self-absorption and artificiality. However, there is a facet of the latter worry that is still outstanding. Recall the point that when fashioning a pattern in one's life, one ought not create bad so that one has a pattern in which bad causes good. For instance, one ought not intentionally make a mess of one's marriage so that one has the opportunity to improve things down the road and obtain redemption. Running with Brännmark's suggestion, and so indicating the proper kind of awareness of the pattern of one's life, does not address *this* respect in which it is self-defeating for one to seek out a pattern.

It is probably this sort of case that is motivating Kamm's proposal, prompting me to return to the issue of aiming for a pattern. What this case shows is that whole-life meaning has a 'non-teleological' or 'deontological' dimension. A teleological account of value is one according to which the proper response to value is always to promote it, either to the maximum available extent or to a satisfactory degree, with the means by which it is promoted not mattering in itself. I claim that certain whole-life facets of a meaningful life are not merely to be promoted. Instead, there are 'side-constraints' on the manner in which one can do so and thereby obtain meaning. So, a pattern of bad causing good in one's life confers meaning on one's life only if one has not intentionally sought out bad so that it would cause good. Instead, if bad has unintentionally arisen, or has arisen intentionally for a reason other than the pattern of it causing good, then, and only then, can it be appropriate for reasons of meaning to intentionally aim to have the bad cause the good. In short, there are *certain ways* of promoting patterns that are meaning-conferring, which means are independent of degree of awareness of the promotion that Brännmark highlights.[13]

Summing up, in this chapter I have argued for a mixed view of the dimensions by which a life can be meaningful or meaningless. I critically explored two major rationales for a pure whole-life perspective, and have argued against them, in part by indicating that life can be meaningful to some degree in virtue of its parts considered in themselves. And I have brought out several distinct respects in which it appears that a pure part-life view would be an inadequate appraisal of a life's meaning. There appear to be at least five ways in which the pattern of a life can plausibly be a bearer of

---

[13] Cf. 10.3-10.5 for discussion of other non-teleological dimensions with regard to part-life meaning.

meaning. In addition to providing a developmental framework in which to place the whole-life facets, I have sought to advance understanding of them by addressing the issues of what the ground of narrative structure is and of how an agent ought to go about trying to add whole-facets to her life while avoiding self-defeat.

The reader should note that I do not say much more about the whole-life dimension of meaning in life in the next several chapters. The overwhelming majority of theories of life's meaning are accounts of the respects in which its parts can make it meaningful, and so I set aside whole-life considerations until the definitive statement of my own theory of what makes a life meaningful (12.4).

4

The Value of Meaning

'(T)hat which befalleth the sons of men befalleth beasts; even one thing befalleth them: as the one dieth, so dieth the other; yea, they have all one breath; so that a man hath no preeminence above a beast: for all is vanity....Wherefore I perceive that there is nothing better, than that a man should rejoice in his own works....There is a vanity which is done upon the earth; that there be just men, unto whom it happeneth according to the work of the wicked; again, there be wicked men, to whom it happeneth according to the work of the righteous: I said that this also is vanity. Then I commended mirth, because a man hath no better thing under the sun, than to eat, and to drink, and to be merry.'

Ecclesiastes 3.19, 3.22, 8.14–8.15 (*The Holy Bible*, King James Version)

4.1 Values in life

These passages from *Ecclesiastes* suggest the view that life is meaningless or 'vanity' because of mortality and injustice, and so one may as well party while one can, or, less glibly, appreciate other God-given goods while one still walks the earth.[1] I ultimately reject the notion that our lives are all meaningless, but I accept the implicit idea that if meaning were not available, another good in life could still be, namely, enjoyment. In previous chapters, I have provided some reason to think that issues of meaning in one's life are distinct from those concerning one's pleasure *qua* pleasure and other facets of one's animal self, such as staying alive and perceiving. While it is controversial to maintain that the latter facets of one's life are desirable for their own sake, nearly everyone agrees that pleasure is one such value. Supposing that meaning and pleasure are distinct respects in which one's life can be worthwhile, it would be of interest to compare them.

In this chapter, I note some similarities between the goods of pleasure and meaning, but do much more to highlight important differences between them. Specifically,

---

[1] There are, of course, competing interpretations of this text, and all I mean to do is focus on one influential reading to draw attention to a difference between two values. For a different reading, made in the context of discussion of life's meaning, see Kushner (1987).

I contrast pleasure and meaning with respect to six value-theoretic factors, among them: what the bearers of these values are; how luck can play a role in their realization; which attitudes are appropriate in response to them; and when they are to be preferred in a life. I aim not only to show that there are several respects in which pleasure and meaning differ as categories of value, but also to bring out some of the logical relationships between the differences, e.g., to note that the value's bearer affects its susceptibility to luck.

I begin by indicating what I take to be the core features of meaning and of pleasure, reminding the reader that pleasure and meaningfulness are both distinct (4.2), which will make it reasonable to compare and contrast the two goods in the rest of the chapter. I first note a few respects in which they are similar in nature (4.3), after which I devote the bulk of the chapter to critically exploring ways in which pleasure and meaning differ (4.4). I conclude by suggesting that there is good reason to identify a pleasant life as a happy one, so that it is ultimately happiness and meaningfulness that form two of the largest and most fundamental values in human life (4.5).[2]

4.2 That pleasure and meaning differ

The reader should by now have a good sense of what I mean by 'meaning'. It is a gradient final good that can be exhibited by an individual's life, where exemplars of its presence are certain kinds of intellectual reflection, moral achievement, and artistic creation, and where key illustrations of its absence include living in an experience machine, being a slave, and rolling a rock up a hill for eternity (Chapter 1). In addition, it is the sort of good that is largely captured by the overlapping concepts of purposiveness beyond one's own pleasure as such, transcendence of one's animal nature, and greatly estimable or admirable conditions (Chapter 2). Finally, I have contended that meaning in life admits of realization along two different dimensions: the parts of a life, on the one hand, and the relationships among the parts, on the other (Chapter 3).

What, though, do I mean by 'pleasure'? Although there are some who think of pleasure as a kind of propositional attitude, such that one can be 'pleased at' something (Feldman 1997), I work with the more traditional understanding of it as a characteristically positive experience. Advancing a full-blown conception of pleasure would require doing some cognitive science, but at a more surface level I can say that pleasure in general is a matter of a feeling that is typically accompanied by a wish that it obtain (and that pain is a feeling normally paired up with a wish that it not obtain). Pleasure is the kind of mental state that most contemporary speakers of English would equate with synonyms such as 'enjoyment', 'gratification', or 'satisfaction'. It is the sensation that is missing when one is depressed or miserable. It is the kind of thing that is

---

[2] Elsewhere (in Metz 2012) I discuss the category of what makes a life worth living and of how it relates to those of happiness and meaningfulness.

accessed with relative ease from the inside, making the best (but not the sole) way to ascertain whether someone is pleased to be a matter of asking her for a considered, sincere report of her state of mind. It is the state of mind that psychologists have shown is not proportional to the degree of one's bodily integrity or the amount of one's wealth (Baumeister 1991: 211-13). It is also what psychologists have shown tends to be higher when one has interpersonal relationships or holds false, overly positive views of oneself and the world, and what tends to be lower when one is isolated or has an accurate picture of reality (Baumeister 1991: 213, 221-5). It is the sort of reaction most people have to being at a party, succeeding on the job, eating a sumptuous dinner with one's beloved, witnessing the flourishing of one's child, winning a competitive sport, listening to powerful music, and having great sex.

I am concerned to compare and contrast meaning with pleasure as such, where this includes both 'lower' and 'higher' kinds. What makes a pleasure a lower one as opposed to higher? In the first instance, a lower pleasure is one available to typical animals, and so includes that associated with eating, orgasming, and sunbathing, which do not require sophisticated cognitive capacities. I do not mean to suggest that it is impossible, or even rare, for higher pleasures to be associated with these activities. The point is that while making and consuming a gourmet meal and having sex to express one's love might well count as higher pleasures, eating a chocolate bar and engaging in sex with a prostitute or a life-like doll would not.

I do not mean to utterly denigrate so-called 'lower' pleasures, in the sense of suggesting that none of them belongs in a life. On the contrary, I believe that the best life would include a decent amount of attention paid to our animal nature, and not merely for instrumental reasons. It is perfectly consistent to assert that tasting chocolate is a lower pleasure but that it is in itself part of a life worth living.

This sketch of what I stipulate talk of 'pleasure' and 'meaningfulness' to connote should be enough to compare the two goods. My aim in the rest of the chapter is to systematically articulate several value-theoretic similarities and differences between pleasant experiences, on the one hand, and meaningful activities such as helping others and creating artworks, on the other. To motivate such investigation, I start by reminding the reader of intuitive evidence to think that the two goods are indeed different.

To begin, note that they are conceptually distinct, in that the question of what makes a life pleasurable does not ask one and the same thing as the question of what makes a life meaningful. Talk of 'pleasure' and of 'meaning' connote different things, for, if they did not, then it would be logically contradictory to speak of an 'unpleasant but meaningful life' or a 'pleasant but meaningless life', but it is not. In addition, the reader will recall that I have gone a large step farther and maintained that part of the concept of meaning includes the idea that we are *not* discussing one's own pleasures as such, which would include the lower ones (2.4.1). Such a view was motivated, recall, by cases such as being subjected to forced passivity, but feeling jolly because of 'soma' (Huxley 1932) and rolling a rock up a hill for eternity à la Sisyphus and enjoying it because of the way the gods have structured one's brain (Taylor 1987: 679-81). However, even setting aside my strong claim that talk of 'meaning' *a priori* rules out a hedonist theory of it, such cases provide strong reason to doubt that whatever makes a life pleasant *just is* whatever makes a life meaningful and vice versa.

Supposing that pleasure and meaningfulness are not one and the same thing, there are nonetheless close links between them that one should acknowledge. For instance, in order to obtain meaning in life, one often must have a certain degree of pleasure, viz., not be so depressed as to be unable to get out of bed. Not only can an absence of pleasure prevent meaning, but a lack of meaning can also impede pleasure; for one could become depressed in the first place from a failure to detect meaning in one's life. I am not concerned to deny that there are intimate causal relationships between pleasure and meaningfulness—indeed, I am keen to affirm that there are, since pleasure and meaningfulness would have to be distinct in order for causal relationships between them to obtain.

In addition, some philosophers have suggested that meaning is *partially* constituted by pleasure, and is not merely a cause of it or caused by it. According to this view, a condition is meaningful for a person only if she enjoys it or is otherwise pleased consequent to it (Wolf 1997a). On this account, even if one were doing something objectively worthwhile, it would not confer any meaning on one's life if one were pained doing it.

I take damning counterexamples to this view to be many cases of 'unpleasant meaningfulness', e.g., undergoing boredom or pain so that others can avoid it (Metz 2007a; cf. 1.2) as well as sundry actual lives such as those of John Stuart Mill and Vincent van Gogh.[3] My own view is that pleasure, particularly when accompanying certain kinds of actions, can enhance meaningfulness, but is not necessary for it (10.2, 12.3). Note that if satisfaction is either a contributory or even essential *part* of what it is for a condition to confer meaning on life, the latter would be far from exhausted by it, and my goal in this chapter is to bring out the differences between the two that remain.

4.3 How pleasure and meaning are similar

One way in which pleasure and meaning, as construed above, are similar is that they are both final goods, i.e., good for their own sake. Supposing that they are indeed distinct values, most people would not consider one to be good merely as a means to the realization of the other one or some third thing. Instead, pleasure and meaning appear to be two different conditions that are each desirable apart from being instrumental for something else.

A second similarity between pleasure and meaning is that they are what T. M. Scanlon would call 'personal goods' (1998: 218-23). They are conditions that make an individual person's existence better or worse and so are in a broad sense 'good for' one.

---

[3] For a longer list of such persons, see Belliotti (2001: 129).

This kind of value contrasts with the 'impersonal' sort, maybe wilderness, ecosystems, species, individual animals, and the bare presence of life.

Third, the goods of pleasure and meaning can be ordered in the sense that some parts of a life are more pleasant and more meaningful than others. There can be a greater or lesser amount of pleasure, on the one hand, and a greater or lesser degree of, say, creativity or beneficence, on the other. There might be other kinds of goods, in contrast, that cannot be instantiated in different degrees. Some clear cases are ones that are sometimes ascribed to God: atemporality (existing never in time), eternity (existing always in time), necessary existence (existing in all possible worlds), immutability (inability to change), impeccability (inability to do wrong), and absolute perfection (the meta-good of having all the perfections to the highest compossible degree). One cannot have better or worse specifications of these conditions.[4]

Fourth, it appears that pleasure and meaning are intrapersonally aggregative, i.e., are amenable to rough judgements of how much of these goods there are in a given life overall. For instance, we can make ballpark estimations of the extent to which some parts of a life are more pleasurable than others in that life, as well as in principle add up the value of the pleasant parts, a procedure that would form the basis for a judgement of whether the life is pleasant on balance. The degree to which the pleasure that accompanies falling in love is greater than the pleasure that comes from having an article accepted for publication is much larger than the degree to which the pleasure that comes from having an article accepted for publication is greater than the pleasure resulting from having given a talk that has been well received. Given these kinds of roughly cardinal measurements of particular times in a life, one could conceivably add them up to inform an estimation of whether the life has enough pleasure in it to count as pleasant overall or period. Similar kinds of claims apply to meaning, even supposing, as I now do (given Chapter 3), that it can include whole-life elements.

Fifth, pleasure and meaning appear to be interpersonally comparative, which means that we can compare different lives with regard to amounts of these goods. For all I know, my life is, so far, more pleasurable than Emily Dickinson's was, but less meaningful than Albert Einstein's.

A sixth commonality, and the last one that I discuss, is that pleasure and meaning are bipolar, as opposed to monopolar; that is, there are both positive and negative scales for these values, as opposed to only a positive scale. This is clear in the case of pleasure, which is well represented with positive numbers, and is opposed by displeasure or pain, aptly characterized by negative numbers. Displeasure is not merely the absence of pleasure that would be well represented with a score of zero, in the way that poverty is the mere absence of wealth or stupidity is the mere absence of intelligence. Instead, pain is a substantial disvalue over and above the mere non-value of the absence of pleasure.

---

[4] Perhaps the same is true of the value of human dignity?

I submit that meaning also has two scales, even though our language might lead us to think otherwise. We lack a singular term in English for the disvaluable opposite of what is meaningful, and instead have terms that suggest a merely monopolar dimension to it. Specifically, we most commonly speak of a condition being 'meaningless', 'insignificant', 'unimportant', 'pointless', or 'nonsensical', which all suggest the mere absence of the positive of meaning, significance, importance, point, or sense, respectively. However, there are longer phrases that suggest the existence of a negative pole to meaning; for example, it does not seem unnatural to speak of behaviour that '*ceteris paribus*, reduces the meaning of the agent's life' (Nozick 1981: 612) or to talk of 'subtracting meanings' (Baumeister 1991: 295).[5] To coin a term, I use 'anti-matter' to designate the concept of the disvaluable scale of meaning that contrasts with those facets of a life that matter. I now advance an argument for thinking that the concept of anti-matter applies to the world, viz., that anti-matter truly exists.

My reason for believing in the reality of anti-matter is that its existence would best explain why some actions are worse than others from the perspective of meaning. If meaning had only a positive dimension, then oversleeping and blowing up the Sphinx for fun would both be well represented with a score of zero on the meaning scale. However, it appears much worse in terms of meaning to do the latter. A person who blasts the Sphinx for the thrill of it has not merely lost an opportunity to acquire some meaning, but rather has done something to weigh *against* whatever meaning he might have had in his life. In evaluating whether this person's life is meaningful on balance or not, one would not merely overlook this action, directing one's attention to the positive instances of meaning, in the way one would with regard to oversleeping; instead, one would consider this action to have set one back with respect to the aim of living meaningfully overall.

Now, if a single action is well represented with a negative number, then a life chock full of anti-Sphinx-like behaviour could have a score lower than zero on the meaning scale. If a meaningful condition is, partially by definition, one that warrants admiration and esteem (2.5), then it is natural to describe the unnecessary destruction of majestic creations as something that warrants revulsion and shame—which conditions are aptly characterized by a negative score. Similarly, if what matters in life is in part analytically achieving goals beyond one's own pleasure that are highly worth pursuing (2.5), then a person who realizes many ends besides his own pain that are extremely worthy of avoidance is naturally understood to have a life that anti-matters.

The existence of anti-matter has been recognized in the field only sporadically and without any depth. Philosophers have merely used language that intimates an awareness of it, with none having provided a characterization of the content of anti-matter, something I aim to help rectify (12.4).

---

[5] For similar talk, see Morris (1992: 49–50) and Munitz (1993: 89–93).

Summing up, pleasure and meaning are similar in that they are both aspects of a person's life that: are desirable for their own sake; have positive and negative dimensions; come in degrees; can be roughly added or subtracted; and can be approximately compared in magnitude between lives. I believe, however, that the similarities end there. In the rest of this chapter, I explore some interesting and important differences between these two goods.

4.4 How pleasure and meaning differ

I now address six value-theoretic differences between pleasure and meaning, supposing that pleasure is a positive experience and that beneficence and creativity are characteristic of meaningfulness.

4.4.1 Bearer: sensation v. action

Pleasure is an experience, something utterly internal to the mind, so that a life that counts as pleasant on balance is one that has had lots of certain feelings or sensations in it. In contrast, meaning is not entirely captured by the content of mental states, and a life is meaningful *principally* in virtue of action. By 'action' I mean in the first instance volition, but not merely that, for I here deem productive thinking, i.e., deliberating, reasoning, or interpreting (as opposed to, say, merely remembering or perceiving), to be a type of 'action'. The least controversial elements of meaning in life such as creativity and beneficence are actions, or otherwise something aptly described as 'productive', and discussions of the topic typically note that it inheres in conditions such as autonomy, authenticity, study, and relationships in which one treats others in positive ways.

Some might object that merely exhibiting certain attitudes can be sufficient for a meaningful life. For instance, in the following, Viktor Frankl can be read as suggesting that some people led meaningful lives in Nazi concentration camps by virtue of something other than their actions.

(T)here is also purpose in that life which is almost barren of both creation and enjoyment and which admits of but one possibility of high moral behavior: namely, in man's attitude to his existence, an existence restricted by external forces....The way in which a man accepts his fate and all the suffering it entails, the way in which he takes up his cross, gives him ample opportunity—even under the most difficult circumstances—to add a deeper meaning to his life. It may remain brave, dignified and unselfish (Frankl 1984: 88).

Now, there are facets of Frankl's discussion indicating that he believes that the relevant mental states are only ones that are under a person's control and that hence that might count as 'actions' (or 'productive') for my purposes. Sometimes Frankl speaks, for instance, of 'the last of the human freedoms—to choose one's attitude in any given set of circumstances' (1984: 86). However, there are some attitudes that are not under much direct control but that plausibly confer meaning on a person's life, perhaps being comparably emotionally affected by others' well-being and woe, viz., feeling sympathy. Is it possible to live a meaningful life simply in virtue of such 'non-active' mental states?

I ultimately contend that a life cannot be *pro tanto* more meaningful for exhibiting attitudes that are completely independent of our control (12.3), but, setting that controversial claim aside, I note here that few would deem a life to be meaningful *on balance* merely in virtue of such dispositions. Most would hesitate to call someone's life 'meaningful overall' just insofar as she wished that others would be helped and were glad when they were helped; a life that is meaningful with regard to help requires doing some helping oneself. Consider that one major reason to hate the prospect of being stuck in a concentration camp is that it would rob one's life of potential for meaning, in particular, would prevent one from engaging in beneficent and creative actions.

Reflection on life in a concentration camp suggests that, while action on one's part is necessary in order to have a meaningful life, no action on one's part is necessary in order to have a meaningless one. Indeed, an effective way to prevent someone from having a meaningful life is to undercut her ability to reason and to engage in sophisticated, skilled activity. This claim helps explain why we abhor the prospect of becoming an Alzheimer's patient; for although we might be able to exhibit certain caring emotions about others, we would be incapable of doing much and hence would lose out on substantial opportunities for meaning in life.

4.4.2 Source: intrinsic v. relational

It seems apt to distinguish between the *bearer* of pleasure or meaning,[6] on the one hand, and their *source*, on the other. I have claimed that a pleasant life consists of certain experiences that are good for their own sake, while a meaningful life is (substantially) made up of certain actions that are good for their own sake. Experiences and actions are in what these values respectively inhere, and they are to be contrasted with the source of these values, i.e., on what the values logically depend in order to inhere.[7] For a value X to logically depend on Y is either for Y to be necessary or sufficient for X, or for Y to enhance some degree of X. Again, Y is a source of X insofar as either X cannot exist without Y or has to exist once Y does, or some amount of X is realized upon Y.

The source of pleasure appears to be intrinsic to its bearer, so that pleasure co-varies with positive experiences alone, whereas the analogous structure does not hold with regard to meaning. Let me spell this out.

---

[6] In Chapter 3 I used the term 'bearer' of meaning to distinguish between whether meaning inheres in the parts of a life or the life as a whole (or both). Here, the same term is apt, as now the issue is in which aspects of human living (whether in part or as a whole) meaning inheres.

[7] This distinction is implicit in much of the discussion about 'non-intrinsic final value'. See, e.g., Korsgaard (1983); Kagan (1998); Rabinowicz and Rønnow-Rasmussen (2000).

In the case of pleasure, its source and its bearer are one and the same thing—namely, positive experiences. Pleasure inheres in positive experiences, cannot exist without them, must exist once they do, and its magnitude is determined solely by them. In other words, positive experiences constitute pleasure regardless of their relationship to other things, and do so purely by virtue of their nature *qua* experience. In particular, neither what has caused an experience nor what it will cause affects the respect in which the experience is a pleasurable one. The cause of an experience can affect whether the pleasure is *appropriate* in some way or not—for example, joy upon a successful theft—but not whether the experience confers some *pleasure* on a person's life.

One might object that the distinction between higher and lower pleasures indicates a respect in which the source of pleasure is relational, i.e., in which pleasure logically depends for its existence on something other than its bearer, positive experience. If lower pleasures are 'purely' bodily ones such as eating doughnuts while sitting in a jacuzzi, and if higher pleasures are ones that have been caused in certain, mental ways, such as the gratification that results from doing supererogatory deeds or listening to music, then pleasure is not strictly intrinsic to its bearer; for the cause of an experience is extrinsic to the experience itself.

However, I submit that the distinction between higher and lower pleasures is not best understood in terms of any cause of the pleasures. It can instead be well captured in terms of differential *qualia*, i.e., what the content of the pleasures is, or how the pleasures feel. The pleasant sensations of listening to music and of going beyond the call of duty would constitute pleasure—and perhaps a higher form of it—even if there were no actual music or good deed. As the experience machine thought experiment indicates, in principle any sensation can be caused by the manipulation of brain states and need not follow from any particular engagement with the world. There is no difference in *pleasure* when the sensation of listening to music is caused by eardrums or by electrodes.

In contrast, it appears that the source of meaning is often relational, i.e., logically depends on something beyond an action, which is (typically) the bearer of meaning. An action can be more or less meaningful because of something outside of it, and, especially, what has caused it and what it will cause.[8] For example, consider creative behaviour. Imagine in one case that it is the result of substantial education, training, and effort, whereas in another case it is the consequence of taking a pill. Or imagine in one case that creative behaviour results in a novel art-object that others appreciate, whereas in another one it does not. In both pairs of cases, it is natural to say that we could have the same creative activity but differential meaning, because of how it was brought about and what its results were.

---

[8] It is somewhat common to note that the effects of an action can affect its degree of meaning. However, Brogaard and Smith (2005: 450–3) are two of the few to discuss the way that an action's cause can affect it as well.

Similar remarks go for the meaningfulness of helping others, although here more argument is needed to establish the point. I claim that help is meaningful at least in part when it actually makes someone better off, and not merely in virtue of the helping behaviour, which I take to be a matter of intending to confer a benefit on someone and acting on the justified belief that one's action will probably confer the benefit. The strongest argument for thinking otherwise, viz., for thinking that it is only a good will, and not any good result, that confers meaning on a life, is a thought experiment in which luck prevents one's action from achieving its aim. Suppose, for example, that digging a drainage ditch does not achieve the intended aim because of an unforeseeable flood. Many think that, because of help's good will elements, it would have been worth digging the ditch despite fortuitous circumstances having prevented good outcomes (e.g., Schmidtz 2001: 180).

However, it is worth noting that, even if one has 'Kantian' intuitions about this thought experiment, one can plausibly hold that the meaningfulness of help is constituted in part by its results. One can grant that one has reason to help others independent of the results, but deny that the final value of help is utterly independent of them. Specifically, I submit that the *moral* worth of engaging in helpful behaviour might well obtain regardless of the results, but that the value of help with respect to *meaning in life* is at least partly contingent on its results. The case of helping obtains much of its force from its association with morality, which I am prepared to accept is largely immune from luck. Even if one grants that the worth of, say, digging a ditch with regard to virtue or rightness is not affected by whether it has in fact benefited the community, its worth with regard to meaningfulness plausibly increases when helping succeeds in improving others' quality of life. I conclude, then, that the cases of creativity and beneficence give us reason to think that the bearer of meaning is not identical to its source, a structure that differs from pleasure, where the bearer and the source are one and the same thing.

4.4.3 Role of luck: total v. partial

Given that the bearer of pleasure is largely experiential and that of meaning is largely a matter of action (4.4.1), there are significant differences in the role that luck can play in realizing these goods. By 'luck', I mean factors over which one has little or no control, and I presume that one has the most control over the actions one performs, and somewhat less control over other things, such as what one feels or how one's actions have affected the world. If so, then luck can *conceivably* be what is *completely* responsible for a person's pleasure, or at least it could affect it to a much greater degree than it could affect meaningfulness.

To clarify the point, I note that luck can of course play a role in whether one's life is meaningful or not, since, as I have said (4.4.2), the consequences of one's actions can affect their significance. Furthermore, whether one's life is utterly meaningless could entirely be a function of luck, e.g., if one were so unlucky as to have had an accident in which one's capacity for action were impaired. However, it appears that luck cannot on its own ever bring about the positive good of meaning to one's life; one has to perform certain actions under one's control. That differs from the positive good of pleasure, which in principle could come about entirely by virtue of factors beyond one's control (even if, in the real world, it often requires a lot of effort).

One might object that a mad scientist could be entirely responsible for making you perform meaningful actions, just as a mad scientist could thrust you into an 'orgasmatron' and thereby be the sole causal factor that has made you feel intense pleasure. However, the cases are disanalogous. Even if there were a sense in which the mad scientist could make you perform certain 'actions', these actions would not be of the sort necessary for meaning to accrue. So-called 'actions' that you perform simply by virtue of external manipulation are not enough to confer significance on your existence. Instead, there must be some kind of authentic or autonomous action in order to ground meaningfulness (see also 9.4.3).

4.4.4 Appropriate attitude: want to continue v. esteem

There are certain attitudes that are appropriate in response to meaningful conditions that are not apt for merely pleasant ones, and vice versa, it appears. For a wide array of factors in virtue of which a life is meaningful, it is analytically the case (2.5) that it is sensible to feel substantial esteem about them, from the first-person perspective. And from the third-person standpoint, it is reasonable to admire another person by virtue of conditions that make her life meaningful. Furthermore, when it comes to meaninglessness or anti-matter, first-person reactions of shame and third-person reactions of abhorrence are often appropriate.

However, these reactions never fit pleasure as such, as I construe it in this chapter. It makes sense neither to admire people simply because they feel a sensation, nor to take pride in the bare fact of doing so. It could be reasonable to feel esteem because of something (say, hard work or an intense relationship) that has occasioned the pleasure, or what the pleasure signifies, but not, I contend, because of the pleasure itself. Similarly, it is inappropriate to feel shame merely because one experiences pain or to abhor someone who does.

I have found it difficult to ascertain whether there is an attitude that is uniquely appropriate for one's own pleasure. However, perhaps when it comes to your pleasant experiences, it is invariably appropriate for you (*pro tanto*) to desire that they continue, and invariably appropriate for you to desire (to some degree) that unpleasant ones end. Meaningful conditions might not invariably call for a desire that they continue. For instance, if you volunteer to be bored so that others avoid boredom, this might confer meaning on your life and be worthy of substantial esteem or admiration, but perhaps it does not call for a *desire* for this condition to continue (even if it does require an *intention* to do so). Maybe when you adopt the goal of undergoing sacrifice for the sake of others, you need not want the sacrifice to continue and may instead wish the sacrifice to end as soon as possible, so long as the relevant end is realized.[9]

4.4.5 When the value is possible: during life alone v. posthumously

From 4.4.1 and 4.4.2 it follows that pleasure can exist only during a life, while the meaningfulness of a life can be increased after it has ended. If pleasure is a function of positive experiences, then no more pleasure is possible upon death, which I conceive as the permanent cessation of existence and hence of experience. In contrast, since the meaningfulness of a life is partly a function of the consequences of one's actions, and since the consequences of one's actions can occur after one is dead, one's life can become more or less meaningful, even though one is no longer capable of action. Vincent van Gogh's life is meaningful in large part by virtue of recognition and influence that obtained long after his death, but that has not made him posthumously any more pleased thereby.

It is often said of meaningfulness that it is something worth living for, though, as the likes of Albert Camus and Joseph Heller have famously suggested, it can also be something worth dying for. Meaningful conditions are naturally understood to be able to provide reasons to die, since dying might have certain consequences that affect the significance of one's existence. Intuitively, sacrificing one's life could enhance its meaningfulness when done, say, in order to save one's children or to protect one's fellow soldiers and thereby advance a just cause.

In contrast, one's own pleasure could never be worth dying for. Of course, there can be fates worse than death, such that dying might be welcome as a way to avoid one from undergoing substantial pain. My point is rather that, unlike meaningfulness, realization of the positive good of one's own pleasure cannot provide a reason for one to die.

4.4.6 When the value is preferable: bias towards the future v. lack of bias

Derek Parfit concocted a thought experiment that, when applied to pleasure and meaning, reveals a striking difference in when most of us prefer these values to be realized (1984: 165-6). Imagine you have just woken up from a surgery and are suffering from a temporary bout of amnesia. Before you are able to remember who you are, you are told that you could be one of two people. You could be either (A) someone who experienced a great amount of pain yesterday; or (B) someone who will experience a small amount of pain tomorrow. Most would prefer to be (A), even though (A)'s pain is larger. To Parfit, that indicates a 'bias towards the future' in the sense that, from the standpoint of any given time,[10] we want our future to be as good as possible.

---

[9] It is not clear how well this point squares with the principle I offer later, that meaningfulness increases, the more of one's rational self is contoured towards a particular object (12.3).

[10] Parfit's thought experiment abstracts from any whole-life considerations, which would often affect the judgements one is inclined to make. Note, by the way, that the coherence of this thought experiment is further evidence against a pure whole-life view (3.3).

Notice that such a bias also arises with regard to pleasure. Suppose you are told that you could be either (A★) someone who experienced a great amount of pleasure yesterday; or (B★) someone who will experience a smaller amount of pleasure tomorrow. Most would prefer to be (B★), even though (B★)'s pleasure is smaller than (A★)'s.

Some have pointed out that there are goods and bads for which we lack a bias towards the future, but the field lacks a plausible comprehensive principle to account for all those that come with a bias and all those that do not. In light of several thought experiments that I conduct below, I proffer and defend the theory that we have a bias toward the future with regard to pleasure and pain and that we lack such a bias with regard to meaningfulness and anti-matter. Although this account of when we are inclined to have a bias towards the future probably does not *justify* the bias to someone who doubts that we should have it, it will indicate that certain arguments that have been explored as potential justifications of it are non-starters, clearing the way for a more promising attempt in the future.

First, then, Thomas Hurka has noted that we are disinclined to exhibit a bias towards the future with regard to certain non-experiential goods related to self-realization (1993: 60-1). Would you rather be (A#) someone who saved another person's life yesterday; or (B#) someone who will help an old lady cross the street tomorrow? Most would prefer to be (A#) because the value is greater, even though the value of being (B#) is in the future. Also, would you rather be (A¬) someone who murdered someone in the past; or (B¬) someone who will tell a minor lie in the future? Most would rather be (B¬), which again suggests a lack of bias towards the future. As an initial suggestion, then, perhaps we lack a bias towards the future simply with regard to goods and bads of self-realization, i.e., the development or stunting of valuable aspects of our human nature, one of which is the capacity to help others for their sake.

However, consider the following counterexample of a good unrelated to self-realization for which we lack a bias towards the future. Suppose you are told that you could be either (A†) someone who has been widely recognized for having produced a masterpiece in the past; or (B†) someone who will be only mildly recognized for having produced a mediocre poem in the future. Although there is self-realization in both cases, the good of recognition is beyond that of self-realization, and it is part of the explanation of why one would rather be (A†).

One might suggest that the recognition would be desirable only if it were in response to actions that one has in fact performed. And so one might be tempted to suggest that we lack a bias towards the future in cases of goods and bads logically dependent on actions we perform for their value or disvalue. That seems to capture the cases of help and recognition for creativity.

Here, though, is a counterexample to that tempting proposal. Suppose you are informed that you might be either (A^) someone who was systematically humiliated behind your back—for example, by virtue of a long string of romantic affairs that your spouse had unbeknownst to you; or (B^) someone who will be humiliated to only a small degree in the future—for example, your spouse will flirt with someone else at a party. Most would prefer to be (B^), which suggests a lack of bias towards the future. However, since no action would be involved on your part, it cannot be disvalue logically dependent on actions of yours that explains the lack of bias.

Another proposal is that we have a bias towards the future with respect to isolated goods and bads such as our own pleasure and pain, but lack such a bias in the context of interaction with others. That would seem to account for the lack of bias in the cases of helping others, being recognized by them, and being humiliated by them. All three cases are ones involving some kind of interpersonal engagement.

However, it appears that we also lack a bias towards the future with regard to certain isolated goods. Would you rather be (A') someone who created a masterpiece in the past (that went unrecognized by others); or (B') someone who will create a mediocre poem in the future? Most would prefer to be (A') because the value is greater, albeit in the past and without anyone else being involved.

My proposal is that we lack a bias towards the future with respect to goods for which it makes sense to feel great esteem and bads for which it is reasonable to feel shame. Helping, harming, being recognized for great works, being humiliated by others and creating are all conditions for which it is reasonable to feel either esteem or shame. Note that my claim is not that it is the bare fact of *feeling* esteem or shame that explains when we have a bias towards the future. It is rather the *appropriateness* of these emotions in reaction to certain conditions that does the work. To see this, imagine that along with being told you could be one of two people, you were offered a pill to remove any unpleasant feeling you might have about being one or the other. Knowing you would not actually feel shame, you would, I presume, still prefer not to be someone who has done a shameful deed in the past.

Above I noted that meaningful conditions are typically ones that warrant great esteem, whereas anti-matter warrants great shame (2.5, 4.4.4). Hence, my suggestion is that conditions of meaning and anti-matter are those for which we are inclined to lack a bias towards the future, whereas conditions of pleasure and pain are those for which we do have such a bias. Bias towards the future is another, key respect in which these two values differ.

If these claims about when we exhibit or lack a bias towards the future are correct, then a number of arguments philosophers have examined in search of a justification for the bias are clearly misguided. For instance, some have addressed whether it might not be the passage of time or the direction of causation that makes it reasonable to have a bias towards the future (Parfit 1984: 168-86). However, such general metaphysical considerations are out of place, for they will not discriminate between those goods and bads for which we have a bias and those for which we do not. The reason to exhibit a bias must be contoured to the particular dis/values involved. I do not yet have an argument for believing it not only to be reasonable to have a bias with regard to the experientialist good of pleasure and bad of pain, but also to be unreasonable to have it with regard to the estimable good of meaningfulness and shameful bad of anti-matter. However, the search for one would be the next logical step to take, if my argumentation is sound.

4.5 Happiness and meaning as basic goods

I wrap up this discussion with a brief summary, and by drawing the conclusion that it is not merely pleasure, but happiness, that differs structurally from meaning. I began by providing *prima facie* evidence for the idea that pleasure and meaning are conceptually and substantially distinct, and suggested a rough definitional essence of each value, with pleasure being a function of positive experiences and meaning being typified by creativity and beneficence. I noted several similarities between the two goods, but then used the most space to reflect on six major value-theoretic differences between them, arguing that: the bearer of pleasure is sensation, whereas that of meaning is principally action; the logical conditions for pleasure are intrinsic to its bearer (that is, are simply sensation), whereas some sources of meaning are extrinsic (e.g., can be the causes or results of a person's action); the role of luck can be complete when it comes to attaining pleasure, but in principle can never be complete with regard to obtaining a meaningful life (which requires performing actions under one's control); the appropriate attitude to have towards pleasure seems to be to want it (the experience) to continue, whereas that for meaning is rather one of great esteem or admiration (for what has been done); the time when pleasure is possible is only during one's life (when there is the capacity for experience), whereas meaning can obtain after one has died (as the consequences of actions can affect their significance); and the time when pleasure is preferable is invariably in the future, even if it is small, whereas most would prefer a more meaningful past to a less meaningful future.

Now, most modern laypeople and psychologists, as well as probably a majority of philosophers, believe that *happiness* is something substantially mental. Philosophers, for example, tend to hold either hedonism or the desire satisfaction theory. Hedonism is the view that happiness is constituted by pleasure, while the desire satisfaction theory is the view that happiness is a matter of obtaining whatever one wants.

I think stock counterexamples to the standard version of the desire satisfaction theory are conclusive. For example, if a blind man wants a red house, his house is painted red one night, and he is never told of that fact, then he is not any happier for having had his desire satisfied. If desires matter with regard to one's happiness, I submit that they do so most clearly insofar as the objects of a person's desires are positive experiences. So, if desires are relevant, the blind man is probably no happier for having a red house because he has not yet fulfilled a desire for a pleasing visual perception.

Summing up, both of the two dominant, mental theories of happiness, at least in their plausible versions, entail that a happy life is one with a substantial amount of pleasant experiences and a small amount of painful ones. There are some who might have been inclined to think that a meaningful life just is (substantively) a happy one. However, if a happy person is basically just one who is pleased, then the implication of this chapter is that this is a misguided view. Perhaps those who have tended to think that happiness and meaning are more similar in content than I have suggested might now change their minds; given the six value-theoretic differences between pleasant experiences, on the one hand, and actions such as creativity and beneficence, on the other, 'happiness' and 'meaningfulness' are two terms with which it would be useful to track them—at least if one was not already content with biblical terms such as 'mirth' and 'vanity'.[11]

---

[11] There is a further question of how the category of meaning relates to that of well-being. By 'well-being' some philosophers just mean happiness, whereas others have something more broad in mind, such as anything that is in a person's interest or would improve her quality of life. By the latter construal, meaning and happiness are two different kinds of well-being.

II

Supernaturalist Theories of Meaning in Life

5

Purpose Theory I: Questioning Motivations

'Our Father which art in heaven, Hallowed be thy name. Thy kingdom come. Thy will be done, as in heaven, so in earth.'

Luke 11.2 (*The Holy Bible*, King James Version)

'Ye shall observe to do therefore as the Lord your God hath commanded you.... And thou shalt love the Lord thy God with all thy heart, and with all thy soul, and with all thy might.'

Deuteronomy 5.29, 6.5 (*The Pentateuch and Haftorahs*)

'I have only created Jinns and men, that they may serve Me.'

The Holy Qur'an 51:56

5.1 Overview

The function of part one was to set the stage for a critical discussion of theories of life's meaning, basic accounts of what the meaningful aspects of a life have in common as distinct from the meaningless ones. I first indicated what is involved in seeking out the most defensible theory and why it is worth undertaking (Chapter 1), after which I analysed the concept of life's meaning that is common to a wide variety of salient conceptions of life's meaning to be found in the philosophical literature (Chapter 2). This concept, largely comprised of ideas about purposiveness beyond one's own pleasure as such, transcendence of one's animal self, and conditions worthy of great pride or admiration, is what makes a theory about life's meaning as opposed to about something else. The next step was to note that talk of meaning in 'life' is vague, in the sense that it could mean that life as a whole bears meaning or that some subset of it does, or both (Chapter 3). I argued for the latter, mixed view, according to which both parts of a life and certain kinds of relationships between its parts are capable of exhibiting meaning, with a complete theory giving an account of both part- and whole-life dimensions. Finally, I indicated several respects in which the category of meaning substantively differs from another value that is often pursued, namely, pleasure, which I identify with happiness (Chapter 4). I pointed out that the two goods warrant different attitudes in response to them, that luck in principle plays a different role in their realization, and that they are preferred at different times in a life, among other contrasts.

By now the reader should have a firm grasp of what a theory of meaning in life is about. The aim of parts two and three is ultimately to provide fresh and compelling reasons to reject all the existing theories of life's meaning that one finds in the literature, and to develop and defend a new one that is more promising. A large majority of the theoretical discussion addresses only the part-life respect in which a life can be meaningful. Despite a willingness to acknowledge, at least upon reflection, that a whole-life can exhibit a kind of meaning different from that in its parts, nearly all theories that have been proposed in the literature neglect to address it. I do address the issue of pattern, when I specify the theory that I find most attractive (section 12.4), but I bracket the issue until then. Another notable facet of the theories from the literature is that they are invariably *pro tanto* accounts, i.e., principles about respects in which a life would be somewhat more meaningful, and not necessarily meaningful on balance, which I also largely set aside (until 8.6 and 12.4).

At the most abstract level, I categorize theories of life's meaning metaphysically, according to the kind of property that is taken to constitute the meaning. Supernaturalism is the general view that one's existence is significant just insofar as one has some kind of relationship with a spiritual realm, which I consider in Chapters 5 to 8. Naturalism is the broad perspective that meaning in life can be a function of living in a purely physical world as known by science, which I address in Chapters 9 to 12. A small handful have also discussed non-naturalism, the view that a life matters if, and only if, it engages with properties that are neither spiritual nor physical, which I take up in only the most brief way (8.7). I ultimately conclude in favour of a new version of naturalism.

In section 5.2, I explain in more detail what makes a theory supernaturalist, differentiate types of supernaturalism, indicate that I am first exploring the dominant version of it, which I call 'purpose theory', and spell out its essential elements. Purpose theory is the view that meaning in life is constituted by doing what God intends one to do with one's life. I briefly consider some salient arguments for purpose theory, which I find fairly easy to rebut (5.3), after which I explicate the most powerful and influential argument for purpose theory, which appeals to the idea that God's purpose is necessary for there to be an objective morality, which, in turn, is necessary for meaning in life (5.4). In the last section, I evaluate that argument, contending that while God's purpose could *entail* an objective morality, it nonetheless could not *explain* it very well, for an underexplored consideration that I advance (5.5).

5.2 Supernaturalism: God- and soul-centred theories

This section is largely a matter of defining terms, analysing concepts, and proffering taxonomy. I begin by spelling out the most general category of supernaturalism, then indicate the major versions of it, and finally articulate the basics of the specific instance that I principally discuss in this chapter, namely, purpose theory. In the course of clarifying purpose theory, I also point out that some criticisms of it in the literature are misguided to the point of being *non-sequiturs*.

Supernaturalism is the general view that what constitutes, or is at least necessary for, meaning in life is a relationship with a spiritual realm. Drawing on the concept of meaning (2.5), a supernaturalist theory implies that the relevant purpose beyond one's pleasure as such to seek out is a relationship with a spiritual realm, or that such a relationship is the relevant way to transcend one's animal nature or to make one's life worthy of great pride or admiration.

Such a characterization begs the question of what counts as 'spiritual', a difficult question, which I do not seek to answer definitively. Most aim to capture the nature of the spiritual by first pointing out that it is beyond space and time, or at least *our* space and time, and that it is not composed of sub-atomic particles. However, this negative analysis fails to differentiate supernaturalism from non-naturalism, for the latter conditions are also deemed to be essentially beyond space and time and not to be composed of strings, or quarks, or the like. In addition to being outside the realm of nature, then, the spiritual, at least in Western monotheism, is thought to be populated by concrete entities or substances, individual things capable of exhibiting properties. More specifically, the standard understanding of the spiritual in this tradition includes thinking that there are persons, self-aware beings, who are beyond (our) space and time and composed of something other than sub-atomic particles.

There are two kinds of persons salient in supernaturalism about meaning in life—namely, God and souls—often taken to be offshoots from Him. The three basic kinds of supernaturalism are a function of which of these types of persons are held to constitute meaning. Perhaps most supernaturalists, following Thomas Aquinas, believe that both God and a soul are jointly necessary and sufficient for meaning in life, a matter of one's soul communing with God in a certain way (e.g., Tolstoy 1884; Morris 1992; Craig 1994). However, it is worth acknowledging simple (or pure) God-centred and soul-centred theories, for two reasons. First, some supernaturalists explicitly reject the composite view, maintaining that only one sort of person beyond the natural world is constitutive of meaning in life. For example, I examine below the views of some important supernaturalists who believe that meaning requires one to relate to God in a certain way, but not to have a soul. Second, some rationales ostensibly offered for the composite view in fact provide support for no more than a simple one.

I begin by examining pure God-centred theories (Chapters 5 and 6), according to which a person's existence is significant solely by virtue of a certain relationship with a non-physical person who is all-knowing, all-powerful, and all-good, and who is the ground of the physical universe. The literature offers a variety of reasons why a relationship with God alone might constitute meaning in life, but, in the present chapter, I focus strictly on the dominant, pure, God-centred view, the purpose theory, according to which one's life is more meaningful, the better one fulfils a purpose that God has assigned to one. This theory implies that if God did not exist, or if He did but one failed to fulfil His purpose to an adequate degree, then one's life would be utterly meaningless. And on the simple version of purpose theory that I explore, all that counts is doing God's bidding in the physical world, without any sort of spiritual afterlife being required for one's existence to be significant. By the present, pure theory, as per a plain reading of the Hebrew Bible, what matters is a relationship with God while on earth, before the death of one's body and hence of oneself. Such a theory is explicitly held by, among others, Brown (1971), Hartshorne (1984), Levine (1987), and Cottingham (2003, 2005).

Before considering what there is to be said for and against purpose theory, I explicate it in some detail. I distinguish what is essential to the view from the contingent versions of it that may be held, and I also clear up some common misconceptions and dispatch objections to the view that plainly misfire.

Purpose theory as such is the view that one's life is meaningful just insofar as one fulfils a purpose that God has assigned to one. Note that purpose theory implies nothing about whether God in fact has a purpose or whether God even exists. Of course, many theists do hold purpose theory, but it would be possible for, say, atheistic existentialists to hold it as well (see 13.2). Supernaturalism is a view about what *would* confer meaning on life, and does not imply that these conditions obtain.

I therefore dismiss straightaway one objection to purpose theory, namely, the charge that the existence of evil in the world, or some other consideration, shows that there is no God. It has been common to maintain that the purpose theory is false, since God does not exist, or that it is at least unjustified, since we do not know that God exists.[1] It has not been appreciated that, strictly speaking, this objection is a *non sequitur*, for purpose theory is a thesis about the conditions for a meaningful life, not about whether these conditions obtain. The view is that life would be meaningful just to the extent that God existed and gave us a purpose that we then fulfilled; it makes no claim about whether life is in fact meaningful. As a thesis about what can alone confer meaning upon our lives, purpose theory in combination with atheism would entail the nihilist conclusion that our lives are meaningless.

Key differences among purpose theorists turn on their conceptions of (a) God; (b) God's purpose; (c) the way God assigns it to us; and (d) the way we are to fulfil it. Let me briefly examine some competing interpretations of these elements.

---

[1] Examples include: Dahl (1987: 11–12n); Hanfling (1987: 50); Ayer (1990: 191–2); Singer (1996: 72–3); Gewirth (1998: 176–7); Kekes (2000: 22–6); Mintoff (2008: 67–8).

I consider talk of 'God' essentially to connote a spiritual person who is omnipotent, omniscient, and morally ideal, and who is the ground of the physical world (a). That is the understanding of God common to the three monotheistic traditions of Judaism, Christianity, and Islam. Although most purpose theorists are theists who hold that God is a transcendent being who is engaged with us, it is worth noting that mystical and deist versions of purpose theory are possible, too. Furthermore, while God might have additional properties such as necessity, infinitude, atemporality, immutability, and simplicity, I tend to set such possibilities aside (until 6.5).

With regard to the nature of God's purpose (b), many defenders of purpose theory hold that any end God assigns to an individual would be part of a larger, single plan for the universe. It is often thought that if God created the natural world, it was done with one highest-order goal in mind, whereby all other ends would be necessary components of, or instruments for, its realization. However, it is not clear that this is the only possible or plausible account of God's purposes. God could have brought about the universe with several higher-order ends in mind, at least if the ends did not (greatly) conflict. For example, God arguably could have made the world for the sake of being generous to the creatures in it, maximizing temporal values, glorifying Himself, and enhancing the meaning of His own existence.

Another distinction with regard to the content of God's purpose(s) concerns scope, that is, whether there are multiple purposes that are relative to individuals or groups, or whether God's one or several purposes apply to all human persons. The dominant view, among philosophers anyway, is that any of God's purposes, whether singular or plural, would apply to all human persons. The commands to love God and to love one's neighbour as oneself are good examples of universal purposes. However, there are those who maintain that God would tailor some purposes to particular groups. Some in the Jewish tradition, for instance, have thought that God has assigned aims to the Jewish people that He has not to others. And some have suggested that a meaningful purpose would be unique to various individuals, entailing that God would in effect need to have several billion purposes in mind (Affolter 2007).

A last point about the content of God's purpose relates to whether it would be moral or not. The standard view among purpose theorists is that God's ends are the source of moral rules (5.4), but there is nothing inherent to the view that would require this. In principle, the defender of purpose theory could hold that moral facts obtain independently of God's will, but that meaning facts do not.

Adherents to purpose theory often disagree with one another about the precise content of God's purpose and the way we might come to know it. These are notoriously difficult questions to answer. However, it is a mistake to think, as some critics have, that purpose theory is unacceptable if it fails to conclusively specify our assigned end (Baier 1957: 106-8; Ellin 1995: 322). The idea behind the objection is that a theory of life's meaning should provide some practical guidance. Now, the purpose theorist can maintain that we do have a reasonable amount of insight into what God's purpose might be. After all, the question of why God would create something rather than nothing, or would create us in particular, is amenable to intelligent reply. But even if the defender of purpose theory could provide no indication as to the content of God's purpose for us, I do not think her view would thereby be disqualified. Utilitarianism has been widely deemed to be a good candidate for a moral theory, despite the enormous difficulty of knowing which course of action would actually produce the best results. Similarly, purpose theory could be an acceptable theory of what would make our lives meaningful, even if we did not know how to fulfil God's purpose.

There are further differences among versions of purpose theory regarding the way God might assign a purpose to us (c). For instance, would God command us to realize His end? Might God punish us with eternal damnation if we failed to realize the end assigned to us?

Finally, purpose theorists disagree about how we ought to fulfil God's purpose (d). Most hold that it is possible for us not to realize God's end, viz., that we are not predestined to do what God would like. Typical adherents to purpose theory believe that we are to freely fulfil the end God assigns. Hence, it is incorrect to say that purpose theory entails that everyone's life would be meaningful merely because God assigned us an end (Baier 1957: 105, 106, 115) or because we could not avoid realizing it (Ayer 1947: 224-5, 1990: 193). The standard form of purpose theory is that one must fulfil the end, not merely be assigned one, and most purpose theorists hold that one must fulfil it by means of a free choice.

Obviously, many different versions of purpose theory are possible. In the rest of this chapter, I appeal either to core features of it or to versions of it that are particularly strong and influential. Doing so will enable purpose theory to avoid many objections that have been levelled against it, while at the same time making my criticism of it fair and powerful.

5.3 Weaker arguments for purpose theory

Purpose theory is a *prima facie* attractive account of what could make a life meaningful. It spells out what it would mean to 'exist for a reason' or for 'life to have a point'. It jibes with the fact that talk of 'purpose' is closely associated with that of 'meaning'. It accounts for the widespread intuition that, *contra* Sartre (1946), meaning in one's life cannot be determined by making just any decision. It provides a plausible candidate for what could confer significance on our lives, namely, a holy being. Finally, it squares with the judgement that most (if not all) people are capable of living a meaningful life, but that not everyone in fact does live one.

These are considerations in favour of the theory, but they are not unique to it. The data above provide no less support for many other theories in the field. So, in this section, I begin to search for arguments that would do a better job of supporting purpose theory in particular, i.e., that would explain both why *God's* purpose is key and why God's *purpose* is. I argue that while the following arguments would better single out purpose theory if they were sound, they are not, and are fairly easily shown not to be.

First, some maintain that fulfilling God's purpose is necessary and sufficient for meaning in life because of what God would give to those who do so, perhaps in an afterlife (Davis 1987: 290-1, 295-6, 300-2; Craig 1994: 49-50). Christians tend to think that God would bestow grace on those who try to come close to Him by accepting His son as their saviour, while Jews and Muslims tend to think in terms of the deserved reward that God would mete out.

However, there are damning objections to this proposal (so to speak). For one, this rationale suggests that it would actually be the state of grace or positive desert that constitutes life's meaning, for which the fulfillment of God's purpose would be merely instrumental (as Jacquette 2001: 22-3 has noted). For another, since a natural, impersonal force akin to Karma could conceivably apportion happiness in response to the right beliefs or choices, this rationale cannot even demonstrate that God's existence is instrumentally *necessary* for meaning in life.

A second argument for purpose theory appeals to the idea that having been created by God for a reason would be the only way that our lives could avoid being contingent or accidental. The ideas associated with talk of 'contingency' and 'accident' need to be sorted out, as theorists use the terms in a variety of ways (see, e.g., Gordon 1983: 238-41; Haber 1997; Jacquette 2001: 10-12). I do not undertake that mini-project here, and just aim to tease out what William Lane Craig might have in mind, as he has articulated this reasoning more explicitly than anyone else:

> Without God the universe is the result of a cosmic accident, a chance explosion. There is no reason for which it exists. As for man, he is a freak of nature—a blind product of matter plus time plus chance. Man is just a lump of slime that evolved into rationality. There is no more purpose in life for the human race than for a species of insect; for both are the result of the blind interaction of chance and necessity....What is true of the universe and of the human race is also true of us as individuals. Insofar as we are individual human beings, we are the results of certain combinations of heredity and environment. We are victims of a kind of genetic and environmental roulette....(1994: 45).

One idea the passage suggests is that life would be meaningless without God's purpose because without it no one has intended that we exist, perhaps in a particular way. However, even if, *per argumentum*, our life would be meaningful just insofar as it were created by an agent with an end in mind, no reason is given for thinking that God must be the relevant agent. It might follow that wholes such as the physical universe or the human race would be meaningless in the absence of such a creator, but I am interested strictly in the conditions for an individual's life to be meaningful (1.2). And so long as at least one human person intentionally creates another, Craig's concern about the latter's being the product of mere chance seems to be avoided.

No doubt Craig would reply that being intentionally procreated by our parents is not sufficient to avoid the kind of contingency he has in mind. Our parents might not have existed, had the course of human evolution and society been different. So, perhaps the best way to express Craig's deep concern is that, without God having participated in our creation with a certain goal in mind, our individual existence would not be well grounded in the fabric of reality. A life that need not have arisen, i.e., is ultimately a product of chance (and that will perish in a handful of years), seems accidental in this respect. In contrast, a life springing from (and returning to) a spiritual source of the physical world would seem not to be accidental.

There are three serious problems with this reasoning.[2] First, suppose, as many theists believe, that God's nature did not require Him to create us, and that it was His arbitrary, or at least contingent, decision to do so. If we sprang from the source of the physical world, but if that springing were itself a mere likelihood and not inherent to the source, then our existence would still be accidental.

Second, suppose, in contrast, that God's nature did require Him to create us (or grant for the sake of argument that His contingent choice to create us would prevent our lives from being 'accidental' in the relevant sense). It is still not clear that *only* God could prevent contingency, understood as a condition in which one's life is not deeply rooted in the nature of the world. To see this, imagine that the universe were everlasting. In addition, assume that fundamental physical laws dictated only one path for the universe, a path that deterministically resulted in our coming into existence. One could further suppose that we were somewhat like vampires, able to live indefinitely. Since under these entirely naturalist conditions our lives would be an unavoidable part of the structure of reality, it appears that God is not necessary to avoid contingency. Hence, even if God could prevent our lives from being contingent, we lack an explanation of why He alone could do so.

The third problem with Craig's contingency-based rationale for purpose theory is that it fails to show that *fulfilling* God's purpose is necessary for life to have meaning. The argument focuses entirely on the idea that meaning is a function of having been created in a certain manner. What one does after the creation plays no role in the argument, but nearly all purpose theorists maintain that it is the free decision to participate in God's plan that would confer the meaning on one's life. Another way to put the point is that, if originating from a certain, divine source were sufficient to make one's life meaningful, then everyone's life would be meaningful and to the same degree. However, it is a given among meaning theorists that some lives are more meaningful than others, with some probably not being meaningful at all (1.2, 4.3).[3]

5.4 The strongest argument for purpose theory

It is time to explore the philosophically most powerful rationale for purpose theory. This is the argument that God's purpose could be the sole source of invariant ethical

---

[2] For another angle, according to which chance makes no difference to meaning in life, see Trisel (2012).
[3] This last objection also plagues a third rationale for purpose theory, that God's purpose is necessary and sufficient for the universe to constitute an intelligible aesthetic object (Gordon 1983: 241–6). If one's life is meaningful just insofar as one's environment has meaning in the way that aesthetic objects do, as products of an agent's intentional activity, then freely doing something to realize God's purpose would be irrelevant, and everyone's life would be equally meaningful.

rules, where our lives would obtain meaning by conforming to them, and would be meaningless if such rules did not even exist.

There are a variety of articulations and defences of this argument in the literature (Davis 1987: 296, 304–5; Moreland 1987: 124–9; Craig 1994: 44, 48–9, 54–5; Adams 1999; Cottingham 2003, 2005). To obtain focus, I critically discuss John Cottingham's work, as he has most explicitly and thoroughly sought to draw conclusions about the meaning of life from considerations about God's purpose as the only possible source of an objective morality. In recent years, Cottingham has been the most prominent English-speaking voice defending this (or any other) God-centred approach to meaning in life, doing so with care and rigor.

I argue that Cottingham's version of the appeal to a 'divine command theory' of morality, as philosophers call it, fails, and that the reasons it does apply with equal force to any other version that could be presented. Note that I do not argue against the premise that a universally applicable and binding moral code is necessary for meaning in life, something others have questioned (Margolis 1990; Ellin 1995: 327). I am inclined to agree with Cottingham and the other friends of divine command theory cited above,[4] that at least a necessary condition of a meaningful life is the existence of a suitably non-arbitrary moral system. Living in a world in which Nazism and Stalinism are not wrong, or are wrong only for some people at certain times, would be topsy-turvy, I accept.

The premise of Cottingham's argument that I reject is the claim that only God's purpose could avoid such a world. Now, I do not reject the divine command theory of morality for the old problems traced back to Plato's *Euthyphro* dialogue; I maintain that Cottingham and other philosophers of religion have resolved them. I present a new reason to reject a God-based morality and hence the key premise of the present argument for a God-centred theory of meaning. In the rest of this section, I spell out Cottingham's argument, appraising it only in the following one.

I count four logically distinct features of the invariant or absolute morality that, for Cottingham, is required for meaning in life. First, moral norms of the right sort are *universal* in scope. A meaningful morality must be 'more than a temporary fragile disposition possessed by a percentage (perhaps a minority) of a certain class of anthropoids' (2003: 72), and 'independent of the contingencies of fluctuating human desire and uncertain historical development' (2005: 54n37). For example, it must be true of any person in any place and at any time that it would be wrong for her to torture a baby for fun.

Second, in order to confer meaning on our lives when we fulfil them, moral norms must be *objective* as well. Objectivity is mind-independence; that is, for moral norms to obtain not merely because they are the object of the (variable) mental states of human beings or other finite persons. Here, one often finds Cottingham rejecting

---

[4] As well as many others in the literature, including Kant (1790); Murphy (1982: 12–17); Tännsjö (1988: 258–60); Jacquette (2001: 12–16).

the importance of a morality that has 'no reality beyond the localized and temporary desires and conventions of humans' (2003: 33) and favouring an ideal of 'attuning ourselves to a creative order that is inherently good' (2003: 62). In short, the fact that it is wrong to torture babies for fun must not be constituted by us, but instead must be part of 'the ultimate nature of the cosmos' (2003: 66; see also Cottingham 2005: 57).

Third, in addition to universality and objectivity, Cottingham believes that meaningful moral norms have *necessity*. He argues that if moral norms were a product of the particular evolutionary history of our species, then they would be merely 'contingent facts' (2003: 71) and a 'cosmic accident' (2003: 62; see also Cottingham 2005: 54n37, 55-7), unable to confer meaning on our lives when we act in accordance with them. Important norms are instead 'eternal and necessary moral verities' such that 'cruelty is wrong in all possible worlds' (2005: 55).

Fourth, and finally, Cottingham believes that the relevant kind of moral norms are *normative* in the sense of providing a categorical reason for action that is (at least often) conclusive. A categorical reason for action is a reason that is binding on an agent regardless of her desires and interests, where a conclusive reason is one that outweighs all other reasons for action. So, the wrongness of torturing babies for fun must be such that it provides an all-things-considered reason not to do it, regardless of whether the agent wants to do so or whether it would benefit her.

Consider, now, why it is reasonable for Cottingham to believe that God could most plausibly ground a morality with these four features of universality, objectivity, necessity, and normativity. A God who stood apart from and above finite moral agents could straightforwardly provide a set of norms that applies to all of them. If God were to command all persons to perform certain acts and to refrain from others, He could create universally applicable moral norms. Furthermore, these commands would obtain independent of the mental states of any of these agents and hence would be objective. In addition, supposing that God would exist necessarily and could not change His mind (perhaps because He would always already be perfect), then the command not to torture babies for fun would be necessary, 'a timeless moral truth or principle held in the mind of God, an inseparable part of the structure of the divine mind' (2005: 47). Finally, the commands of a perfect being could straightforwardly ground a reason for us to perform an act, regardless of its effect on our desires or interests (2005: 46-57).

Although it is difficult to conceive of a supernatural realm and how it might operate, Cottingham provides some reason to think that an invariant morality not only can be grounded in it, but also could *alone* be grounded in it. In particular, *contra* the naturalist, who is on the most firm ontological ground for positing a relationship of constitution between natural and moral properties, Cottingham appeals to the familiar Humean rationale that, upon apprehending values, 'we humans are plainly recognizing something that goes beyond the observed facts of the natural world' (2005: 48).[5]

---

[5] As another commentator has remarked about the suggestion that the ethical could be constituted by the physical: '(O)ne will search in vain for any physics textbook describing moral value as one of matter's properties!' (Copan 2004: 299-300).

5.5 How to account for an invariant morality

Does God's will best account for the possibility of an invariant moral system? In this section, I explore two kinds of reasons for thinking not. First, I consider traditional *Euthyphro* problems about the alleged inability of divine commands to entail non-arbitrary moral rules, and I explain how Cottingham has plausibly dealt with them (5.5.1). In the next section, I raise a fresh problem for Cottingham's invocation of God's will to ground an invariant morality, that even if an appeal to God could *entail* invariant moral rules, doing so would not best *explain* them (5.5.2), after which I respond to replies from Cottingham (5.5.3).

5.5.1 Euthyphro problems

Cottingham is well aware of the *Euthyphro* problems facing an attempt to ground morality in God's will, and he adopts promising strategies to resolve them. Here are three of the traditional reasons for doubting that God's commands could entail the key features of an invariant moral system. First off, they seem unable to support an intuitively attractive morality, for if the bare fact of a command were right-making, then torturing babies for fun would be right if God commanded it. Second, although God's commands would be objective (in the sense of independent of our minds), they would be neither universal nor necessary if God's commands changed, something major monotheistic traditions often agree has happened. Third, for some it is hard to see how the bare fact of being commanded to do something could be normative, that is, provide an overriding, categorical reason for action.

Following Aquinas and some other contemporary theists (MacDonald 1991; Adams 1999), Cottingham convincingly resolves all three problems in one fell swoop, by grounding God's commands in God's perfect nature. Suppose that God's essence just were perfection, where perfection is creative, knowledgeable, and benevolent personhood. Suppose, further, that perfection is unchanging.[6] If God's immutable nature just were creative, knowledgeable, and benevolent personhood, then God's commands would be fixed by it. Hence, God could not command us to torture babies, for an essentially benevolent agent could not issue such a command. In addition, God's commands could not change, since His nature would be unchanging. And, lastly, God's commands would ground conclusive, non-instrumental reasons for action in that they would spring from perfection itself; normativity would, roughly, be a matter of becoming like God, 'a source that is generative of truth, beauty and goodness' (Cottingham 2003: 90), as much as we can, or striving to 'participate, however dimly, in the divine nature' (Cottingham 2005: 49).

---

[6] Perhaps because any alteration would be a matter of 'going downhill' from an apex, or because atemporality is a higher state than the 'feebleness of division' (Plotinus) inherent to temporal extension, or because an utterly simple being incapable of changing for lack of parts would be more independent and hence higher than a complex being dependent on parts for its existence (cf. 6.5).

Summing up, Cottingham's appeal to God's perfect nature enables him to avoid many of the standard *Euthyphro* concerns about the ability of a God-based ethic to entail an invariant morality. In the following section, I raise a different problem with Cottingham's God-based meta-ethic.

5.5.2 Explaining an invariant morality

Granting Cottingham that rules with the right sort of content that are universal, objective, necessary, and normative could be grounded in God, I question whether appealing to God provides the best theoretical account of them. I first argue that Cottingham's appeal to a God-based meta-ethical view evinces a logical incoherence, and then note that most readers would manifest a similar incoherence if they were to adopt such a view.

If there were conclusive evidence for a God-based ethic, then the evidence for God's existence would be comparable in strength to the evidence for the existence of wrongness—but Cottingham's works indicate that it is not. Cottingham's writings make it clear that he thinks he knows that certain acts are wrong but does not think he knows that God exists, which discrepancy entails that he is not justified in claiming to know that wrongness is fundamentally a function of God.

To clarify the nature of the incoherence, consider that for any entailment 'If X, then Y', it would be inconsistent to claim the following three things: I know 'If X, then Y' is true; I know X obtains; I do not know whether Y obtains. Now, Cottingham claims to know that 'If wrongness exists, then God exists' is true and also to know that wrongness exists, but he denies knowing that God exists. That set of claims is incoherent.

If Cottingham wants to retain the idea that he knows that a God-based ethic is true, then, to avoid incoherence, he must either claim to know that God exists or deny knowing that wrongness exists. However, as I show below, Cottingham explicitly rejects both of these positions. Furthermore, I maintain that Cottingham and the rest of us *would be correct to reject these two claims*, meaning that, to avoid incoherence, we must reject a God-based ethic.

First, Cottingham could try to argue that there is conclusive evidence of God's existence, as much evidence as there is that certain acts are wrong. If that were true, then it would be coherent to claim that there is conclusive evidence that wrongness is a function of God. However, Cottingham expressly denies that we have any conclusive evidence that God exists. He maintains that, at best, the evidence does not indicate that God does not exist, such that theism is at most consistent with, but not positively defended by, the evidence. Invoking the tradition of Pascal and Kant (Cottingham 2005: 6-8, 18), Cottingham's conclusion is that 'there is at least the possibility of a religious interpretation of reality' (2003: 62), one that the evidence neither indicates is actual nor indicates is non-actual, thereby permitting one to have faith in God's existence without contradicting one's rational nature. Summarizing his analysis of arguments for atheism, Cottingham concludes, 'the evidence from the observable world was at best compatible with a claim about its ultimate divine source: although not ruling it out, it was not such as to support it either'.[7]

Now, the qualification that evidence 'from the observable world' favours neither theism nor atheism might suggest that Cottingham believes that there is some other kind of knowledge of God's existence available. And Cottingham does at times suggest that there is a way to find 'knowledge of God' (2005: 12), to access religious 'truths that are made manifest' (2005: 139, 16), and to have 'some form of support for the theistic interpretation of reality' (2005: 136).

The epistemic reason Cottingham discusses in this context is a certain kind of religious experience. Adamant that he is not making an abductive argument that would posit God as the best explanation of religious experience, Cottingham instead maintains that sometimes the world is 'seen as carrying traces of the transcendent divine world that is its ultimate source' (2005: 123), an experience that provides non-inferential warrant for theism. These experiences include 'glimpses...of a world transfigured by overwhelming goodness and beauty' and 'transformations wrought in our lives by prayer and meditation' (2005: 133; see also Cottingham 2003: 61). These are supposed to be apprehensions of the world as participating in the divine, such that, for example, one does not merely perceive beauty, but rather a beauty *that has God as its source*. Not everyone has these experiences, which cannot be replicated willy-nilly in a scientific experiment. But for those who have engaged in spiritual practices over some time— that is, those who are most likely to have these experiences—they provide an immediate epistemic justification for theism. Call this the 'religious experience' defence of God's existence.

Cottingham's texts themselves indicate two strong reasons for denying that religious experience ultimately entails that any of us has conclusive evidence of God. First off, recall that Cottingham says that the evidence 'from the observable world' is equivocal with regard to God's existence, and notice that Cottingham explicitly characterizes religious experiences as '*observational* modes of inquiry' (2005: 131-2). Second, immediately after noting that religious experiences have traditionally been regarded as 'signs of the divine presence' (2003: 61), Cottingham points out that there are other, equally weighty indications of God's non-existence (for example, the quality and quantity of evil) and concludes that there is a 'stand-off when it comes to evaluating the nature of the cosmos we inhabit...that leaves the door open for the theist (as it does for the atheist, or the agnostic)' (2003: 62). These two passages render consistent the tension that initially seemed to exist between Cottingham claiming that God is no more than consistent with the evidence and claiming that religious experience provides epistemic reason to believe in God. The tension is reconciled by noting that this epistemic reason to believe in God is merely *pro tanto* and is not all things considered. And if the epistemic reason is not conclusive, then Cottingham still faces the objection that the

---

[7] Cottingham (2003: 92); see also Cottingham (2005: 6–8, 13, 24–5, 47–8, 57–8, 61–2, 118–19, 122–4, 133).

evidence for God is inconclusive and the evidence for wrongness is conclusive, where the evidence for both must be comparable if we are to have conclusive evidence that wrongness is a function of God.[8]

Hence, Cottingham is committed to the view that there is no conclusive evidence that God exists. And he therefore cannot coherently think that wrongness is a function of God, if he also thinks, as he does, that there is conclusive evidence that wrongness exists. If there were knowledge that wrongness logically depends on God, then there would have to be knowledge of the existence of both, but there is not, for Cottingham and, as I discuss below, for us.

This brings me to the second major way that Cottingham could object to my claim of incoherence, namely, by maintaining that our knowledge of the existence of wrongness is equivocal in the way that our knowledge of God is. However, Cottingham explicitly—and plausibly—denies that there is merely inconclusive evidence that wrongness exists. Commenting on the fact that some people enjoy being greedy, harmful, and arrogant, Cottingham says (2005: 55):

> (D)espite the grizzly satisfactions so described, such actions are wrong, indeed necessarily wrong: cruelty is wrong in all possible worlds. (Those who doubt this are invited to try to construct a coherent scenario of a possible world in which such behaviour is good or right.)

This quote expresses not the weak view that if wrongness existed, cruelty would be wrong. It rather indicates the strong view that cruelty is in fact wrong. And rightly so. It would be poorly motivated for Cottingham to backtrack by saying that we lack conclusive evidence that anything is wrong. *For all we know*, torturing babies for fun, enslaving others so that one can become rich, raping others to feel a sense of power, and intentionally shooting innocents for target practice are indeed wrong. Most (though, of course, not all) of the debate in contemporary moral philosophy is not about whether wrongness exists, but rather about what its nature is and how it is known.

I conclude that there is an incoherence in Cottingham's views, one that is unavoidable so long as he adopts a God-based meta-ethic. On the one hand, Cottingham maintains that wrongness is constituted by God, but, on the other hand, he claims to

---

[8] Cottingham's work suggests another move he might make. He distinguishes between two sorts of evidence, namely, that which is 'discursive' (2005: 122, 133), 'demonstrative' (2005: 118), 'propositional' (2005: 124), 'empirically testable' (2005: 136), 'impartial' (2005: 138), and 'argumentative' (2005: 133), on the one hand, and (roughly) non-inferential, non-propositional, and private, on the other. Cottingham's *oeuvre* could be read as saying that the former kind of evidence is inconclusive as to the existence of God, but that the latter sort is conclusive, tipping the scales in favour of theism for those who have had religious experiences. On this way of reading Cottingham, he would say that some of us do in fact have knowledge of God comparable to knowledge of wrongness.

However, not many of this book's readers will have had such religious experiences. In addition, the appeal to them is probably not enough to show that the evidence of God is so strong as to confer knowledge. When seeking evidence of something external to oneself, one ought to encounter some kind of consensus; if many different minds converge on the same opinion about what is beyond them, often the best explanation of the convergence is that there really is something beyond them that they are all tracking. But Cottingham does not provide any reason to believe that religious experiences across the world have a substantially similar content, and I doubt that such evidence is available.

know that wrongness exists and claims not to know that God exists. And it is urgent to see that this argument is *not merely an* ad hominem *against Cottingham*: a large majority of readers will think that they know that certain actions are wrong, but not that they know that God is real. If one encounters this discrepancy, then one cannot coherently claim to know that whether actions are wrong or not logically depends on God. To be coherent, one should hold that wrongness is a function of something other than God, since, I suggest, one is not likely to find either more evidence that God exists or less evidence that wrongness does.

In light of the conclusive evidence that wrongness exists and the inconclusive evidence that God does, a more coherent meta-ethical position would be that wrongness is a function of natural properties. There is substantial evidence that there are natural—that is, physical—properties, where this evidence is comparable in strength to the evidence that wrongness exists. Hence, if a naturalist meta-ethic can be shown to entail an invariant morality, then it should be favoured over a supernaturalist one for reasons of explanatory strength.[9] Given that we do not know that a spiritual ream exists, and given that we do know that matter exists, a naturalist absolute morality would fit much better with what else we (think we) know about the world. What I do in the rest of this sub-section is to articulate the way a naturalist metaphysics could plausibly underwrite an absolute ethical system.

Cottingham does briefly address naturalist meta-ethical views (2005: 54n37), but not the one that I find most easily able to account for an invariant morality, namely, a kind of moral realism that is analogous to scientific realism, the hallmark of Cornell meta-ethics for some time.[10] Consider briefly how realism captures invariance in the scientific realm, before applying it to the moral.

Let us suppose that the claim 'Water is H_2O' is universally true, objectively true, and necessarily true—that is, it is true for everyone, as someone who did not believe it would be mistaken; it is true in virtue of something independent of our beliefs about it, for it took a lot of empirical discovery to ascertain that it is true; and it is true in all possible worlds, for if something were encountered that seemed a lot like water but were composed of XYZ rather than H_2O, it would not be water. Of course, there are those who reject these assertions about the claim that water is H_2O, but my aim here is not to defend them. It is rather to point out the way that realists capture these facets of invariance in science.[11]

The invariance is accounted for in terms of synthetic *a posteriori* necessities and a causal theory of reference. When we claim that water is H_2O, we are expressing a real property identity such that the term 'water', which is associated with features such as

---

[9] I favour a naturalist meta-ethic over a non-naturalist one, since it is clear that there are physical properties, but not at all clear that there are ones that are neither physical nor spiritual (see 8.7 for further discussion).

[10] See, for example, Boyd (1988); Sturgeon (1988); Brink (1989); Gilbert (1990); and Miller (1992), who put the semantics of Saul Kripke and Hilary Putnam to work in defending the view that there are mind-independent, physical moral properties such as injustice.

[11] A way that Cottingham happens to accept (2005: 29–30).

being a clear, odourless liquid found in the ocean and in the rain, picks out the same thing in the world as 'H_2O', a term associated with a certain chemical composition. The co-reference of the two terms is thought to be necessary because the terms 'rigidly designate' one and the same existent, once a certain dubbing process has taken place. This co-reference is not ascertained *a priori* in the way that the claim 'A bachelor is an unmarried male' is, but rather through *a posteriori* methods of perception, induction, and abduction. Relatedly, the co-reference is not analytically true in the way 'A bachelor is an unmarried male' is, for the sense of the term 'water' does not include the chemical composition associated with 'H_2O'. Summing up, we have learned empirically over time that our terms 'water' and 'H_2O' essentially refer to one and the same thing.

Moral realists account for the invariance of morality in the same way. They view moral principles as synthetic *a posteriori* necessities, so that, by a Kantian ethical theory, 'Wrongness is degradation of persons' would be analogous to 'Water is H_2O'. The term 'wrongness', which is, say, associated with actions that *pro tanto* warrant guilt or blame, picks out the same thing in the world as 'degradation of persons', behaviour that fails to treat rational creatures as having a superlative final value. The co-reference of the two terms would be necessary, since the terms rigidly designate one and the same property, again, once a certain dubbing process has taken place. The co-reference has not been ascertained *a priori*, or at least *a posteriori* methods have played by far the dominant role in supporting it (particularly in the past 40 years' expansion of normative ethical theorization). And it is, of course, not analytically true, since it would not be logically contradictory to reject it in favour of, say, the consequentialist claim that wrongness is failure to maximize welfare. If one believes the evidence favours Kantianism, then the moral realist would say that we have learned empirically over time that our terms 'wrongness' and 'degradation of persons' essentially refer to one and the same class of actions. Hence, the claim 'Wrongness is degradation of persons' would be universally, objectively, and necessarily true in the same way that 'Water is H_2O' is.

So far, I have explained how one might account for the invariance of moral norms on a naturalist metaphysics, basically in the same way that realists account for laws in science. However, one feature of an absolute morality has so far been unaccounted for, namely, normativity. Here is where the analogy between science and morality might seem to break down. The claim 'Water is H_2O' is not normative; that is, it does not include a conclusive, categorical reason to do anything, not even to believe something. In contrast, claiming that it is wrong to torture babies for fun does include an overriding reason not to do something, which reason obtains regardless of one's desires and interests.

In fact, I submit that the analogy between scientific realism and moral realism can be extended to account for the normativity of morality.[12] The way that a realist ought to account for normativity is by asserting another sort of property identity, this time

---

[12] In a manner that the Cornell realists have been reluctant to do, for, invariably, they are instrumentalists about practical reason.

between wrongness and (practical) rationality. Consider the claim, 'One (typically) has overriding, categorical reason not to perform wrong acts.' There is nothing stopping the moral realist from maintaining that we have empirically learned that the terms 'overriding, categorical reason not to perform actions' and 'wrongness' essentially (or largely) co-refer. Here, it would not be that the term 'wrongness' inherently *connotes* the idea of a conclusive, non-instrumental reason not to perform a certain act (which, as is well known, would oddly make amoralism a logically contradictory view). Instead, 'wrongness' would *denote* the property of having an all-things-considered consideration not to act, a consideration independent of one's desires or interests. There would be a synthetic *a posteriori* connection between wrongness and normativity, such that it is universally, objectively, and necessarily true that one has overriding, categorical reason not to perform wrong actions such as torturing babies for fun. Hence, it is incorrect to think that the naturalist is committed to the view that any reasons that exist for people to act must be 'relative to their desires or inclinations' (Cottingham: 2005: 53).

Consider, finally, the way to refute the Humean reason Cottingham proposes for rejecting naturalism. That objection to naturalism, recall, is that when one apprehends nature, one does not sense any moral properties; ethical norms are not immediately 'observed'. Normativity, especially, is not part of our sense-data, e.g., not something that we see, smell, hear, etc. This rationale supposes that naturalism must be reductive in the sense of holding that moral language is reducible to the language of physics or some sense-based enquiry. However, at the core, naturalism is a metaphysical thesis about what exists (only the physical) and, in the meta-ethical realm, about the nature of ethical properties (they are physical). It is not essentially a view that the language of physics or of what is apprehended through one of the five senses must (or even can) be used to express what exists, a view that most self-described 'naturalists' reject these days. Hence, in the same way that contemporary naturalists in the philosophy of science are happy to grant that we do not literally or immediately see causation or democracy but may infer that they are physical relations, so a naturalist in meta-ethics can maintain that we do not apprehend normativity directly through any one of the five senses, but can reasonably conclude that it is a physical relation.

Summing up, as there is another, naturalist way to capture the invariance or absoluteness of morality besides an appeal to the supernatural, and since the former is more coherent than the latter, fits with our background ontology better than the latter, and is not vulnerable to Cottingham's objections, I find weak Cottingham's key claim that an invariant morality (and hence meaning) is possible only if God exists and assigns us a purpose.

5.5.3 Replies and rejoinders

Cottingham has replied in two major ways to this objection.[13] For one, he has questioned the ability of naturalistic moral realism to ground invariance of the sort he

---

[13] Which I initially made in Metz (2008).

thinks is relevant. For another, he has argued that my charge of logical incoherence is fallacious. I take up both replies in turn, aiming to provide conclusive reason to believe that my objection is sound.

Although I may have shown that realism could account for certain kinds of moral universality, objectivity, necessity, and normativity, I have perhaps failed to capture the precise kinds that Cottingham believes are not only necessary for meaning in life, but could also be grounded by God alone. First off, as discussion of 'Moral Twin-Earth' has suggested,[14] moral realism is at best able to account for truths that are universal across (and normative for) *the human race, not all species or from a God's-eye point of view.* Suppose that another species used the term 'wrongness' to refer to some property other than degradation of persons (or whatever feature one most plausibly thinks constitutes impermissible action on earth). Then what counts as 'wrong' for them will differ from what counts as 'wrong' for us. Cottingham, in contrast, believes that the important sort of universality is one that avoids the 'unacceptably relativistic conclusion that rightness or wrongness depend on the contingences of species development'. (2005: 54n37; see also Cottingham 2003: 71). In response to me, Cottingham remarks:

> Without a divinely based teleology, something like an ideal pattern of life or goal for human existence, I cannot see how we can distinguish among which of the 'natural' inclinations and satisfactions of our species have ultimate normative force. The systematic indulgence by a powerful tribe of their impulses towards conquest and genocide might if sufficiently successful, usher in a millennium of stability, prosperity and personal flourishing for the winners, and I can see no ultimate conclusive reason which, for the naturalist, would count against such a course of action (2008: 267).

To run with Cottingham's reply still farther, when it comes to objectivity, the realist is committed to thinking that the content of wrongness is fixed by a linguistic dubbing process dependent on human behaviour. Of course, the essential nature of what it is that gets referred to is a mind-independent matter, and so there is a much stronger objectivity than forms of conventionalism or social relativism. However, what our species denotes with a certain term is a subjective issue, making the objectivity weaker than a God-based ethic, which removes the content of morality altogether from being a function of human disposition.

Finally, the strength of necessity when it comes to a realist approach is admittedly also weaker than a God-based one. Realism grounds a less robust kind of necessity in that it entails that truths in all possible worlds obtain by virtue of facts in only one world. Specifically, necessary truths about which actions are wrong are fixed when our species has rigidly designated certain properties 'wrong'. In contrast, on Cottingham's model, there can be necessary truths about wrongness that do not obtain by virtue of facts in only one world, but rather facts in all possible worlds, namely, the contents of God's mind.[15]

[14] E.g., Horgan and Timmons (1990–1991).
[15] For more on this distinction between types of necessity, see Nozick (2001: 120–68).

There are three ways that I believe the moral realist can plausibly respond to the charge of being able to account for only a weak invariance. First, with respect to the scope of universality (and normativity), she can point out that even if another species used 'wrongness' (and 'conclusive, categorical reason') to denote properties different from the ones we pick out with these terms, *we* could still say *of that group* that its members are acting wrongly when they degrade persons (an act that they have all-things-considered and non-instrumental reason not to do). Just as the claim 'Water is H_2O' is true for any species *from within our language*, so would the claim 'Wrongness is degradation of persons' be universally true, too.

Second, and also regarding the scope of universality and normativity, the naturalistic moral realist can tell a compelling story about why all persons on earth would have a common history leading them to dub the same properties with the same terms. Familiar socio-biological rationales about the evolutionary advantages of cooperation among finite agents provide some reason for thinking that natural selection would lead human societies to refer to the same essential property with the term 'wrong' in the way they presumably do with 'water'. Perhaps a similar evolutionary force would apply to those beyond the human race.

What the realist can say, thirdly, in response to all four respects in which her invariance is more limited and weak than a God-based invariance is that the former is nonetheless sufficient for meaning in life. Although a number of theorists have suggested that life would be meaningless if there were no invariant morality, no one, so far as I know, has provided an account of precisely which sort of invariance is key.[16] It would be useful for me to provide some reason for thinking that if an invariant morality were necessary for meaning, it would suffice to be an invariance with the scope and strength that a physicalist meta-ethic could ground.

My strategy is to draw an analogy between the true and the beautiful, on the one hand, and the good, on the other, suggesting that the meaning-conferring invariance of the former is merely naturalist. Cottingham himself often invokes the classic triad of the good, the true, and the beautiful as largely constitutive of meaning (2003: 33, 90, 103, 2005: 43), and, as I have suggested, they are exemplars of meaning for the field more generally (1.2). So, if I can ascertain that a naturalist invariance grounds meaning in two of these conditions, I can fairly draw a similar conclusion about the remaining one.

Intuitively, meaning can come from making scientific discoveries and creating works of art, and I think it is clear that the kind of invariance sufficient for meaning in these cases is one for which a naturalist could well account; that is, when it comes to discovering scientific laws or the laws of beauty, the laws need not be ones that apply to all species, that are utterly independent of human reference, or that obtain necessarily by virtue of facts in all possible worlds. It is sufficient that these laws are true merely

---

[16] Nor has anyone yet provided an explanation of *why* invariance of some sort or other is required for meaning.

for all human beings, fixed by what we denote with certain terms and necessarily true by virtue of human reference in this world. For example, even if another species theoretically carved up the world in such a way that $E = MC^2$ were not true for it, Einstein's law was a fantastically important discovery. And if another species of intelligent, finite creatures with much different sensibilities and experiences from ours were to find Picasso's works to be ugly or otherwise aesthetically revolting, they conferred a terrific amount of meaning on his life. Similar remarks, I suggest, apply to morality. Even supposing Mother Teresa (or the stereotypical understanding of her) were not morally superior by another species' norms, her actions made her existence significant by virtue of living up to high standards that have a human-wide invariance. The sort of invariant morality that a naturalistic moral realism could underwrite appears sufficient to avoid the meaninglessness of a world in which, for example, Nazi polices are not *really* wrong.

Cottingham's second major reply to my objection is that there is a logical problem with my charge of incoherence. Recall that I maintain that for any conditional 'If X, then Y', it would be inconsistent to claim the following three things: (1) I know 'If X, then Y' is true; (2) I know X obtains; (3) I do not know whether Y obtains. Now, Cottingham claims (1*) to know that 'If wrongness exists, then God exists' is true; and (2*) to know that wrongness exists; but (3*) not to know that God exists. I pointed out that not only does Cottingham commit himself to all three claims, but also that most readers already firmly hold (2*) and (3*), making it incoherent for them to adopt (1*), or the divine command theory of morality that logically implies it. In reply, Cottingham constructs the following counterexamples:

> I can surely maintain that apples are constituted by quarks, and yet be far more confident of the evidence that apples exist than of the evidence that quarks exist. Or I can surely maintain that the properties of the number zero are essentially constituted by complicated properties involving relations between sets of sets, yet be far more confident that the number of coins in my pocket is nil than I am about the existence of sets (I may have just read an ingenious paper refuting their existence, which I cannot see how to get round) (2008: 266).

Do these counterexamples rebut the charge of incoherence in holding a God-based morality?

Consider the apples case first. The parallel would be this: (1~) I know that apples are constituted by quarks, or what is implied: I know that if apples exist, then quarks exist; (2~) I know that apples exist; and (3~) I do not know whether quarks exist. Cottingham is maintaining that if one accepted (2~) and (3~), there would be nothing incoherent about coming to believe (1~). However, I continue to find incoherence in this case. If one has enough evidence to *know* both that if apples exist, then quarks exist, and that apples exist, then I cannot fathom how one could consistently claim not to have enough evidence to know that quarks exist. Similar remarks apply to the other case, which, to be parallel, would be this: (1^) I know that zero is constituted by set relations, or that if zero exists, then set relations exist; (2^) I know that zero exists; and (3^) I do not know whether set relations exist. Again, the incoherence is patent to me.

I suspect where Cottingham's purported counterexamples go awry is the focus on differential 'confidence'. It might be true that it could be coherent to think that if wrongness exists, then God exists, supposing that one were (merely) *more confident* that wrongness exists than that God exists. However, the relevant principle is one that appeals to differential *knowledge*, in other words, differential 'confidence' in the sense that there is *conclusive* evidence that wrongness exists, to the point where one knows that it does, on the one hand, and inconclusive evidence that God exists, where one does not know that He does, on the other. That seems to make a striking difference with regard to whether the incoherence obtains.[17]

To conclude this chapter, recall that it has begun the part of the book in which I systematically appraise theories of meaning in life, searching for the most defensible one. There are two major categories of such theories, supernaturalist and naturalist, and I have begun by critically exploring the former, the view that one's existence can be significant if and only if there exists a spiritual realm with which one interacts in a certain way. I spelled out the most influential version of supernaturalism, purpose theory, which is common to many theologians in the Jewish, Christian, and Muslim traditions. According to this view, one's life is more meaningful the better one fulfils a purpose that God has assigned, and I have identified the most promising argument for purpose theory to be one that John Cottingham has articulated with most care in the context of discussion of life's meaning. According to this rationale, God's purpose constitutes meaning in life because life would be meaningless without an objective system of moral rules for human persons, and because only God could create such a system. In evaluating this argument, I have elected to focus on the latter premise alone. I argued it is incoherent for Cottingham and for a large majority of readers to believe that only God's purpose could be the source of an objective ethic, since we know that wrongness exists but do not know whether God does. And I explained how nature could plausibly ground such an ethic.

Even if my objection to the divine command theory were successful, it would show merely that one of the most influential and powerful reasons for believing in purpose theory should be disbelieved. It would not show that purpose theory itself should be disbelieved, that the heart of monotheistic thought about meaning in life does not merit acceptance. Establishing that is my aim in the next chapter.

---

[17] Roger Crisp in correspondence has suggested another way out, here. He points out that one could claim to apprehend conclusive evidence of God's existence if one were initially sure both that wrongness exists and that if wrongness exists, then God exists. That is true. However, most of us are quite unsure of the latter claim; indeed, Cottingham's work is important precisely because so few in the field believe it and because he marshals resources to defend it that are worth taking seriously.

6

Purpose Theory II: Advancing Objections

'(W)e complain of men and their mutability, and of the mutability of all temporal things; but God is unchangeable, this is our consolation, an entirely comforting thought—so speaks even frivolity....(I)n a manner eternally unchanged, everything is for God eternally present, always equally before Him. No shadow of variation.... changes Him; for Him there is no shadow. If we human beings are mere shadows, as is sometimes said, He is eternal clearness in eternal unchangeableness.'

Søren Kierkegaard ('The Unchangeableness of God')

6.1 Aiming to undermine purpose theory

In the previous chapter, I argued that the motivations proffered for purpose theory are unpersuasive. In particular, I presented reason to reject the strongest rationale for purpose theory, that God's will alone could constitute an objective moral system, conformance to which is necessary for life to be meaningful. However, showing that the arguments for purpose theory are unconvincing does not show that the theory itself is incorrect. The latter is what I seek to establish in the present chapter.

Before presenting my own, novel reason for rejecting purpose theory, I demonstrate that the literature indeed needs a fresh objection to it. I maintain that the various arguments against purpose theory that one currently finds are unsuccessful. The most interesting ones[1] hold that purpose theory, when conjoined with very plausible theses, entails logical contradictions. There are arguments purporting to show that purpose theory entails the absurdities that God is not all-good (6.2), God is not all-powerful (6.3) and God is not eternal (6.4). I argue that these three *reductio* arguments against purpose theory fail, as there are independently attractive versions of purpose theory that do not entail these absurdities.

However, critical discussion of these arguments will point the way to a more telling one (6.5). My argument against purpose theory is that the best explanation of why

---

[1] I have responded to the less interesting ones in 5.2.

God alone might constitute meaning in life is that God has supremely valuable properties that we cannot conceivably exhibit, features such as simplicity, immutability, and atemporality, which Kierkegaard celebrates in the above passage. Unlike Kierkegaard, though, I maintain that such features are incompatible with God having a purpose, as normally conceived. In short, the fundamental motivation for holding a God-centred theory in general provides reason to reject the particular, purpose-oriented instance of it that has held sway in the monotheistic tradition for several centuries, or so I argue. After defending this argument from criticism (6.6), I conclude by indicating a need for God-centred theorists to develop alternative conceptions of how God might be the source of meaning in our lives, which are not centred on purposiveness, some of which I explore in the next chapter (6.7).

6.2 God's purpose v. God's morality

A prominent argument against purpose theory is that it would be immoral for God to *assign* a purpose to other agents, regardless of the *content* of that purpose. It has often been pointed out that God's purpose for us must have a certain content in order to be meaning-conferring; serving as food or entertainment for intergalactic travellers would not be sufficient.[2] The standard reply to make on this score is to invoke the divine command theory of morality, viz., to argue that God's will would not merely conform to ethical ideals, but would constitute them (5.4). The present objection is consistent with the view that the content of God's purposes would not be morally suspect, but is that, even so, His *giving* us a purpose, for example by commanding us, would be.

Purpose theory arose in the context of a teleological conception of normativity, where what is deemed of most value for us is achieving a particular end appropriate for human nature as such. Modern conceptions of normativity famously tend to reject the idea that everyone ought to realize some predefined, specific end; they instead tend to hold that we ought to live according to norms that rational agents have self-legislated in some way. Now, it appears that being assigned an end might conflict with the dictum that rational beings ought to live by their own choice. Hence, the atheist and naturalist Jean-Paul Sartre once said of his subjectivist theory, according to which the meaning of a person's life is a function of whatever choices she has made, that it 'alone is compatible with the dignity of man; it is the only one which does not make man into an object' (1946: 45).

In order to pin down in exactly what respect a God who assigned us a purpose might thereby degrade us, I appeal to Kantian ideas about morality. In characteristically modern fashion, the Kantian standpoint presumes that our most important feature is our capacity to make decisions for ourselves and proposes that the fundamental

---

[2] Nagel (1971: 721); Nozick (1981: 586–7); Murphy (1982: 14); Ellin (1995: 319–20); Sharpe (1999: 15); Thomson (2003: 54); Affolter (2007: 446–50); cf. Nielsen (1981: 184–7); Kekes (2000: 25).

moral norm is to respect this capacity. In the following, I examine the 'disrespect objection', the central claims of which are that it is immoral to treat our capacity for self-determination solely as a means to an end, and that God's assigning us a purpose would do exactly that. These claims together entail that if God assigned us a purpose, God would be immoral, a logical contradiction, since God is by definition morally ideal.

Of course, the purpose theorist could always reject a Kantian ethical perspective, denying that autonomy is the most important value (Walker 1989), and perhaps favouring utilitarianism instead. Or she could grant that Kantianism applies to us, but deny that it applies to God (Hanfling 1987: 45-6). Rather than spend time considering how the Kantian might reply to these two claims, I merely set them aside. For the sake of argument, let us suppose that both human and divine wills are morally obliged not to treat rational beings disrespectfully. It would be interesting if it could be shown that purpose theory is consistent with this Kantian thesis. In the following, I explore four ways in which God's assigning us a purpose might seem to degrade our capacity for free choice. I show that there is nothing inherently disrespectful about God's assigning us a purpose and hence that purpose theory need not absurdly entail that God is immoral.

6.2.1 Coercion

Why hold that God's ascribing us a purpose would be disrespectful? A first reason might be that restricting a person's choice by making threats is a quintessential form of disrespect, and it appears that God would threaten us by making eternal damnation the consequence of not realizing His end (Baier 1957: 107; Kurtz 1974: 86; Murphy 1982: 14-15; Baggini 2004: 13-17).

To begin to reply, consider that a threat is not necessarily disrespectful; it depends on why the threat is made. Specifically, a threat made incidentally in the course of maintaining a retributive punishment system need not be disrespectful. To fix ideas, suppose that a human society instituted a punishment system for the sake of giving violators of just laws the punishment they deserve. While not intending to deter crime with this punishment system, the society would nonetheless be making incidental threats to those who would break just laws. Regardless of its aim, the mere existence of a punishment system threatens citizens in saying, 'If you break a law, you will be intentionally harmed for having done so.' It does not appear that such threats would be disrespectful, on the plausible assumption that retributive punishment is respectful of autonomy.[3]

If it would not be disrespectful for a state to make threats in the course of maintaining a retributive punishment system (particularly within a Kantian moral framework), then it would not be disrespectful for God to make threats in doing the same. Supposing, as per the dominant account of purpose theory (5.4), that God's purpose

---

[3] On which see Metz (2006).

would constitute morality, or at least that it would be God's purpose for us to be moral, then our failing to fulfil that purpose would warrant punishment, and any threats God would make incidental to imposing that punishment would be respectful.

Now, I cannot rest content with this response, as it stands, since it appears impossible for a human to deserve eternal damnation. It is unlikely that any finite action can warrant an infinite reaction, meaning that even monstrosities such as Hitler and Stalin do not deserve to be in hell forever (cf. 7.4.1). Hence, I believe the purpose theorist must reject the idea that God would impose eternal damnation upon those who do not fulfil the purpose He assigns. The purpose theorist can accept that we have souls that live forever, and that God would impose a finite punishment upon souls who have rejected His (moral) end. However, to avoid the charge that God's assigning us a purpose would be disrespectfully coercive, I suspect that the purpose theorist can and should reject the postulate that God would send recalcitrants to hell forever, as that would be a disproportionate penalty flouting retributive principles.[4]

6.2.2 Exploitation

Even if the purpose theorist rejects the notion of eternal damnation, charges of disrespect may still arise. In fact, God's offering the reward of heaven for realizing His purpose might seem to be disrespectful. Some might deem this to be a 'coercive offer', or, in terms that I find more applicable, a form of 'exploitation'. It seems exploitive, and hence disrespectful, to offer a starving person food in exchange for doing what you like. What choice would she have but to conform to your will? By analogy, a critic could maintain that it would be exploitative for God to offer finite creatures an eternity of bliss in exchange for doing His bidding. The God-Father would be making an offer we can't refuse.

Of course, one option for the purpose theorist would be to reject the reward of eternal heaven as I have argued she must reject the punishment of eternal hell. She could hold that God would provide either no reward or a moderate reward for realizing His end. That conception of God's response to purpose fulfilment would easily resolve the problem.

However, I think the purpose theorist might be able to maintain that God could offer us the reward of eternal bliss for acting according to His will. The charge of exploitation arises most clearly when the purpose of the person making the offer involves degrading or harming the recipient. Most would not hesitate in calling 'exploitive' the offer of food to a starving person in exchange for sex or a kidney. It is not so clear, though, that it would be exploitive to offer such a person food in exchange for, say, working part-time at a soup kitchen. If this would not be exploitive, then perhaps it would likewise not be exploitive for God to offer us heaven forever in exchange for accomplishing His aim that we act morally.

---

[4] The lay conception of hell is largely out of fashion these days among philosophers of religion. See, e.g., Adams (1975); Kershnar (2005).

6.2.3 Condescension

There is yet a third version of the disrespect objection, namely, one most forcefully stated by Kurt Baier. Baier's rendition interestingly does not turn on God's imposing any scheme of punishment or reward. Since Baier's remarks are important and influential, I quote in full the relevant passage:

> We do not disparage a dog when we say that it has no purpose, is not a sheep dog or a watch dog.... Man is in a different category, however. To attribute to a human being a purpose in that sense is not neutral, let alone complimentary: it is offensive. It is degrading for a man to be regarded as merely serving a purpose. If, at a garden party, I ask a man in a livery, 'What is your purpose?' I am insulting him. I might as well have asked, 'What are you for?' Such questions reduce him to the level of a gadget, a domestic animal, or perhaps a slave. I imply that we allot to him the tasks, the goals, the aims which he is to pursue; that his wishes and desires and aspirations and purposes are to count for little or nothing. We are treating him, in Kant's phrase, merely as a means to our ends, not as an end in himself....[5]

Baier's claim is not that God's purpose would be 'selfish', i.e., in His best interest but not ours. Therefore, it will not suffice to point out that God's purpose would be in our objective interest (as Levine 1987: 461n does). Baier's concern about God's assigning us a purpose is that it would *degrade* us, not that it would *harm* us. Baier objects that being assigned a purpose would treat one's capacity for rational choice as a mere tool to be used for the realization of a purpose one does not share. It is irrelevant that realizing the purpose would be good for oneself; that would merely add a paternalistic aspect to the degradation.

It is also worth noting that Baier's claim is not that it is disrespectful ever to view someone as being useful. Hence, one cannot respond to Baier by noting cases in which it is not disrespectful to view people as having a use-value. If we ask a stranger what he does for a living, we are in effect asking how he contributes to society, and there is nothing disrespectful about this. Such a case does not tell against Baier, however (*contra* Brown 1971: 20), since enquiring about someone's job need not involve treating the person *merely* as a means, which is Baier's concern.

If God were to assign us a purpose, then God would clearly have to regard us as a means, i.e., as being useful for the realization of his end. The question is whether God must thereby regard us *solely* as a means, and that is not so clear. If God did not coerce, exploit, or deceive humans to get them to fulfil a purpose, then God would engage in no *manipulation*, the central form of treating a person merely as a means. However, might God *insult* us insofar as He assigns us a purpose? It seems so. If God has a purpose He wants us to fulfil, He would have to inform us of it, directly or indirectly saying to each of us, 'There is something I would like you to do with your life, and this is the reason that you exist.' As it stands, this statement does sound patronizing.

---

[5] Baier (1957: 104); see also Hepburn (1966: 262–3); Joske (1974: 259); Singer (1996: 29).

However, if we reflect some more on what God's purpose might be and how God might seek to promote it, the statement can be seen to be part of a respectful address. For example, suppose that the end God assigned us were to exercise our free will in a moral way. Informing us of such an end need not be condescending. Suppose that we filled out God's statement as follows:

Dear Thaddeus, There is something I would like you to do with your life, and this is the reason that you exist. Specifically, I would like you to be a moral person. Your free will is such that I cannot cajole you into exercising it morally, and your moral choice would be valuable only if it were made freely. Therefore, I ask you to pursue the fundamental end of pursuing moral ends. Yours sincerely, God.

We could even imagine that a 'please' were thrown in. I, for one, would not find this condescending.

In short, being assigned a purpose could be a matter of divine request, rather than divine command. If being assigned an end can be a matter of being asked to adopt the end voluntarily, then there need not be anything insulting about being assigned an end.

6.2.4 Creation for the wrong reason

Baier makes some additional remarks suggesting a fourth way that purpose theory might be thought to entail that God is disrespectful. We often think that it is possible for parents to be immoral insofar as they create offspring for the wrong reasons, and the same might go for God's creation of us. Baier claims that the purpose theorist 'sees man as a creature, a divine artefact, something halfway between a robot (manufactured) and an animal (alive), a homunculus, or perhaps Frankenstein, made in God's laboratory, with a purpose or task assigned him by his Maker' (1957: 104).

To illustrate the problem, suppose that a couple decides to have a child fundamentally because they would like a musician in the family. Merely acting on the maxim of creating a child in order to have a musician might constitute disrespect, so the argument goes. The same apparently goes for creating a child with the aim of having a worker on the farm. In contrast, it would not be disrespectful to make a baby for the sake of promoting a being who will set its own ends. Baier might therefore suggest this principle to govern the creation of rational beings: it is disrespectful to create a person for any purpose other than to pursue its own purposes. Irving Singer, who echoes many of Baier's concerns, approximates this principle when he objects, 'If humanity, or life in general, was created to serve a purpose beyond itself, our being would be analogous to that of a manufactured artifact' (1996: 29). Now, God's purpose for us would presumably involve living morally, and it therefore appears that God would create us for a purpose other than setting our own, arbitrary ends.

Obviously, the purpose theorist must cast doubt on the principle I have culled from Baier's and Singer's remarks. In particular, the purpose theorist must contend that the principle is too broad, i.e., that it can be respectful to create a person for a given purpose beyond that of adopting her own purposes. It is clear that someone—whether God or a parent—who created a person in order to have another moral agent on earth would be treating that person as a means, but it is not obvious that he would be thereby treating her *merely* as a means. If, as we assume, such a creator did not coerce the created into being moral, did not take advantage of the created's weakness to get her to be moral, and did not condescendingly tell her to be moral, then creating her so that she would become virtuous would not dishonour her autonomy. Again, if God the creator pursued His end of having the created become ethically upright merely by reasoning and requesting, then the fact that He made the created in order for her to be moral would not appear to treat her capacity for free choice solely as a means to His end.

I conclude that while some versions of purpose theory, particularly those involving eternal damnation, are vulnerable to the charge of disrespect, purpose theory as such is not. If we can imagine a God whose end for us includes moral action, who does not threaten us with an eternity in hell to get His way, who does not make offers we can't refuse, and who does not address us in a patronizing way (but rather argues and asks), then we can conceive of a purpose theory that does not entail the absurdity that God is disrespectful and hence immoral.

6.3 God's purpose v. God's omnipotence

The second major argument against purpose theory in the literature is that it entails that God is not omnipotent, which is a contradiction. If it were possible for us not to fulfil the purpose assigned to us, and if the purpose assigned to us were necessary to realize God's plan, then God would need our help. But if God needed our help, God would not be omnipotent. Therefore, what follows from purpose theory is the absurdity that God would not be omnipotent, something that He is by definition (Britton 1969: 31, 34–5; Hanfling 1987: 48).

A standard move for the purpose theorist at this point is to begin by maintaining that God would bestow on us libertarian free will, i.e., an agency that is free from causal determination, and hence to accept that we could avoid realizing the end God assigns us. The purpose theorist also grants that realizing the end God assigns us would be necessary to fulfil God's own, higher-order end(s); assuming that God is the source of the universe, we must presume that our existence is an integral part of the reason(s) for which it was created. That means the purpose theorist in effect admits that there is indeed a sense in which 'God would need our help': the realization of God's higher-order end(s) would depend on our realizing the purpose He assigns us, something we could freely choose to do or not. Hence, the purpose theorist responds to the objection by refuting the claim that God's needing help implies a lack of omnipotence. That can be fairly done, as follows.

It is widely accepted among philosophers of religion that we need not conceive of God's omnipotence as implying that God could do what is logically impossible; for example, God's 'omnipotence' is not undercut by virtue of His inability to make a circle square. It is also commonly held that it would be logically impossible for God to bring about by His own power the superlative value of a person freely choosing to act morally. Therefore, if one of God's higher-order ends required us to act morally on the basis of our free choice, then we would have a sense of God 'needing our help' that does not impugn God's omnipotence. God could need our help in that God conceivably could not *entirely* of His own accord fulfil one of His valuable higher-order purposes.

However, this response might raise a new worry. Even if God's omnipotence were not called into question by virtue of God's being logically unable to bring about one of His higher-order ends, one might find such a God to be irrational. It seems incredible that a person would create the universe for the sake of an end that might not be realized. How could it be sensible to go to the trouble of making a physical world for a goal the attainment of which depends on the contingent decisions of members of our undependable species?

There are two strong replies to this new concern. First, recall that one can conceive of God making the universe for several higher-order ends (5.2). It would seem rational to create nature knowing that, say, two higher-order ends would certainly be realized but that one would only probably be realized. Second, it could even be rational to bring about a world for the sake of a single highest-order goal that God could not ensure would be achieved. At least if free moral choice were extremely desirable, then the expected value of creating a world for the sake of free moral choice, which may or may not materialize, could be greater than that of creating a world in which human beings lacked free will but that unavoidably achieved some less valuable goal.

I conclude that God's needing our help to attain His higher-order goal(s) threatens neither God's omnipotence nor His rationality. Let me turn now to the third *reductio* against purpose theory that one encounters in the literature.

6.4 God's purpose v. God's eternality

The last charge of absurdity in purpose theory is that there is a tension between claiming that God is beyond space and time, on the one hand, and claiming that God sets an end for us, on the other. Irving Singer clearly voices this worry about speaking of 'God's purpose':

(T)o talk in this way is to assume that one can refer to an intentionality outside of time and space comparable to what occurs within. That is the basic flaw in the analogy.... It is not a question of determining whether we can fathom the cosmic plan, or prove that a cosmic planner exists, or manage to fulfill his purposive program. It is a question of knowing whether our mind is able to formulate these notions with any degree of clarity (1996: 31, 32; see also Hepburn 1966: 223).

In short, Singer finds the traditional concern about the compatibility of eternality and personality in God sufficient to disqualify purpose theory.

The most straightforward reply, of course, is to deny that eternality must be interpreted in atemporal terms. It is open to the purpose theorist to maintain that God would be everlasting, i.e., exist *always* in time, rather than exist *never* in time.

However, to start to develop my own line of criticism, I suggest that it would be interesting to maintain that a merely everlasting God would not be special enough to be a unique source of life's meaning. If a temporal being could not have the right sort of status to be the only source of life's meaning, and if an atemporal being could not be conceived to have a purpose, then the purpose theorist would be caught in a dilemma: either God would be atemporal and could be the sole source of meaning, but God could not have a purpose, or God would be temporal and could have a purpose, but God could not be the sole source of meaning. I believe this is the most important argument against purpose theory, one that I advance in the following section.

6.5 A new argument against purpose theory

The arguments in the previous three sections contend that purpose theory entails logical absurdities. I have replied that there are independently plausible versions of purpose theory that can avoid these implications. However, I now argue that there is a criticism of purpose theory that no version can avoid. Specifically, I maintain that the most compelling motivation for God-centred theories in general is in tension with purpose theory, the most influential instance.

6.5.1 A fresh strategy

The argument I develop can be understood in terms of the answers that one might offer to two questions. First, why think that the significance of our lives essentially depends on some kind of relationship with God? Second, why think that the relevant relationship with God is a matter of realizing His purpose? I argue that the most promising answer to the first question entails that no satisfactory answer to the second question is available.

In particular, I submit that the best explanation of why God could be the sole source of life's meaning must appeal to features that are utterly unique to God. I contend that the most promising account of why a relationship with God alone could make our existence significant is that significance comes from orienting our lives towards a unique being with perfections such as atemporality, simplicity, and immutability, qualities that humans cannot even be imagined to manifest. And since, as Singer notes, these qualities appear to be incompatible with purposiveness, God is unlikely to be a purposive agent if He alone could confer meaning on our lives.

Before presenting and defending this argument in the rest of the chapter, I respond to a concern about its form. Some might be inclined to ask, 'So what if purpose theory cannot be the correct version of God-centred theory? Whoever supposed that God-centred theory is the proper way to articulate a religious thinker's perspective on the meaning of life?' A God-centred theory, as I construe it, maintains not just that the better one's relationship with God, the more meaningful one's life, but also that the existence of God is necessary for one's life to be *at all* meaningful (or at least meaningful on balance). A God-centred theory of meaning entails that, if the physical world did not spring from God, then there would be no way to acquire a meaningful life in it. The question is whether such a theory is indeed most attractive to supernaturalists. Why not hold instead that life could be somewhat meaningful in the absence of God, but that it would be a lot more meaningful with His presence?[6] If Western religious scholars denied the view that God's existence is necessary for a meaningful life, or if such a view were *prima facie* implausible, then it would not substantially discredit purpose theory to show that it is incompatible with the underpinnings of this view.

I have three reasons for thinking that God-centred theory, as I construe it, is the relevant standard for appraising purpose theory, at least given a Western context. First, the most influential statements on the meaning of life in the Western religious tradition are clear instances of God-centred theory (usually combined with a soul-centred one). Staying within the past 150 years or so, consider Søren Kierkegaard's writings on God, where he expresses the judgement that life would be without significance if He did not exist (1850). For another instance, think of Leo Tolstoy's *My Confession* (1884), the most widely read religious discussion of the meaning of life, in which he acutely expresses the worry that life would be meaningless without God. Finally, recall Martin Buber's view that meaning is a function of an I–Thou relationship, where at least the possibility of such a relationship with God is what enables it with other beings (1923). And, more recently, the most prominent supernaturalists about meaning, including Craig (1984), Hartshorne (1984), Adams (1999), and Cottingham (2005), maintain that without God, our lives would be meaningless.

In addition to historical prominence, there are strong theoretical reasons for using God-centred theory as a base from which to evaluate purpose theory. The concept of meaning implies the idea of final value, and one would expect a religious theory of meaning to accord with religious theories of other goods. Consider, then, that in moral theory a religious view is standardly understood to maintain that moral rules are identical to God's commands and hence that 'if God does not exist, everything is permitted'.[7] For another example, consider the theory of human excellence. Here, many religious thinkers follow Aquinas, holding that God is the unique source of perfection and that other things can obtain excellence only by virtue of participating in God's (e.g., MacDonald 1991; Adams 1999). And even when it comes to happiness, more than a few follow Pascal: 'All men seek happiness...(and) the infinite abyss can be filled only by an infinite and immutable object, that is to say, only by God Himself. He alone is our true good...' (1669: #425; Theron 1985). By analogy, then, a religious theory of meaning that appeals to God ought to hold that He is the sole source of a

---

[6] This sort of view can be ascribed to Quinn (2000).
[7] A catchy line often misattributed to Dostoyevsky.

meaningful life and that, without God, there would be no meaning, just as there would be no morality, excellence, or well-being.

Finally, God-centred theory cuts out philosophically interesting territory. In order for the dispute between supernaturalists and their critics to be substantive, the former must hold that a relationship with God is necessary to make our lives meaningful. To see this, suppose that a religious thinker instead maintained the weaker view that while a relationship with God would make our lives more meaningful, it would not be necessary for meaning. The problem is that virtually no naturalist would dispute this claim. Few naturalists have said or would contend that, if God existed, relating to him would fail to enhance the meaning of one's life, and any potential arguments for this contention are unpersuasive (cf. 8.3). Naturalists emphasize that God's existence is irrelevant for meaning, that for a life to be meaningful there need not exist a spiritual person who grounds the universe. Therefore, for religious thinking to carve out a distinct and interesting position, it must maintain that God's existence and a certain relationship with Him are necessary for life to acquire significance.

I conclude that for reasons of tradition, coherence, and relevance, it makes sense to appraise purpose theory in light of God-centred theory, as I characterize it. If, as I contend, purpose theory is not consistent with the best motivation for God-centred theory, then there is strong reason to reject purpose theory and to seek out some other account of meaning that is grounded on God.

My thesis with regard to purpose theory is that if a relationship with God were essential for a meaningful life, then the meaning could not come by accomplishing a purpose He sets. To defend this thesis, I now argue for two premises. Premise 1 contends that the best explanation of God-centred theory includes the claim that God has certain properties such as simplicity, immutability, atemporality, or perhaps infinitude. Premise 2 maintains that these properties are incompatible with a purposive God. These two premises entail that purpose theory probably cannot be the correct version of God-centred theory.

6.5.2 Premise 1: The qualitative properties

In looking for an acceptable explanation of why God alone could make our lives meaningful we must, of course, appeal to features that cannot be found anywhere but in God. Again, if our lives acquire significance just to the extent that we have a proper relationship with God, then to explain why God is central to life's meaning we must invoke features that only God could manifest. The more God's proposed meaning-conferring features were like properties we or other, lower creatures could exhibit, the more reason there would be to think that a relationship with them would be sufficient for meaning in life. Keeping this in mind, I quickly canvass some common explanations of why life might be meaningless without God, in search of the *most promising* one.

I begin by briefly canvassing the several arguments for purpose theory already criticized (5.3–5.5). First, I noted that some take God to be central to meaning in life because only He could give people the appropriate sort of afterlife. However, it appeared that a physical, Karmic-like force could conceivably do the same. Second, I pointed out that some deem God to be essential, since only He could prevent our lives from being accidental, in the sense of not being deeply rooted in the structure of the universe. However, it looked as though a deterministic universe would be sufficient to avoid accidentalness. Third, I devoted much space to considering the idea that only God could create an objective ethic, without which our lives would be meaningless. However, I argued that that is greater reason to think that the kind of morality necessary for meaning in our lives is something that could obtain independently of God's will.

For a few as yet unexplored suggestions about why God might be essential for meaning in life, consider, fourth, the idea that only God could make our lives part of a grand scheme, one that encompasses the universe. If God existed, then our lives would be important for, say, helping to constitute the temporal life of an all-good being who is the ground of the physical world as a whole (Hartshorne 1996: 16-18; Smith 2000: 255-9). However, God's existence is not necessary to avoid this condition, which a brief reflection on the television series *Star Trek* might indicate. If we could travel extraordinarily quickly, then we could enact a plan with an enormous range. Therefore, it cannot be mere scope that explains how God could be the sole source of life's significance.

Fifth, the famous process theologian, Charles Hartshorne, has proposed a God-centred theory that appeals to God's love for all creatures (1984; see also Runzo 2000). For Hartshorne, love involves making another's experience one's own, that is, paying attention to her life, remembering it, and rejoicing when it goes well. According to Hartshorne, God is the key to a meaningful life because 'our welfare is appreciated and immortalized in God, along with the welfare of all those we care about' (1984: 160). The meaning of our lives, on this view, consists in our experiences being assimilated into an everlasting unity with those of others, something possible only in God's eternal recollection of our existence. However, surely actually living together in loving relationships would confer more meaning on life than a loving fond remembrance, which suggests that an eternal world filled with vampires or other immortals would be sufficient for meaning.

Sixth, and finally for now, there is Robert Nozick's influential argument for God-centred theory (Nozick 1981: 594-619; cf. Cooper 2005). His basic idea is that for a finite condition to be meaningful, it must obtain its meaning from another condition that has meaning. So, if one's life is meaningful, it might be so in virtue of being married to a person, who is important. And, being finite, the spouse must obtain her importance from elsewhere, perhaps from the sort of work she does. And this work must obtain its meaning by being related to something else that is meaningful, and so on. A regress on meaningful finite conditions is present, and Nozick's suggestion is that the regress can terminate only in something infinite, a being so all-encompassing that it need not (indeed, cannot) go beyond itself to obtain meaning from anything else. And that is God.

The field is aware of two major objections to Nozick's rationale. First, one can deny its 'trickle-down theory of importance', to use Daniel Dennett's pithy phrase (cited in Baggini 2004: 14; see also Thomson 2003: 25, 48). For instance, perhaps an altruistic action and its proximate effects are meaningful in themselves, not needing to obtain their meaning from anything else, such as the fact that the helpful result will have some *other* significant effect. A second criticism is to accept that meaning is relational, but to deny that a condition must obtain its meaning from another meaningful condition. Nozick himself ultimately concedes that a condition could obtain meaning by virtue of being related to something finally valuable that is not necessarily meaningful, such as a person or artwork (1981: 610; see also Thomson 2003: 25-6).

There is a common theme present in the objections to all six suggestions about why God might be essential for meaning in life. In all the cases considered, I have objected that, for all we know, *nature*, independently of God, could perform the functions of which God alone has been thought capable. There are, of course, replies that religious thinkers have made and could make to my objections, but it is not my concern to address them. My point is that there is a *more auspicious* strategy for explaining why God's existence and a relationship with Him might be necessary for our lives to acquire significance. The best way to avoid the objection that God is not necessary for meaning since life in a material world could fulfil the relevant role is to look for something *utterly* supernatural, viz., something that nature simply *could not* exhibit.[8]

Here, I take a cue from perfect being theology (e.g., Rogers 2000), according to which, roughly, God would have an absolutely unique essence, one that could not be found in the physical world and that alone has the kind of final value towards which it would be worthwhile contouring one's life. What is it about God that might make Him qualitatively both different from and more valuable than anything in nature? I do not think that God's being all-good, all-powerful, or all-knowing are very plausible candidates, for, again, we find goodness, power, and knowledge in our world to some degree, and we can imagine these traits being realized to a much greater degree in the physical universe. To be sure, God would have these to a superlative degree, but this appears to be more of a quantitative difference from human beings or other possible, natural creatures, not a qualitative one.

It would make most sense to look for features of God that other beings such as humans or greater persons could not exhibit, indeed, cannot even be *conceived* to exhibit; such features would most clearly indicate in what respect God's nature and value would be unique. Four properties come to mind: atemporality, immutability, simplicity, and infinitude, which I call the 'qualitative properties'. If God alone had a certain combination of the qualitative properties, if a being with this combination of

---

[8] Morris (1992) approximates this strategy when he suggests that if both permanent and impermanent persons exist, only a permanent person who could exist beyond the physical universe would be worth loving and hence confer meaning on our lives. However, this argument rests on the unfortunate dialectical assumption that a permanent person in fact exists.

these properties had an exceptional excellence, and if we gained significance by contouring our lives towards such a being, then we would have a satisfying explanation of how God alone could be the source of significance. I now fill out this strategy.

First off, then, it is clear that human beings cannot have any of the qualitative properties. We essentially are spatiotemporal, changeable, decomposable, and limited. Angels, too, are typically understood to be limited and, at least in thought, decomposable. Note that simplicity, a condition of being unable even to be conceived as being composed of separate parts, implies atemporality and immutability. A being without parts obviously cannot change, while a being in time implies that it has extension, viz., stretches over moments, and hence has parts. That is one reason to focus on simplicity as one of the deeper explanations of why God might be unique.

One might object that it is possible for there to be parts of nature that are simple; perhaps certain sub-atomic particles are in fact simple. A fair response is that this is not the right sort of simplicity, since such particles would still be in time and thereby have a kind of extension. However, rather than delve into tricky metaphysics any more than I already have, consider the response that it is not *mere* simplicity that counts, but rather personhood combined with some or other qualitative property. To explain, consider that rarity is a final value, but one that obtains only when it co-varies with something that is independently good. A rare species or Stradivarius is to be prized, whereas a rare moral monster such as Stalin, or a rare form of cancer, is not. Similarly, simplicity can reasonably be deemed a final value, and a superior one, that obtains when conjoined with the independent good of personhood.

Next, it is reasonable to think that the qualitative properties of simplicity and infinity exhibit important sorts of final value. Classical theists provided many arguments purporting to show that the two values of unity and independence are (partially) constituted by the qualitative properties, which I tightly sketch here.[9] Consider the value of independence, which is a matter of not being confined by or dependent on anything else. All four of the qualitative properties may be deemed instances of independence *par excellence*. A simple being, viz., one without parts, would be free from dependence on these parts for its existence. Such a being would be completely unto itself or *a se*. A being beyond space and time (implied by simplicity) would be free from the limits of these forms. Such a being would be free not only from decay and death, but also from a point of view restricted to now and here. An immutable being (implied by atemporality) would similarly be something that utterly determines its own nature; if a being that exists in a certain mode can neither have begun to exist nor cease to exist in this mode, then it is free from any influences save itself. Finally, by an 'infinite' being I do not merely mean one that is immeasurable or exhibits the classic triad of omnipotence, omniscience, and omnibenevolence. Instead, I mainly have in

---

[9] Many of the following arguments can be found in Plotinus, *The Enneads*; Anselm, *Monologion* and *Proslogion*; and Aquinas, *Summa Contra Gentiles* and *Summa Theologica*.

mind Nozick's idea of an unlimited and all-encompassing being.[10] Such a thing would by definition be free of any restrictions.

Now think about the value of unity. Integrity and oneness are better than disintegration and fragmentation, and the qualitative properties are plausibly manifestations of the former. A simple being, having no parts, forms the ultimate unity in that it cannot even be conceived to dissolve. A being beyond space and time would lack extension or the 'feebleness of division' (Plotinus). An immutable being cannot help but remain what it is. Finally, an unlimited being would be utterly whole.

These have been sketches, not full-blown arguments; one could spend a lot of work developing just one of them in detail. My goal is not really to convince the reader that unity and independence are superlative final values that the qualitative properties exhibit. My aim is more modestly to show that the *most promising* explanation of why relating to God would be essential for meaning in our lives is that God would exhibit certain features incapable of being found in the physical world and that these features would have an incomparable value (that confers significance on our existence when we orient ourselves towards it). Even without a complete analysis of the qualitative properties and of the values of unity and independence, my hope is that the reader will agree that, particularly given the problems facing the several other rationales for a God-centred theory above, it is fair to say that this is a *more plausible* rationale for holding any sort of God-centred theory.

6.5.3 Premise 2: The qualitative properties v. purposiveness

Premise 1 of my objection to purpose theory is that if God alone could confer meaning on life, this would have to be because He has utterly unique features, ones that have a superior worth and that cannot even be imagined to obtain in a physical world, roughly, simple or infinite personhood. Premise two, which I now defend, is the claim that God's having such qualitative properties is incompatible with central tenets of purpose theory.

As Singer pointed out (6.4), it is hard to see how a being with the qualitative properties could play the role that purpose theory requires. In fact, this is a fairly common concern among philosophers of religion;[11] my aim is not to add anything new to this particular issue, but rather to articulate it. So, first off, how could there be a simple and hence unchangeable being beyond time that is purposive? The worry here is not that it is difficult to conceptualize an immutable, timeless being. While it is difficult, it is not impossible, for, as has been noted, states do not seem essentially to be temporal or to involve change. The concern is rather that, to the extent that we can conceive of an immutable being beyond time, such a being appears unable to engage in goal-directed

---

[10] Those in the perfect being theological tradition tend to interpret an infinite being as one not limited by space, time, and matter. Such a notion of infinity is not much distinct, then, from considerations of simplicity.

[11] For contemporary statements of the tension between God's personality and some of the qualitative properties, see Gale (1991: 37–56) and Swinburne (1993: 217–38).

activity. Activities are events, and events fundamentally involve change and time. For God to adopt an end for humans might presuppose some deliberation, and deliberation would appear to be a temporal event involving alteration in God. And even if God adopted an end without prior deliberation, the adoption alone would seem to be something that takes time and that forms something new in God. Still more, creating a world according to a plan seems hard to understand as something that does not cover a span of time. Finally, purpose theorists standardly hold that God would inform us of the end He has set and that He would respond to our free decision of whether to fulfil it or not. Again, these are activities that seem irreducibly to take time to perform and to involve change on God's part.

For another instance of the problem, how could there be an absolutely simple agent who has multiple ends, one for humans and one for animals (or perhaps multiple ones for humans)? God presumably would have created bees for a purpose that they do not fail to fulfil. In order to avoid the counterintuitive implication that animal lives can be as meaningful as human lives, the purpose theorist must maintain that we would be assigned a purpose different from and better than those assigned to animals.[12] Now, even if human and animal purposes were components of a single plan for the universe, the fact of there being components would seem to imply a lack of simplicity. The same goes for the different acts mentioned above; more than one act would appear to compromise God's simplicity, and it is difficult to see how a single, grand act could ground purpose theory's conception of what God does.

Finally, how could a being that is unlimited be the sort of entity that has a purpose? As Nozick has put it, 'To be one way and not another is to have limits. It seems, then, that no terms can describe something unlimited, no human terms can truly apply to it' (1981: 608). I doubt that analogical reasoning will get the purpose theorist very far. While I have argued earlier that purpose theory need not provide much practical guidance to be viable (5.2), it must at least be theoretically comprehensible.

Summing up, if indeed a God with the qualitative properties cannot be purposive (Premise 2), and if God must have the qualitative properties to be the sole source of meaning (Premise 1), then it follows that we probably could not acquire meaning in our lives by realizing a purpose that God assigns us. I submit that this is the most significant problem facing purpose theory. Before considering objections to this argument, I first clarify it by indicating how it differs from two other rationales with which it is liable to be confused.

For one, consider that my argument is not that God, *qua* perfect, by definition has the qualitative properties and hence cannot be purposive. My criticism is not easily reduced to this reasoning, since God probably does not, merely *qua* perfect being, have all the perfections. To see this, imagine a quite valuable being who created the universe but who is short of being Anselm's 'thing of which none greater can be

---

[12] For more discussion of human and animal meaning in the context of purpose theory, see Nozick (1981: 586–7) and Hanfling (1987: 48–9).

conceived'. Such an entity would sensibly be called 'God'. Hence, insofar as the concept of God does not include having all the perfections, God does not (or at least does not obviously) by definition have the qualitative properties (see Nozick 1989: 46-54; Nagasawa 2008).

Instead, my claim is that the most plausible reason for holding a God-centred theory of life's meaning of some kind or other is inconsistent with the traditional version of God-centred theory that appeals to God's purpose. God alone could make our lives significant just insofar as He has some combination of the qualitative properties, and it is difficult to conceive of a purposive agent who has no limits or is absolutely simple and therefore can neither change nor act in time.

For another, notice that I am not merely reiterating points from the tradition of wondering whether and how God's otherness might be consistent with God's personality, for I am not making a claim about inconsistency within the concept of God. My argument is that if God alone could be the source of meaning, this must be because God would have the qualitative properties, perfections that are incompatible with purposiveness. Unlike a thesis about incoherence in the very idea of God, my position does not threaten God-centred theory as such. It instead suggests that, if one is sympathetic to God-centred theory, one should consider versions of it other than the one that has been central to Western religious thinking about the meaning of life.

6.6 Reconsidering the qualitative properties and purposiveness

There are two ways that the purpose theorist might object to my argument, namely, by questioning its two major premises. I focus on criticism that purpose theorists have actually made,[13] aiming to provide conclusive replies to them.

6.6.1 Are the qualitative properties key?

First, some might seek to knock down premise one, the idea that personhood imbued with the qualitative properties, possible only in a supernatural realm, should be deemed central to meaning by a God-centred theorist. Perhaps I was wrong to suggest above that nature could perform a certain function, such as grounding an objective morality or ensuring that reward is apportioned to virtue, of which God has often been supposed alone capable. In the final analysis, that, of course, could be true. But I claim that, for all we can tell, nature appears much more metaphysically able to perform those functions than to produce a person with the qualitative properties.

Another possibility is that I missed a salient function. Perhaps of most temptation, Philip Quinn (in correspondence) once suggested to me that it is God's being the creator of the universe in virtue of which He has the unique ability to ground life's

---

[13] To the initial statement articulated in Metz (2000).

meaning. It is inconceivable, after all, that we or any other spatio-temporal creature could have created the universe, and it is plausible to think that such creative agency is a superlative final value.

However, since the universe is essentially spatio-temporal, God must be an atemporal being to have been its creator, meaning that my initial objection about the conflict between atemporality and purposive activity resurfaces. How could a person who is beyond time create a spatio-temporal world, when doing so would appear to require time?

6.6.2 Can the qualitative properties and purposiveness be reconciled?

The purpose theorist might grant, for the sake of argument, that God could confer meaning on our lives only if He had the qualitative properties, but try to show that it is conceivable that a being with them could be purposive. Aquinas is a promising source of ideas needed to flesh out this strategy. For example, one could contend that simple, atemporal and immutable knowing is possible, that willing and knowing are not really distinct in God, and that having a purpose is part of the concept of willing.[14]

To take the debate into a fresh direction, I do not pursue that approach, and instead address a strategy that Jacob Affolter (2007) has explicitly directed towards my argument. He maintains that, upon considered reflection on the precise kind of purpose that would confer meaning on our lives were we to fulfil it, one can see that a knowing and willing God without extension could ground such a purpose. Affolter also suggests, more tentatively, that one finds reason to think that *only* such a God could ground it.

The purpose necessary for meaning in life, according to Affolter, has two salient characteristics. First, it is an end that has what Affolter calls 'absoluteness', meaning that it is constitutive of our nature such that failure to comprehend it implies a failure to grasp what we essentially are. Drawing on largely pre-modern, teleological views of ontology and explanation, Affolter claims that if realizing a purpose were to make our lives significant, it would have to be a final cause that is inherent to us. What we are just is (at least in part) to be such that we ought to realize a certain state of affairs. It follows from this kind of ontology that an acceptable understanding of what we are must refer to this prescribed end. Now, when we are asking whether our life is meaningful, we are often asking whether it has a point and whether it makes sense. Affolter captures these intuitions well by suggesting that meaning in life comes from realizing a purpose that renders intelligible the essential nature of our existence.

In addition to absoluteness, Affolter suggests that the purpose necessary for meaning in life also has 'specificity'. By the latter, Affolter means that the purpose is unique to a given individual. What makes one's life meaningful cannot come from realizing an end that applies to all human beings, so he suggests, but rather from realizing a purpose that

---

[14] For interesting work along this line, see Kretzmann and Stump (1985); Rogers (1996); Lodzinski (1998).

defines one in contrast to everyone else. Perhaps Affolter believes that only a specific purpose would be special enough to accord with the dignity of being an individual.

Supposing that the kind of purpose the realization of which alone constitutes meaning has both absoluteness and specificity, Affolter maintains that it could come from an atemporal God and offers *prima facie* considerations for thinking it could come only from such a God. First, nature apparently cannot ground specific purposes. Although, arguably, there could be natural ends for species that are absolute, such that a caterpillar just is the sort of thing that ought to achieve the end of becoming a butterfly, we are much less familiar with the idea of natural ends for each member of a species. Ends in nature seem unable to provide an answer to the question of what would make my particular life meaningful.

Second, although human beings are able to provide specific purposes, they are unable to ground absolute ones. The ends adopted or prescribed by human beings are too contingent and limited to a particular standpoint to be able to constitute the essence of a human being. For instance, parents might assign a purpose to their child, but that purpose is hardly one that makes the child who she essentially is.

If neither non-human nature nor human beings can ground a purpose that is both absolute and specific, then what else could beside God, whose viewpoint is alone independent and unconstrained? And while Affolter grants that an atemporal God might not be able to *have* such a purpose, he maintains that a God who is beyond time could *assign* one. Affolter claims that the purpose would not be in God's mind so much as in the created object, thereby avoiding concerns about God's mental acts taking time. Affolter sums up his view this way: 'The idea is that God designs a person in such a way that the need to fulfil some purpose is built into the person's very nature, such that his life makes no sense apart from that goal' (2007: 453).

To reply, I begin by questioning whether a meaningful purpose must be specific, a claim Affolter tentatively puts forth to provide some reason to think that only an atemporal God, and not nature, could ground the relevant purpose. Traditional and sensible understandings of purpose theory conflict with the claim that, for a purpose to confer meaning on life, it must be unique to a given individual. First, in terms of tradition, standard conceptions of purpose theory envision God ascribing the same purposes to all human beings. For example, many Christians think of it as loving God and one's neighbour, while most Muslims think of it as upholding the five pillars of faith, prayer, charity, pilgrimage, and fasting.

Second, and more deeply, there is good reason to think that the sort of purpose that confers meaning on a person's life would not be unique to that person. What the monotheistic traditions have in common is the view that at least one of God's central purposes for human beings would be to live a moral life. Intuitively from a religious point of view, a purpose could confer meaning on a life if and only if it were ethically sound. Assuming that Affolter does not wish to embrace relativism about moral norms, and more plausibly holds that they are universal in scope, he cannot easily hold the notion that a purpose confers meaning only if it varies from person to person. Indeed, it is precisely because morality is an invariant end among human beings that religious thinkers have been inclined to find it—and hence God—relevant to meaning in life (see 5.4). Sartre (1946) might deem his dignity to be compromised by the fact of invariant norms prescribing mutual aid, but virtually no supernaturalists about meaning do.

Suppose, now, for the sake of argument that Affolter's conception of the purpose relevant to meaning in life is correct. I argue that, even granting this conception, a God who is simple (and hence atemporal and immutable) could not ground it. Affolter's claim, recall, is that although a God without extension might not be able to *have* absolute and specific purposes for human beings, He could *assign* one to each of us. '[P]urpose₁ belongs in a strong sense to the object as much or more than to the person who confers it. In other words, what people think that God alone can assign is not something that is a purpose for Him that we adopt' (2007: 453). Now, even if there is a distinction to be drawn between what is in God's mind, on the one hand, and what God has put into His creation, on the other, I submit that *God could put into creation only what was first in His mind*. If assigning a purpose to human beings requires the prior condition of God having had it in mind, and if (as Affolter is inclined to agree) it is correct that having a purpose is impossible for a God without extension, then the distinction between having and assigning does not help to resolve the tension between grounding purpose and lacking extension.

Recall key instances of this tension. It is hard to conceive of a spiritual person who creates a physical universe in accordance with an end without the creation being an event with a certain temporal duration. It is difficult to imagine an unchangeable God beyond time and without parts who has multiple purposes, one for each individual person, and still other purposes for the animal, vegetable, and mineral kingdoms. I am not sure whether these are tensions within God's 'having' purposes or 'assigning' them; but that does not matter, since, however these are plausibly understood, assigning presupposes having. I think this is the key respect in which Affolter's attempt to reconcile the tension between God's otherness and His personality falls apart.[15]

6.7 The end of purpose theory

This completes a lengthy critical analysis of the purpose theory of meaning in life, according to which one's life is more meaningful just insofar as one fulfils a purpose that God has assigned. In this chapter, I have explored arguments against the theory, and I began by arguing that three standard criticisms in the literature are unsuccessful. Specifically, I refuted the charges that purpose theory absurdly entails that God would treat us immorally, lack omnipotence, and fail to be eternal. Then, I raised a new problem for purpose theory, that it does not square with the best explanation of

---

[15] For another reply to Affolter, namely, an account of how a purely material world could ground the sort of absolute and specific purpose he deems essential for meaning, see Metz (2007c: 462–3).

God-centred theory. After rejecting several accounts of why a relationship with God might be necessary for life to have meaning, since it appears that nature could do the requisite work, I advanced what I take to be the most promising account. I suggested that God could be the unique source of meaning just insofar as he exhibits the qualitative properties of being infinite or simple, superlatively valuable properties that cannot even be imagined to apply to a person in the natural world. And then, appealing to long-standing concerns about how a God who is radically other could be a person, I argued that God with properties such as infinity and simplicity could not be purposive, at least not in the way that purpose theory standardly maintains.

It is, of course, open to purpose theorists to pursue further objections to this argument, perhaps by taking up Thomistic perspectives. I suggest, however, that it would also be reasonable, given the objection made here, to develop a God-centred alternative to purpose theory. Suppose God's having some combination of the qualitative properties is part of the best explanation of what would enable Him alone to confer significance on our lives. Precisely how would we have to relate to a being with the relevant qualitative properties in order for our lives to acquire meaning in them? This question, too, is worth considering at this point, in order to develop the supernaturalist perspective, and it is among those that I seek to answer in the next chapter.

7

Non-purposive Supernaturalism

'And yesterday I saw you kissing tiny flowers,
But all that lives is born to die;
And so I say to you that nothing really matters,
And all you do is stand and cry.'

Led Zeppelin ('That's the Way')

7.1 How to be a supernaturalist

In the previous chapter, I argued that the most influential God-centred conception of meaning is in tension with the underlying rationale for holding a God-centred theory as such. Specifically, I argued that the best explanation for thinking that a relationship with God is alone what can make life meaningful is that only He could exhibit the 'qualitative properties' of being a person who has superlatively valuable features such as simplicity or infinity. And I then suggested that a being with such perfections could not be purposive, at least not in the way characteristically understood by purpose theorists. In order to appraise supernaturalism, what remains to be done is, first, to spell out some God-centred conceptions of meaning in life consistent with conceiving of God as having the qualitative properties (7.2). These are the most defensible God-centred theories of meaning, on the supposition that the argument against purpose theory is successful; but even if it is not, it is worth considering other God-centred accounts, perhaps as additions to the purpose theory, if not alternatives to it.

Next, I articulate a pure soul-centred view of meaning in life, one that does not appeal to God as being necessary for meaning in life, but rather to the conditions of being constituted by an immortal, spiritual substance that is in a certain state (7.3). I aim to advance discussion of soul-centred theory in a variety of ways. One major thesis is that the existing arguments for soul-centred theory are weak in that their premises, even if true, provide little support for the intended conclusion (7.4). I work to reconstruct these arguments, and I demonstrate that once they provide reason to accept soul-centred theory, they provide equal reason to believe God-centred theory as well. Then, I defend soul-centred theory and what it implies from extant critics, particularly those who maintain that immortality would be sufficient for a meaningless life (7.5). And I conclude by highlighting the single, fundamental motivation common to the supernaturalist positions that I address in this chapter: all are grounded on the 'perfection thesis', the idea that meaning in life requires engaging with a maximally conceivable value (7.6). It is our nature as beings able to idealize, to imagine flowers that live forever, that makes us wistful.

7.2 God-centred theory grounded on the qualitative properties

I have argued that the most promising reason to believe any God-centred theory is that God would have perfections such as simplicity or infinity and also that a being with such properties could not have a purpose (6.5). Supposing those claims are worth taking seriously, it invites the question of how meaning in our lives could be constituted by relating to such a being, which I take up in this section.

I have not engaged in the kind of thorough metaphysical analysis that would prove that features such as simplicity and infinity are incompatible with being purposive; all I did was provide some decent reason to think so. If I am wrong, so that a simple or infinite God could in fact be purposive, then the following accounts of how to acquire meaning in life from God should be seen as ways to supplement purpose theory, not to supplant it. The key point is that the field needs to go beyond the idea that the principal, let alone sole, way that God could confer meaning on life would be by fulfilling a purpose He has assigned. In addition, if not instead, there are several ways in which one's existence could plausibly become more significant by relating to God (one with or without the qualitative properties). I discuss three: respect (7.2.1), love (7.2.2), and communion (section 7.2.3), and then articulate what I take to be the most defensible God-centred theory, supposing my argument against purpose theory is attractive (7.2.4).

7.2.1 Respect

The least intense of the three relationships that one could have with a higher being that may not be purposive is to respect it. Often talk of 'respecting' a being is a matter of saying that one ought to treat it as having the final value that it does, with the opposite, 'degrading', connoting the idea of treating a being as having less than its final value. In the first instance, respect involves an awareness of the being and its value; respecting a perfect being at a minimum requires not ignoring it, not treating other things as though they are more worthy of one's attention (idolatry). It also might require setting time aside to reflect on it and its superlative worth, where one does not 'contaminate' this time with lower trifles such as playing video games and surfing the web (keeping the Sabbath). It could also involve studying the being's nature, becoming as intimately aware of its functioning as one can (reading scripture, studying theology). If the being is aware, then it would plausibly include capturing its attention, and if it is self-aware, then it would no doubt further involve communicating with it (praying). Respect also forbids insults, symbolic expressions to the effect that the being is worth less than it is (blasphemy), and it would probably also require taking care to protect likenesses of the being where they may be encountered (not harming those made in its image). Finally, if the being is the highest one conceivable, it could warrant making sacrifices on its behalf, spending the time, labour, wealth, and other resources needed to construct, say, elaborate buildings in which to pay one's respects (temples, churches, mosques).

7.2.2 Love

A more intense relationship to have with a perfect being that might not assign a purpose would be to love it. The standard conception of love among English-speaking philosophers is a matter of two jointly necessary and sufficient conditions, namely, sharing a way of life with a being, on the one hand, and caring for its quality of life, on the other. The former is a matter of ideally living with, or at least spending a certain, attentive amount of time with, a person, whereas the latter consists of helping her for her own sake.

With regard to the latter, it is initially difficult to see how one could aid a perfect being, given that it would be omnipotent and, *ex hypothesi*, beyond space and time. However, such a being, despite not acting according to purposes, probably could still exhibit states such as wishes or desires, and it might be that, by fulfilling them, we would be acting for its sake, even if not improving its condition. In addition, it is fairly clear that one could share a way of life with such a being. One facet of this would include identifying with it, in the sense of thinking of oneself as a member of a common relationship with it, viz., as a 'we', and another would be habitually spending time in its company, not merely, say, once a week at an assigned time.

Furthermore, it is likely that the standard conception of love is misleading or incomplete, as David Velleman (1999) has persuasively argued. Velleman believes that essential to love is neither of the two conditions of 'to benefit and be with' (1999: 353), as he puts it. Instead, according to him, the essence of love is nothing more than an arresting awareness of another's nature and value that tends to break down emotional barriers. Without haggling over what is merely a contingent form of love and what is not, it seems correct to think that at least a desirable form of love includes the kind of emotional engagement that Velleman highlights. And this, too, we could exhibit towards a God with the qualitative properties.

7.2.3 Communion

More intense than respect and love is what I have in mind by the word 'commune'. Most strongly, there are those in the supernaturalist tradition who would construe communion in terms of actual union with God, some kind of merger between our soul and God's nature in a spiritual realm. Less strongly, but still more powerfully than the previous two relationships, would be imitating God, becoming 'birds of a feather' with Him or sharing his point of view by, say, shedding the physical aspect of one's nature and entering an atemporal realm, even if one would not thereby literally 'become one' with Him. And less strongly than this, but still more robustly than respect and love, would be to remain on earth, or to otherwise be embodied, but seek to dissolve the distinction between the profane and the sacred, as in the Hasidic tradition (for discussion, see Scholem 1991). Here, one aims to impart mundane tasks with a sense of the holy, or to imitate the holy as much as possible. This would characteristically involve cleaving to God in the sense of constantly having him in mind and perhaps seeing him in, or bringing him into, everything that one does.

7.2.4 *The most defensible God-centred theory*

Of course, one could exhibit all three relationships at the same time, and, furthermore, implicit in these analyses is the idea that later forms of relationship include earlier ones; that is, there is no love, or proper love, without respect, and there is no communion, or truly desirable communion, without respect and love. If indeed God must have the qualitative properties of either simplicity or infinity to be the sole source of meaning in our lives, a plausible form of God-centred theory would be that our lives would be more meaningful, the more intensely we related to Him by virtue of respecting, loving, and communing.

An additional facet of God-centred theory would include the idea that our lives can become more meaningful by virtue of God engaging in these kinds of relationships with us in turn. A non-purposive person could constantly keep us in mind in such a way that He would attend to whom we really are, experience an arresting awareness of our value, and perhaps even lack emotional defences, e.g., be open to being disappointed. Particularly if Velleman is correct that love 'can have an object but no aim' (1999: 355), that it is 'essentially an attitude toward the beloved himself but not toward any result at all' (1999: 354), and that it is specifically 'a state of attentive suspension, similar to wonder or amazement' (1999: 360), then a non-purposive God could easily love us, and it would be reasonable to think that our lives would be more meaningful thereby.

In short, at the core of what I take to be the most defensible God-centred view, or at least an underexplored one in the context of philosophical debate about meaning in life, is the idea that our lives have more meaning in them, the more intense our positive relationships to a perfect being and the more intense His positive relationships to us. More specifically, the more we respect, love, and commune with a (non-purposive) being with the qualitative properties, and the more it does so with us, the more meaningful our lives.

7.3 An analysis of soul-centred theory

I now set aside God-centred theory, and turn to soul-centred theory, the general view that one's life is meaningful by virtue of having a certain spiritual and eternal existence, regardless of whether God exists. In this section, I spell out what is essential to such a view and what is not.

Soul-centred theory is the view that a significant existence is nothing but being constituted by a soul that lives forever in a certain way, where a soul is an indestructible, spiritual substance. A soul-centred theory implies that immortality, a life that never ends, is necessary for meaning in life, but the claim that immortality is necessary (and sufficient) for meaning in life does not imply a soul-centred theory. After all, a life that never ends could conceivably be a physical life that forever remains in this world. Think of a vampire in an infinitely expanding universe, or of being able to upload one's consciousness into an infinite succession of different bodies.

For many religious thinkers, a soul is an offshoot of God, but that is not true by definition of the word 'soul' (*vide* Hinduism). Furthermore, even if it were substantively true, it would not necessarily follow that God is what constitutes meaning in life; it could be that a soul (necessarily) comes from God, but that it is *merely* a certain condition of one's soul, utterly apart from God, in virtue of which one's life matters or not. On this latter view, God would be instrumental for meaning, but a relationship with Him would not be meaningful for its own sake. I have explained that one can conceptually accept a pure God-centred theory; for example, some think that fulfilling God's purpose would be sufficient for meaning, even if one were not granted eternal life (5.2). Conversely, there is no logical contradiction in holding the other pure or simple form of supernaturalism, that a soul but not God is constitutive of meaning in life.

Soul-centred theory is not merely the claim that a spiritual, eternal life could *enhance* the meaning of one's life; it is instead the strong thesis that life would be downright meaningless if it were to end or if one lacked a spiritual nature. However, soul-centred theory is not so strong as to claim that spiritual immortality is, as such, sufficient for a meaningful life. Such a thesis would be counterintuitive in at least two respects. First, presuming that everyone is spiritually immortal if anyone is, this thesis would entail that either everyone's life is meaningful or no one's life is meaningful. But most think that some people have meaningful lives while others do not, or at least differential degrees of meaning in their lives. Second, if a spiritual immortality were sufficient for meaning, then an eternity in hell would be meaningful, which is surely incorrect. Hence, immortality and spirituality are plausibly proposed to constitute merely necessary conditions for meaning, not sufficient ones; to acquire meaning, one's soul must live in a certain way.

Notice that soul-centred theory differs from the claim that people need to believe in a spiritual, eternal life for their lives to be meaningful (one major claim in Swenson 1949). It is also different from the view that one can learn about the meaning of life only by getting answers from God in an eternal afterlife (Levine 1987). Soul-centred theory is rather the view that a soul itself (not the belief in it) is a necessary metaphysical (not epistemic) condition of life's meaning.

A few different instances of soul-centred theory are possible, depending on the way immortality is understood. The concept of immortality at the core is that of a life that will never end, but this could be realized in a variety of ways. First, one may conceive of eternal life in temporal or atemporal terms. An immortal life could be one that will never cease to be in time or one that will transcend time altogether. Second, there are sundry ways of thinking about the modal status of immortality. Some think of immortality as a matter of being unable to die, while others think of it merely as a matter of being able to live forever while also being able to die, usefully called 'immortability' (Hocking 1957: 74, 154).

Just as God-centred theory does not imply that God exists, no version of soul-centred theory implies anything about whether we in fact have a soul. Of course, many of those who believe that they have a soul do hold a soul-centred theory, but it would be possible to hold the latter and think that we will perish along with the inevitable deaths of our bodies. Hence, soul-centred theory also does not imply anything about whether our lives are in fact meaningful; it is compatible with nihilism, the view that our lives are meaningless (13.2).

7.4 Arguments for soul-centred theory

There are three major arguments for soul-centred theory in the literature, in catchwords appealing to realizing justice (7.4.1), making a permanent difference (7.4.2), and transcending limits (7.4.3). In this section I spell them out and maintain: that they are all weak in the sense that their premises do not support a soul-centred conclusion; that they can be reconstructed so that this fault is remedied; and that, once they are reconstructed so that a soul-centred conclusion is supported, the logic of the arguments also supports the conclusion that one must engage with a perfect being in order to obtain meaning. In short, what I demonstrate is that the fundamental motivation for holding a God-centred theory is implicit in the most promising arguments for holding a soul-centred one. By the end, unity amongst the most plausible supernaturalist positions will have emerged.

7.4.1 Realizing justice

One common argument for soul-centred theory is that life would be meaningless if the injustices of this world were not rectified in another world. As noted above (Chapter 4), the Biblical text *Ecclesiastes* acutely expresses the concern that there is no afterlife and hence that life is 'vanity', since both good and evil people share the same fate of death, and contemporary philosophers continue to express a similar sentiment.[1]

There are two ways in which justice requires different conditions to befall the righteous and the wicked. First, those who have suffered from wrongdoing are owed compensation for their losses. Second, those who have been evil deserve punishment for

---

[1] *Ecclesiastes* 3: 19-20 and 9: 2-3 in *The Holy Bible*, New King James Version. For recent statements, see Davis (1987); Craig (1994); Quinn (2000); Suckiel (2003: 32–3).

their wrongness and vice, and those who have been upright deserve reward for their rightness and virtue. One is indeed inclined to question whether one's life could be meaningful were one's good deeds to go unacknowledged or, worse, to be punished, and were tyrants to get away scot-free with embezzled billions, Rolls-Royces, and seaside chalets. Obviously, neither compensatory nor retributive justice is completely achieved in this world, which means that, for our lives not to be non-sensical, they must extend beyond the death of our bodies.[2]

The straightforward problem with the present argument is that while perfect justice might require an afterlife, at least given the way our bodies on earth are or must be, it is not clear that it requires an *eternal* afterlife that would be concomitant with having a soul (Perrett 1986: 220). Soul-centred theory implies the robust claim that, for our lives to be significant, we must possess a spiritual nature that either continues infinitely into the future or enters an atemporal realm where there is no distinction between past and future. It seems that humans would deserve an eternity in heaven only if they did something infinitely good, or an eternity in hell only if they did something infinitely bad, and we may reasonably doubt that infinite (dis)values are possible in a finite world (Kershnar 2005). Furthermore, even if they were possible, it would not follow that eternity is needed to give people what they deserve; for supposing that one could do something infinitely (dis)valuable in a finite amount of time here on earth, it would seem that a response proportionate to this deed would require merely a finite amount of time. If infinitely good or bad deeds were possible in a finite timespan, then so would punishments and rewards matching these deeds. Hence, soul-centred theory apparently gains no support from the view that meaning requires complete justice to be done.

How might the friend of the justice rationale reply to this problem? One way would be to appeal to metaphysical considerations about what would make an afterlife possible. One might argue that if a person were able to survive the death of his body at all, then he would have to be both essentially spiritual and immortal. Perhaps the only way to separate from one's physical self would be to have a spiritual nature that is indestructible. If that were true, then imposing any posthumous scheme of reward and punishment would require a soul.

There are two serious problems with this suggestion. For one, many thinkers view personal identity in terms of a causal and propositional chain of conscious states, which chain could conceivably outlast a given body and yet not last forever. Most philosophers believe that your mental life is what makes you who you are, not any *thing* (such as a soul, or a brain) in which your mental life contingently inheres. For another, even

---

[2] Note that God *qua* ideal judge, i.e., a perfectly impartial, powerful, and omniscient personal being, does not seem to be necessary for completely just conditions to obtain. An impersonal, natural force akin to Karma, or a personal being whose powers are not quite as robust as God's, would appear metaphysically sufficient to mete out desert. Cottingham, though, appears to suppose that only God could advance goodness such as desert (2003: 64–73).

if one's personal identity were constituted by a spiritual substance (not merely a chain of awareness), it is still conceivable that one could have a spirit that outlasts one's body but dissolves at some point. There is no compelling reason to think that there must be an utterly incorruptible aspect of one's identity in order for one to survive the death of one's body.

Let me therefore examine a second reply on behalf of the perfect justice rationale for soul-centred theory. So far, I have considered the argument that life would be meaningless without perfectly just responses to *imperfect behaviour*. The perfect justice theorist might do better if she claimed that life would be meaningless without perfectly just responses to *perfect behaviour*. On this view, immortality would be necessary not for rewarding relatively good people, but primarily for enabling people to become absolutely good and then to receive a fitting reward. This view is inspired by some of Immanuel Kant's remarks:

> The achievement of the highest good in the world is the necessary object of a will determinable by moral law. In such a will, however, the complete fitness of dispositions to the moral law is the supreme condition of a highest good.... But the perfect fit of the will to moral law is holiness, which is a perfection of which no rational being in the world of sense is at any time capable. But since it is required as practically necessary, it can be found only in an endless progress to that perfect fitness.... This infinite progress is possible, however, only under the presupposition of an infinitely enduring existence and personality of the same rational being; this is called the immortality of the soul (1788: 128-9).

We can avoid Kant's technical terminology and theoretical baggage and still find something worth discussing. Kant himself does not speak in terms of life's 'meaning' (any more than the author of *Ecclesiastes* does), but his remarks are relevant for a conception of meaning that promises to ground soul-centred theory. When Kant speaks of the 'highest good', he is referring to the best state of affairs for finite rational beings. For Kant, the highest good is our *final end*, not only in the sense that it must be our foremost goal, but also in that we may conceive of the world as having been created for such a state of affairs (for discussion, see Pogge 1997). Now, the purpose that we must above all pursue and that is grounded in the order of the universe, according to Kant, is moral perfection and happiness fitting that condition. And since moral perfection is possible only if we have a soul, a soul is necessary for the highest good, for the purpose the fulfilment of which alone confers meaning on our lives.

The problem with the Kantian response is that it is hard to see why one should think that moral perfection requires immortality, let alone a soul. Talk of 'perfection' suggests a maximal state, a condition in which the best has been achieved. Although there would arguably be *more* moral perfection possible if one lived forever, it is not clear that living forever is necessary for moral perfection itself. We seem able to conceive of a morally ideal agent who eventually dies, perhaps an unresurrected Jesus.

How might we motivate Kant's view that a soul is necessary for moral perfection? One possibility is that one's physical nature is incompatible with moral perfection.

Some philosophers have held that our sensuous nature interferes with the functioning of our rational, moral nature so much that the latter cannot be perfected until it is free of the former (see, e.g., Plato's *Phaedo*). Others have suggested that an ideal moral disposition is one that cannot be corrupted, with Kant's conception of moral perfection including the idea of unchangeable purity (1793: 46, 57, 61, 65).

In fact, though, neither view entails that a soul is necessary for moral perfection. Even if it were true that pristine virtue required a nature that is spiritual or cannot become bad, it would not follow that such a nature must never come to an end. Hence, there is still no reason for thinking that doing complete justice to a moral agent (even a saint) requires that agent to have both an immortal and spiritual nature.

Let me address yet another, quite different reason why one might think that doing perfect justice to a perfectly upright agent requires her to have a soul, based on ideas about what is involved in rewarding someone. To reward a person, at least in the right way, is to give her something that one expects her to like and to appreciate, at least if she were functioning normally and aware of what she could have. One does not reward someone, or at least not properly, by giving her just whatever she happens to want, for she might want to suffer physical torment out of a warped sense of guilt or to satisfy a cocaine addiction. She might also have adapted her preferences to the perceived limits of what is available, or simply not thought clearly about various possibilities. Instead, reward, of the most desirable sort, involves pleasing someone in a way she would like, on the supposition that she is mentally healthy and knows her options. Now, since people who are sane and autonomous would invariably want eternal bliss once the idea occurred to them, eternal life is necessary to reward the perfectly upright. And supposing it is true that life's meaning depends on being perfectly upright and receiving reward in proportion to it, we have an argument that supports a soul-centred theory.

Or do we? The argument presented, as it stands, would support the conclusion that immortality is necessary for meaning, supposing its premises were true. But I have been at pains to indicate that such a conclusion is not identical to soul-centred theory, which is the view that, to have a life that matters, it must be not only eternal, but also spiritual in nature. So, I need to say more about why sane and autonomous people would invariably want not merely eternal bliss, but the sort that could come only with a soul.

The natural explanation is that normally functioning people aware of possible states of the world would want to be able to come as close to being like and with God as they could (while remaining themselves). A physical form of immortality would preclude one from entering a spiritual realm in which one could be most akin to and in touch with a perfect being who is the source of everything. One need not be terribly religious to admit that one would like, say, to commune with a perfect being and to live in a universe that is oriented towards a spiritual end. A person with a maximally excellent nature, however conceived, would count as a V.I.P. worth meeting. And if mentally healthy people familiar with the concept of God would strongly want the pleasure of communing with Him, then a God-centred view also follows from the idea that a person must be rewarded consequent to ideal moral behaviour for her life to be meaningful.

In short, the present rationale for soul-centred theory has led towards an impure form of it, i.e., one that is conjoined with God-centred theory. I now demonstrate that similar results obtain in the context of the two other major rationales for soul-centred theory.

7.4.2 Making a permanent difference

In the most widely read text advocating a religious approach to the meaning of life, Leo Tolstoy argues that something can be worth striving for only if one faces no prospect of death. He says:

Sooner or later there would come diseases and death (they had come already) to my dear ones and to me, and there would be nothing left but stench and worms. All my affairs, no matter what they might be, would sooner or later be forgotten, and I myself should not exist. So why should I worry about all these things (1884: 11)?

One way of putting Tolstoy's point is that life would be meaningless if nothing were worth pursuing and that nothing would be worth pursuing if it would not have an 'ultimate consequence'.[3] Since one could apparently make a permanent difference only if one's life did not end with the death of one's body, this rationale appears to support the idea that one must have a soul for one's existence to matter. Tolstoy's rationale continues to have many supporters.[4]

One central criticism of this argument is that death intuitively cannot undercut the worth of performing certain constructive actions. For example, Anthony Flew remarks that it would be odd to 'think of a doctor despising his profession on the Keynesian grounds that in the long run we are all dead' (1966: 105). Such a case strongly suggests that helping others can be worth doing, even though the helping agent will die and the helping action will have no infinite ramifications (cf. 8.4).

It is open to Tolstoy and those with similar views to deny intuitions of the sort Flew invokes, and indeed they do. For instance, with regard to helping his family, whom he supposes are, of course, also mortal, Tolstoy asks, 'Why should they live? Why should I love them, why guard, raise, and watch them?' (1884: 12).

Tolstoy would have a stronger response to Flew if he could explain why it at first seems as though it is worthwhile for a mortal to help other mortals and why this judgement is false, upon further reflection. Probably the strongest explanation is that while such activities seem to merit performance from an everyday perspective, *from a broader perspective* nothing is worth doing unless it will have an ultimate consequence

---

[3] For someone who speaks this way, see Clark (1958: 30).
[4] Including Hanfling (1987: 22–4); Morris (1992); Craig (1994); Suckiel (2003: 32).

(see Hanfling 1987: 22-4; Craig 1994; Mawson 2010: 32-3). Although the claim that our lives must have an ultimate consequence for them to be choice-worthy is eminently questionable, I grant it, for the sake of moving on to a deeper criticism.[5] There remains a serious problem with the inferential structure of Tolstoy's argument, as it is neither deductively valid nor strongly inductive. We can accede the premise that an ultimate consequence is necessary for meaning and still deny the conclusion that a soul is necessary for it.

The glaring problem is that immortality is not the only way for a life to have an ultimate consequence. One's life could make a permanent difference if it made a lasting impression on *other* infinite things (as Levine 1987: 462 has pointed out). For instance, suppose that you made a substantial contribution to God's plan and that God fondly remembered you forever for having done it. Or imagine that angels eternally sung your praises. Or envision generations of mortal humans recounting tales of your great deeds successively without end. Tolstoy seems particularly worried that his life will, in his terms, 'sooner or later be forgotten', that it will seem as though he never existed or added anything to the world, but this condition could be prevented in several ways without Tolstoy being immortal.

For the friend of Tolstoy to resolve this problem, she must contend that not just any ultimate consequence is needed for constructive actions to be worthwhile. Instead, a particular kind of ultimate consequence is needed for a life to be choiceworthy, namely, one bearing on *oneself*. Now, such a view needs to be motivated, not merely asserted. What independent reason is there for thinking that for a given project to be worth doing, it must have some eternal ramification for the person doing the project?

Here is an answer worth considering. Meaning is a function of not just any sort of final value (e.g., bodily pleasure), but a special sort. Specifically, suppose that meaning depends on an infinite value. Now, if interacting with an infinite value required an immortal condition, then it would follow that meaning requires immortality. Some remarks of another immortality theorist, William Ernest Hocking, suggest this sort of view:

> The best of our experiences are normally long looked-forward-to and long remembered....Without this natural time dimension we know we have not 'done justice' to the event: meanings may be seen instantly, but they are not 'realized' (by beings with our time-extended mode of thinking) except with a certain amplitude of the process of pondering. Deprived of their due aftergrowth they fail to attain their proper value....And if there were such a thing as 'eternal value' accessible to us mortals, it would rightly call for unlimited time for its realizing (1957: 68, 141).

The suggestion is that actions are worth doing only if they give due consideration to their objects. Giving due consideration to an 'eternal value' requires an infinite amount of time in which to honour it, part of which will involve thinking about

---

[5] Below I return to the issue of appraising life from an 'objective' or 'external' standpoint (13.3).

one's involvement with it. Now, if the only object able to confer meaning on our lives had 'eternal value', then it would follow that for the actions relevant to meaning to be worth doing, one must have an eternal life. This line of thought provides a reasonable explanation of why, for example, an infinite chain of mortal humans who remember you would not be an ultimate consequence sufficient to make your actions worthwhile; *you* must not fail to engage with eternal values if you want your life to be meaningful, and this requires that *you* live forever.

So far, so good for the soul-centred theorist, but she has another major step to take in order to defend her view. The above is a rationale worth taking seriously for thinking that immortality is necessary for meaning in life, but it fails to indicate why a soul in particular might be necessary. Why think that an eternal life that is *spiritual* is necessary in order to make the relevant sort of permanent difference to the world? Why would a physical immortality not suffice?

To support a conclusion that immortality in the form of a soul is necessary for meaning in life, I revisit what it is that might count as an 'infinite value' and of how we plausibly ought to engage with it in order for our lives to be important. It is natural, of course, to identify God as that which has an infinite or eternal value, something that would require a life that never ends in order to give it due consideration. What the friend of Tolstoy's view ought to suggest next is that merely respecting or even loving such a value for an eternity in the physical world would not be enough for a genuinely meaningful life. Communing with God, in the most intense ways, would plausibly require a soul (7.2.3), a spiritual nature that is similar to His, if not springs from Him. Positively relating from a distance, while embodied on earth, would not be as desirable as doing so up close and personal, so a defender of the present argument might suggest. And this is indeed what Tolstoy concludes: 'No matter how I may put the question, "How must I live?" the answer is, "According to God's law." "What real result will there be from my life?"—"Eternal torment or eternal bliss." "What is the meaning which is not destroyed by death?"—"The union with the infinite God: paradise"' (1884: 18).

7.4.3 Transcending limits

The third major rationale for soul-centred theory, particularly inspired by Nozick, also provides reason to accept God-centred theory. Recall that Nozick suggests that the meaning of something in general appears to be a matter of asking about its relationship with other things (2.4). A life is meaningful, then, insofar as it relates to something beyond it in the right way. And Nozick suggests that mortality is a boundary that, if not crossed, renders a life meaningless.

A significant life is, in some sense, permanent; it makes a permanent difference to the world—it leaves traces. To be wiped out completely, traces and all, goes a long way toward destroying the meaning of one's life.... Attempts to find meaning in life seek to transcend the limits of an individual life. The narrower the limits of a life, the less meaningful it is.... Mortality is a temporal limit and traces are a way of going or seeping beyond that limit. To be puzzled about why death seems to undercut meaning is to fail to see the temporal limit itself as a limit (Nozick 1981: 582, 594, 595).

Many conditions that intuitively confer meaning on a life do seem to be instances of transcending limits. For example, finding a cure for cancer is a way of going beyond one's narrow interests, creating a great work of art is a way of connecting with complexity or beauty, and discovering the basic laws of the universe is a way of linking to reality. And it also seems true that immortality would be an instance of transcending a substantial limit, namely, the limit of an immediate perspective restricted to the now. Hence, Nozick may reasonably think that mortality is a limit the crossing of which is central to meaning.

However, this argument obviously needs to be tightened up if it is to provide strong support for soul-centred theory. Exactly which kinds of limits must one transcend in order to acquire meaning? As I have noted (2.4.2), breaking the speed limit and pinching a stranger are ways of 'crossing boundaries', but these are not *prima facie* candidates for a meaningful life. Furthermore, why believe that the limit of time is a boundary that specifically must be crossed in order for one's life to meaningful? Why would loving another person or creating a work of art not suffice?

Remember that Nozick's main strategy for specifying the relevant limits involves thinking of meaning as transcending limits that keep one from something finally valuable. '(M)eaning is a transcending of the limits of your own value, a transcending of your own limited value' (Nozick 1981: 601). On this view, one must protect, produce, or respect objects' good for their own sake that are beyond one's person. Now, both love and creativity can constitute 'a connection with an external value' in the absence of immortality. Clearly, a Nozickian needs a careful specification of the final values with which a person must connect, and of how to connect with them, in order for the transcendence rationale to entail soul-centred theory.

Let me reformulate the transcendence rationale this way: a meaningful life is one that connects in the strongest possible way with final value farthest beyond the animal self. The animal self is constituted by those capacities that we share with (lower) animals. These include our being alive, experiencing pleasures and pains, and exercising perceptual capacities. These conditions might be finally valuable, but they do not seem to have the sort of value with which one must connect to acquire significance; a life is not meaningful merely for being alive or feeling pleasure, probably by definition (see 2.4). Instead, on this view, a life is meaningful for intensely linking up with final values that are qualitatively superior to our animal natures. And there are degrees to which one can positively orient one's life to these values, ranging along various dimensions. For instance, *ceteris paribus* it is more intense: to intentionally promote a value than to knowingly do so but not aim to do so; to promote a value by engaging in labour rather than just by donating inherited funds; and to love or devote oneself to a value than to respect it or to leave it alone. The current transcendence conception of meaning is the view that meaning can come only from an intense positive relationship with the final value most beyond our animal nature.

Such a conception of meaning gives direct support not only to soul-centred theory, but God-centred theory as well. God, a perfect being, would, of course, be the highest final value with which a person could relate. One could get no farther away from one's physical, sensual nature than by relating to the divine. And, as I have noted above, the most intense relation for a person to have with the divine would be to commune with it, which would require a soul. So, I have found a version of the transcendence rationale that entails soul-centred theory, supposing its premises are true.

To bring this section together, what I have done is to examine three central arguments for soul-centred theory from the literature. In each case, I presented the basic rationales, showed that all of them, as they stand, actually fail to provide much, if any, reason to believe a soul-centred conclusion, and then reconstructed them. Once the three arguments were revised so that they would support soul-centred theory if we accepted the truth of their premises, they turned out to support God-centred theory as well. Although I have not addressed every possible argument for soul-centred theory, I submit that there is a broad lesson to be learned here. I conclude this section, not by evaluating the truth of the premises (see Chapter 8), but instead by bringing out the fundamental reason why each of the major arguments for soul-centred theory has also been, in effect, an argument for God-centred theory and by suggesting that this common denominator is strong evidence for expecting other arguments for soul-centred theory to have the same implication.

Recall that the realizing justice rationale, inspired by the author of *Ecclesiastes* and Kant, is that soul-centred theory follows from the view that compensatory or retributive justice is necessary for a meaningful life. I initially found it unreasonable to think that giving people what they deserve would require immortality, as opposed to a finite afterlife. This claim seemed reasonable, however, once I supposed that rewarding those who have been perfectly moral would require giving them something that they would like to have if they were functioning normally and aware of their options. One such desire would be for eternal bliss of a sort that includes a relationship with God, making God just as necessary for positive desert. The permanent difference rationale, springing from Tolstoy's classic text, maintains that a meaningful life depends on one doing something that is not only desirable, but also either continues to exist indefinitely or to positively influence the world without end. It was at first difficult to grasp why one would have to be immortal in order to make a permanent difference, but this notion was easier to accept when I considered that central to meaning might be making a particular sort of permanent difference, namely, intimately engaging with an infinite value. And since communing with God is the most straightforward answer to the question of what constitutes an intimate engagement with an infinite value, God-centred theory also follows from this rationale for soul-centred theory. Finally, the transcending limits rationale, articulated with care by Nozick, holds that a meaningful life is one that overcomes certain boundaries in the right way. It was *prima facie* implausible to think that one would have to be immortal to transcend the kinds of limits relevant to meaning. Yet this became plausible upon taking the relevant limits to be ones that keep a person from final value that is the farthest away from her animal nature, and deeming the relevant way to transcend them to require an intense, positive relationship. Again, a prescription to commune with God appears to follow.

In all three cases, the rationales support soul-centred theory in a straightforward way once, and only once, an idealized evaluative claim is conjoined with them. The justice rationale makes use of a claim about what is *best* for a morally *ideal* human person, the permanent difference rationale invokes a claim about *infinite* value, and the transcending limits rationale appeals to a claim about the *highest* final value, the one *farthest* from our animal self. Since meaning is a final good, linking meaning with an eternal spiritual life naturally requires an intermediate judgement about *value*. In addition, it requires a judgement about *superlative* value, since it must be of a sort that cannot obtain in any finite lifespan or even in the physical world. And any judgement about superlative value beyond the spatio-temporal world of sub-atomic particles that must be made in order to ground soul-centred theory will in all likelihood also ground God-centred theory, the view that a perfect being is central to life's meaning. That is the deep, logical reason that underwrites the tight, historical association between God and soul in supernaturalist conceptions of meaning.

7.5 Arguments against an immortality requirement

At this point, I turn away from arguments for soul-centred theory and towards arguments against it. One common criticism I mention here only to set aside, namely, the idea that if a finite life would not be meaningful, then neither would an infinite life. The claim, often ascribed to Ludwig Wittgenstein (1921: 149), but routinely advanced by others,[6] is that there is nothing about the *bare fact* of more time that could make life meaningful. This point is probably true, but it is irrelevant, as soul-centred theorists do not maintain that *mere* infinite extension or atemporal existence would be sufficient for meaning in life. As is clear from the discussion above, they instead hold that immortality is a partially constitutive condition of other states that together are sufficient for a meaningful life. For example, soul-centred theory is particularly well motivated by the idea that meaning is a function of communing with God.

There are three other prominent objections in the literature that are on the face of it more promising, and all three target not soul-centred theory as such, but rather a logical implication of it, namely, the claim that a life that never ends (whether in the

---

[6] Griffin (1981: 61); Perrett (1985: 237); Solomon (1986: 241); Levine (1987: 458); Singer (1996: 134–5); Schmidtz (2001: 175).

physical universe or in a spiritual realm) is a necessary condition for it to be meaningful. Call this the 'immortality requirement'. All three objections are 'bold' in the sense that they maintain, in stark contrast, that a life that never ends is a sufficient condition for it to be meaningless. One criticism is that an immortal life could not avoid boredom (7.5.1), a second is that it could not include sufficiently energetic goal-directed activity (7.5.2), and a third is that it could not exhibit certain moral goods (7.5.3). I maintain that these objections are unsound, and for reasons that the literature has not often recognized; some other consideration will have to be advanced in order to reject the immortality requirement with good reason (something I aim to provide in Chapter 8).

7.5.1 Boredom

About 40 years ago, Bernard Williams (1973) presented a powerful reason to doubt that immortality is necessary for a meaningful life and is rather sufficient for a meaningless one. His idea is simple: living forever would get boring, and a boring life cannot be a meaningful one, which instead requires zest and fulfilment. Some might suggest not merely that boredom would prevent meaning from being exhibited, but also that it would constitute anti-matter; i.e., would reduce the amount of meaning in one's life (4.3). This argument against immortality continues to be advanced, or at least taken seriously, by recent commentators.[7]

Roy Perrett (1986) has argued, with Williams, that some key responses to the boredom problem resolve it at the prohibitive cost of losing one's identity. For example, one might respond that a reincarnated immortal life—that is, one that included forgetting large portions of one's past—would not get boring. However, Perrett suggests that continuity of soul here would not be sufficient to maintain a person's identity through reincarnation, that the loss of memory and other parts of one's mental life would constitute a loss of *oneself* (1986: 224-7). One might also respond that immortality could take an atemporal form; presumably the concern about boredom arises only in the temporal world in which remembrance is possible. But Perrett also maintains that entering an atemporal realm would in effect amount to suicide (1986: 231-2). Since we are essentially beings who reflect and act, events that take time, any entity that continues in an atemporal realm would not be us, so he maintains.

Rather than take up the difficult metaphysical issue of whether *we* could enter an atemporal realm, I note that some philosophers maintain that one need not go reaching beyond time in order to resolve the boredom problem. It might be that an infinitely long life in the physical universe need not get boring. For instance, John Martin Fischer (1994) draws a distinction between repeatable and non-repeatable pleasures. Repeatable pleasures are those that we have some desire to experience again, at least after a certain amount of time has passed and we have undergone other experiences.

---

[7] Murphy (1982: 14); Perrett (1986: 223-4); Hanfling (1987: 54); Ellin (1995: 311-12); Kurtz (2000: 18); Belshaw (2005: 82-91).

Fischer argues that, given the existence of repeatable pleasures, we could find an immortal life to be exciting enough. Many others agree with Fischer that immortality need not get boring.[8]

Many of the pleasures Fischer points to are physical ones, such as the taste of food, but, assuming that the soul could remain embodied, this is not a problem for soul-centred theory. A deeper worry is that, well, forever is a very long time, and that repeatable pleasures might not be able to repeat that long. I find it difficult to make a firm judgement about whether an immortal life could avoid boredom to the extent that a mortal life can, and so I now aim to advance the debate by pursuing a different track.

I question the idea that boredom necessarily undercuts meaning. Both Williams and even many of his critics (e.g., Wolf 1997c) assume that boredom of any degree is incompatible with life's meaning, so that the more bored one is, the less meaningful one's life. What follow are two reasons to doubt that claim.

First, imagine that Mother Teresa had been bored by her work (at least in the stereotypical understanding of it). I submit that her life would have been significant, at least to some substantial degree, simply by virtue of having substantially helped so many needy people. One way to capture this case would be to suggest that boredom prevents a meaningful life only to the extent that it makes us incapable of constructive action. Then the question would be whether immortality would bore us to such a degree that we just stay in bed in the morning—whether it would bore us to death, so to speak. Even if immortality would get boring, it might not get *that* boring.

For a second reply, imagine someone who *were that* bored, bored to the point of being unable to do anything constructive. Suppose that this person had volunteered to be bored stiff so that many others would not be bored stiff. Consider, say, someone who volunteers to be head of an academic department, taking on administrative burdens, and attending dull meetings, so that his colleagues can avoid doing so. I submit (indeed, hope) that having done so would confer some meaning on his life, which means that boredom as such is not sufficient for absence of meaning.

7.5.2 Passivity

Another kind of argument for the view that immortality would be sufficient for a meaningless life is that our lives would lack certain kinds of intensity or engagement, which go beyond finding things interesting and avoiding boredom. I provide examples of this reasoning, before showing that they can all be defeated in one fell swoop.

A common suggestion is that being immortal would unavoidably result in a lack of a sense of preciousness and urgency, and perhaps the absence of motivation altogether (Baier 1957: 128; Nussbaum 1989: 339; Lenman 1995; Kass 2001: 21; James

---

[8] For those who agree with Fischer's reasoning, see Harris (2002: 26–7); Horrobin (2005: 15–17); Quigley and Harris (2009: 75–8); for related, but distinct, attempts to explain why immortality need not get boring, see Wisnewski (2005); Bortolotti and Nagasawa (2009); Chappell (2009).

2009). One would not prize a loved one or bother to strive for love so vigorously, if at all, so the argument goes, if one could always come back to her or find another beloved down the road. 'Why not leave for tomorrow what you might do today, if there are endless tomorrows before you?' (Leon Kass, cited in Bortolotti 2010: 43). A related argument is that an immortal person's decision-making would become determined more by external than internal factors (Wollheim 1984: 265-7). To do well in a finite life, a person must figure out what she wants deep down and then ascertain which actions would maximize her (expected) desire satisfaction within a limited timeframe. In contrast, to do well in an infinite life, the suggestion is that a person would not have to choose which actions to perform on the basis of introspection, but, instead, only when to do them in light of external circumstances. 'Weather reports would gain an immense importance in our lives' (Wollheim 1984: 266), which would make them senseless.

Here is the common problem with these suggestions: strictly speaking, they do not tell against soul-centred theory and its immortality requirement, for it is the *belief* in immortality and not immortality *itself* that is the potential culprit. If the logic of these arguments were correct, then a mortal who believed he were immortal would suffer these passive conditions, but an immortal who did not believe he were immortal would not suffer them. And there is no reason to think that being immortal must entail an awareness of that fact, or at the very least not one of the sort that would involve passivity.

7.5.3 Lack of moral goods

A third interesting objection to soul-centred theory is that being immortal would be sufficient to make our lives insignificant because persons who could not die could not exhibit important virtues (Nussbaum 1989: 338-9; Kass 2001: 21-2). For instance, they could not display a beneficent character to any significant degree, or courage of any kind that matters, since life and death issues would not be at stake. They would have no opportunity to become a *hero*, or, as Leon Kass puts it, 'The immortals cannot be noble' (2001: 22).

In one major respect, this argument appears to face the same problem as the previous one, namely, that it is largely a function of an agent's *beliefs* about whether she is immortal or not, and not whether she is in fact immortal. Suppose people were immortal but did not know it. Imagine, say, that when a fireman runs into a burning building to rescue children trapped inside, he thinks that he is putting his life at risk to rescue their lives that are at risk. In fact, though, if his body or the children's bodies were to die, their mental states would be transferred to other bodies or would remain contained within souls that survive. Such a fireman, despite being immortal, would surely exhibit the virtues of courage and beneficence.

Some might bicker over my characterization of these virtues, but let me avoid that debate here. Instead, on behalf of the current argument against soul-centred theory, I revise it by suggesting that certain *actions* would have much less moral worth if we were immortal than if we were not. Nussbaum's best case, here, is of justice, where she suggests,

> The profound seriousness and urgency of human thought about justice arises from the awareness that we all really *need* the things that justice distributes, and need them for life itself. If that need were removed, or made non-absolute, distribution would not matter, or not matter in the same way and to the same extent (1989: 338).

Nussbaum characterizes justice as a 'virtue' (1989: 338), but I do not focus on that. Instead, the strongest way of making her point is that promoting a just distribution of liberties and wealth would not confer much, if any, meaning on one's life, if these goods were not crucial to avoid death.

It might be true that Mandela's and Mother Teresa's lives would be not *as* meaningful as we may have thought, supposing that those whom they 'saved' did not ultimately need saving. However, they would plausibly have obtained substantial meaning by virtue of alleviating intense pain and preventing frustrated goals. And, remaining within the realm of the good (as action supportive of others), childrearing would confer great significance on one's existence, even if one's children could not die. When I think about what I have accomplished in rearing my sons, I do not feel great esteem particularly for having kept them alive. Instead, substantial pride has come largely from having helped them to do such things as avoid neurosis and physical injury, cultivate their intelligence and compassion, and develop self-love and confidence. Still more, most of us would judge meaning in life to be possible in other realms apart from the good, viz., the lives of Darwin and Einstein, as well as Dostoyevsky and Picasso, would have been meaningful for their work in the true and the beautiful, respectively.

A final way to reply to the argument that immortality would be sufficient for meaninglessness, since we could not then exhibit important moral goods, is that its rationale turns on a specific kind of immortality that is not essential. At best, the current rationale applies to persons who cannot die. However, another form of immortality is a condition of being able to die and also being able to live forever (as per vampires). The friend of immortality could seek to defend 'immortability' as the relevant form; under that condition, death might be even more grave (so to speak) than it presently is, for its victims would have otherwise lived forever.

7.6 The perfection thesis as the fundamental motivation for supernaturalism

In this chapter I have fully evaluated neither the supernaturalist theories explored in it nor the arguments made in their defence. I have not raised the question of whether the premises of the reconstructed arguments for soul-centred theory are true, having rested content simply to identify premises that would entail it on the supposition that they are true. I also have not appraised soul-centred theory itself, beyond rebutting some extant objections to it. Furthermore, I have not yet judged the God-centred theories articulated in this chapter, or the motivation for them that I have presented. I save such tasks for Chapter 8.

I conclude this chapter by pointing out that the most defensible supernaturalist positions have appealed to what I call the 'perfection thesis', the claim that *meaning in one's life requires engaging with a maximally conceivable value*, whether it is a perfect being with the qualitative properties, an unsurpassable reward to enjoy upon having achieved moral purity, or an infinite value that must be given its due or intensely related to, or the highest final good beyond one's animal self with which one must connect most intimately. The argumentation in this chapter strongly suggests that supernaturalism is best motivated by the perfection thesis, and, so, appraising supernaturalism requires refuting that claim, something I do in the next chapter. The deep reason why the existing criticisms of soul-centred theory from the literature are unpersuasive is probably that none of them tackles the perfection thesis head on.

8

Rejecting Supernaturalism

'Eternity is in love with the productions of time.'

William Blake (*Proverbs of Hell*)

8.1 How to be a naturalist

In the previous chapter, I presented what I take to be the most defensible forms of supernaturalism, ones that appear truly to follow from reasonable premises offered in their support and to be able to avoid criticisms. At the end of that chapter, I pointed out that these supernaturalist positions share a common thread that I call the 'perfection thesis', the claim that meaning in life requires engaging with a maximally conceivable value or superlative condition. In this chapter, I critically discuss reasons to doubt the perfection thesis and the supernaturalist theories motivated by it.

I begin by considering two objections already to be found in the literature that I think are unsuccessful (8.2), after which I advance three other objections that I maintain are more promising. The first objection is that some lives appear to have meaning *because of* the absence of anything perfect or supernatural (8.3). The second objection is that some lives appear to have meaning *despite* the absence of anything perfect or supernatural (8.4). The third objection parallels one that I developed in the earlier context of divine command morality (5.5), namely, that it is incoherent to believe that meaning in life is a function of anything perfect or supernatural, given that we know that the former exists but do not know that the latter exists (8.5). What one finds in the spatio-temporal universe is enough on its own to make our lives meaningful and, were God to exist, to be worthy of His admiration, as per William Blake.

After presenting and defending these arguments against the supernaturalist position, I present a principled alternative to the perfection thesis insofar as it applies to meaningfulness on balance. The perfection thesis, when applied to whether a life is on balance meaningful, is the view that, even if a life could be *somewhat* meaningful for engaging with something other than a maximally conceivable value, it could not be meaningful *period* unless it engaged with one. This view could appeal to those supernaturalists who concede that *some* meaning would be possible in a purely physical world, but who maintain that this would not be enough for a 'truly' meaningful life.

In the absence of any principled specification of what amount of less than perfect value is sufficient for a meaningful life on balance, the naturalist view is poorly developed. Furthermore, the supernaturalist can at least claim the virtue of simplicity, for a maximally conceivable condition is a simpler ground than some degree of value in between it and zero (Swinburne 1991). It would be useful for me, as a naturalist, to specify in a principled way the less than ideal degree of meaning that determines when a life on balance counts as a meaningful one, that is, to present the most attractive version of the 'imperfection thesis'.

I argue that the major instances of the imperfection thesis that naturalists have offered (or at least suggested) in the literature are implausible (8.6). None fits the considered judgements that naturalists themselves would, upon reflection, have about which degree of meaning is sufficient for a person's existence to be meaningful on balance. Working within a naturalist mindset, I demonstrate that a new principle needs to be developed, and I also articulate the one that I deem to be most justified, defending it from objections. I conclude by pointing out some broader implications of the argumentation in this chapter, specifically, that the objections made to supernaturalism probably apply to non-naturalist theories as well, and that they are consistent with thinking that a supernatural realm could at least enhance meaning in life (8.7). The latter point is important, for while naturalists up to now would rightly have us deem meaning to be possible without anything spiritual, they have tended not to acknowledge that the spiritual is worth imagining as another, possible source of meaning.

8.2 Weak arguments against supernaturalism

The basic line among the most prominent naturalists in the literature is that supernaturalism posits too high a standard for the appraisal of life's meaning. I ultimately agree with this conclusion, but not for two major reasons that naturalists have proffered.

8.2.1 Futilitarianism

Brooke Alan Trisel has done the most of late to try to refute the idea that perfection, or one of a sort that includes a spiritual realm, is essential for meaning in life. Trisel primarily aims to rebut people he calls 'futilitarians', those who claim that a mortal life would not be worthwhile—and hence, for my purposes, would be meaningless—since everything we do is unavoidably futile. Drawing on intuitive notions of futile treatment in a medical context, Trisel provides a revealing analysis of what a futile life is: roughly, one that repeatedly fails to achieve valued ends (2002: 68-70; see also Trisel 2004). Futilitarians, Trisel argues, are incorrect that immortality is necessary for worthwhileness (meaning), since their desires for perfection are unrealistic. Life would indeed be futile if one strove for something that, for all we know, does not exist and cannot be brought about. However, Trisel suggests that we could and should then simply change our desires to avoid futility and consequent meaninglessness. If we were to desire something less than perfect, then life would not be futile, since we could then have a good chance of obtaining our ends. 'When futilitarians wonder why striving is futile, they need not look much further than to their own towering expectations' (Trisel 2002: 79).

The problem with Trisel's rationale against the perfection thesis and the supernaturalist perspective it grounds is that, even if the latter posit too high a standard for meaning in life, Trisel posits one that is too low. He is, by and large, a subjectivist about value, such that a condition would confer meaning on a person's life just because she has desired it and it has been realized. As I argue below, this has seriously counterintuitive implications about what can and cannot confer meaning on a life (9.5). For example, it appears that a person could avoid futility and meaninglessness, on Trisel's view, simply by eating her own excrement, supposing she had that odd desire (Wielenberg 2005: 22). It would be of more interest if naturalists could object to the perfection thesis while granting the supernaturalist that meaning in life is not merely a function of whatever desires or other contingent mental states an individual happens to have.

8.2.2 Ought implies can

Supposing, then, that meaning in human life is something that is not merely subjective, how might a naturalist plausibly contend that the perfection thesis posits too high a standard? One idea, suggested by the work of not only Trisel, but also Kurt Baier, is that if a state of affairs is beyond the control of an individual, it can neither add nor detract from the meaning in her life. The intuitive point here is that the clearest bearer of meaning is an action (as per 4.4.1 and 4.4.2), and that meaning is a practical matter, one that we should take into account when deciding how to live. Recall that most contemporary philosophers writing on life's meaning believe that meaning in an individual's life can come in degrees, and they are interested in the ways an individual could have more meaning in her life. This personal and action-guiding interest might suggest that the relevant sort of meaning judgements are fundamentally 'normative', that is, governed by the dictum 'ought implies can'. If we are unable to do something or to bring something about, then we can have no reason to do it and so it is not relevant to whether our lives are meaningful or not. Baier can be read as hinting at this sort of reasoning in the following:

(T)he standard by which a thing's ability to satisfy certain criteria of excellence is rightly judged depends on the ability...to meet a given criterion....(W)hen I was young, the life of an eagle seemed to me vastly superior to my own life and that of other people. But, as I mentioned, this question has not greatly exercised many people, and that for good reason, since we have no choice in the matter (1997: 63, 63-4).

Baier's attitude appears to be not to cry over spilled milk and more generally not to worry about things that are beyond one's control. If something cannot be prevented, changed, or brought about, then it should make a difference neither to what one chooses to do nor to the way one appraises one's life.[1] This entails that if perfection of whatever sort did not exist, or were unavailable, its absence could not affect the extent to which our existence is significant.

I have two replies to this argument against the perfection thesis. First, the inference from meaning being of practical importance, which is undoubtedly true, to the centrality of 'ought implies can' is weak. Although it is correct that meaning comes in degrees, that it varies within a life and also between lives, and that we have *pro tanto* reason to seek more of it rather than less, these claims are consistent with the view that meaning is at bottom *evaluative* rather than *normative*. Previously, I analysed the concept of life's meaning as including the idea that meaning is a final good, something valuable for its own sake; I did not include the idea that meaning is essentially an action-guiding category (1.2 and Chapter 2). And supposing that meaning is basically a *good* rather than a *should*, we can still be interested in it for the difference it can make to our lives—and indeed through our actions—to no less a degree than we are with regard to the good of happiness. When we ask how to make a life meaningful (or happy), we need not suppose that the only coherent answer implies that meaningfulness (or happiness) is a function of only those conditions within a person's control. Instead, the answer could plausibly be that a life is meaningful (or happy) insofar as a person has realized a certain kind of value, which may or may not be in her control.

Second, there are powerful counterexamples to the claim that conditions capable of producing or reducing meaning in a person's life are alone those within her control. If a state of affairs could affect the meaning of my life only if I could elect to bring it about or not, then, supposing I could not change the fact that my wife and children will die before me, their deaths could not affect the degree of meaning in my life. However, that is absurd (and only some kind of Zen master could think of claiming otherwise). Although I accept that what makes a life matter is often a function of the actions one has taken, there are some conditions beyond a person's control that can affect the extent to which her existence is significant, and, for all that has been said so far, perfection could be among them.

8.3 Meaning because there is nothing supernatural

The first promising criticism of supernaturalism is bold, and while, in the final analysis, it does not tell against supernaturalism as such, it is philosophically interesting for the extent to which it might work against soul-centred theory. It is the suggestion that at least some of the clearest cases in which we are inclined to find meaningfulness are

---

[1] See also Trisel 2002, 2004, as well as adherents to the initial purpose analysis of the concept of meaning (section 2.4.1).

best explained by the absence of anything supernatural. Whereas an argument below is that certain conditions would be meaningful *even if* there were nothing perfect or supernatural (8.4), the present one is that certain conditions would be meaningful *only if* there were no spiritual realm that one could enter upon bodily death.

The clearest sort of this case is one in which people's lives are at stake. Imagine you are driving a bus and must choose between running over a human person or a flower, where the former is immortal and presumably would enter some kind of heaven once her body died. On the face of it, there is good reason to strike the human, and even to do so out of love, as one would thereby maximize her chances of leaving this, imperfect world and sending her to a much better place. Having definite reason to run over the flower is best explained, so the argument goes, by the hypothesis that there is no afterlife, that the physical universe is the only world available to the person, of which she would be permanently deprived upon striking her.

Friends of soul-centred theory naturally reply that the human person, exhibiting a soul made in God's image, would have to be treated with respect, meaning that there would be strong reason not to strike her and instead to mow down the flower, which is worth less than she. And they can also take a cue from an earlier point, that the right sort of love implies respectful treatment (section 7.2.4).

However, the logic of this reply suggests that the respectful and loving thing to do would be to get permission from people to kill them, so that they could go to Heaven sooner rather than later. At this point, friends of soul-centred theory typically suggest that this would be to 'play God', who may alone rightly end the earthly lives of those He has created. But not even that will resolve the problem that, by the present reasoning, the respectful and loving thing to do would be to let people die. Even if there were reason not to kill a human person, there would appear to be no reason of respect or love to save one, supposing she would live on in a much more desirable afterlife. In any event, I do not pursue this dialectic any farther, but merely submit to the reader that there would be *clearest conclusive* reason both to save human persons and not to kill them, if they were mortal.

Even if this criticism were sound, it would not tell against the perfection thesis and supernaturalism in their entirety. The objection applies most readily to the claim that a perfect state of one's spiritual nature is necessary for life to be meaningful, but it does not as easily apply to the claim that a perfect being must exist for it to be meaningful. For the latter to work, there would have to be some plausible reason for thinking that certain conditions would be meaningful only if God did not exist. There are strands of thought, especially in the existentialist tradition, to this effect, for example, that we would be free in an important sense or have a higher status in some way if God did not exist.[2] However, I leave it to others to try to find a plausible kernel in these ideas.

---

[2] For examples, see Sartre (1946); Barnes (1967); Singer (1996: 29); Klemke (2000: 196); cf. Williams (1985: 128).

8.4 Meaning even if there were nothing supernatural

My second objection to supernaturalism and the perfection thesis that underwrites it is already in the literature. I aim to buttress it by bringing out just how counterintuitive the views of supernaturalists would have to be in order to reject it.

The objection invites the reader to suppose, for the sake of argument, that there are no spiritual persons. Imagine that only the physical universe, as best known by the scientific method, exists. Now consider whether certain lives could be on balance meaningful, say, those repeatedly invoked here, such as Einstein, Darwin, Dostoyevsky, Picasso, Mandela, and Mother Teresa. Many will respond that they would find these lives to be meaningful in the absence of anything perfect or supernatural.

Similar remarks go for actions that confer some meaning on a life, such as rearing children with love, patience, and insight, caring for students, parents, or hospital patients, having deep and long-lasting friendships, sustaining a vibrant and close marriage, engaging in charity work, creating artworks, making intellectual discoveries. These, too, appear capable of making a person's existence somewhat more significant in the absence of a perfect or supernatural condition. If so, then the best theoretical explanation of what confers meaning on a life cannot include the claim that a life would be utterly meaningless in the absence of a spiritual realm.

There are some inclined towards theism and supernaturalism who feel the force of these examples, which are common in the literature, and who, in light of them, grant that at least *some* meaning would be possible in a purely natural world. There are, however, some supernaturalists willing to say that nothing would be meaningful if indeed there were neither God nor a soul. Of course, people can *say* anything, but the real question is whether they would truly *believe* what they are saying. And this is open to doubt, for at least many who would say such things. It would require viewing human persons, and their capacities for self-reflective thought and behaviour and for love, as worth no more than rocks and plants, as exhibiting no relevant qualities beyond being a 'fermenting globule', as Tolstoy at one point characterizes mortal life from an extremely alienated perspective (1884: 15). From this standpoint, a supernaturalist would have to view engagement with human persons as utterly unimportant because there is no God in whose plan we could play a role, or because there is nothing with a perfect nature around which we could orient our lives, or because we cannot become morally ideal and consequently experience eternal bliss, or because we cannot commune with an eternal value and thereby make a permanent difference to the world, or because we could do nothing to greatly transcend our animal natures. *Alleviating the pain of a young child from, say, having been burnt,* would, from this standpoint, be incapable of conferring any meaning on a person's life, as doing so would be unrelated to a perfect being and would be a mere momentary effort to care for an entity that does not matter because it and its states are imperfect and will eventually disintegrate.

Tolstoy and Craig do appear to have viewed the human condition from this perspective, but I presume, with several religiously oriented philosophers, that it would be *incredible* (Audi 2005: 334, 341-2) and false *beyond a reasonable doubt* (Quinn 2000: 58; see also Trisel 2004: 384-5). In any event, I presume it would be too much to swallow for a large majority of readers, even those religiously inclined. The cases suggest that the supernatural is not *necessary* for our lives to be at all meaningful or even meaningful on balance, although it could still be that a supernatural realm could enhance the extent to which they are (as I discuss in 8.7).

8.5 Meaning without knowledge of the supernatural

The third argument against supernaturalism that I defend exhibits a structure similar to an objection that I made earlier to the divine command theory of morality (5.5). There, I argued that although we know that wrongness exists, we do not know whether God exists, making it incoherent to claim to know that wrongness is a function of God's will. A similar problem faces the perfection thesis and supernaturalism in general. If we had conclusive evidence for supernaturalism and its perfection thesis, then we would know both that perfection exists and that meaning in life exists, but we do not.

Recall the general form of the incoherence. For any entailment 'If X, then Y', it would be inconsistent to claim the following three things: (1) I know 'If X, then Y' is true; (2) I know X obtains; (3) I do not know whether Y obtains. Now, the supernaturalist must hold the following claims: (1+) I know 'If meaning exists, then perfection (and a spiritual realm) exists' is true; (2+) I know meaning exists; (3+) I do not know whether perfection (a spiritual realm) exists. Since this collection of propositions is inconsistent, the supernaturalist must jettison one, and I claim that it would be most reasonable to let go of (1+), i.e., the perfection thesis and supernaturalism.

Of course, the initial reply will be that the relevant proposition to drop is (3+). After all, many supernaturalists are theists. Perhaps most of those who believe that God or a soul is necessary for meaning in life believe that those spiritual substances actually exist.

However, *to believe* they exist is not necessarily *to know* that they do. Pointing out that many people, and probably a majority of supernaturalists, have faith in God and a soul is not to establish that there is conclusive epistemic reason for them to do so, the only relevant standard.

One avenue at this point would be to canvass arguments for and against the existence of a spiritual realm, but I do not want what is intended to be a book on life's meaning to turn into one about whether there is evidence to warrant belief in the existence of a substance beyond the physical universe. There are plenty of books oriented towards the professional philosopher that are devoted to, say, the existence of God. There are very few devoted to meaning in life. So, I rest content, here, with pointing out the fact that a large majority of professional philosophers are unconvinced by arguments for thinking that God or a soul exists, even if many in their private lives have faith in them.[3]

The only other option to avoid the rejection of supernaturalism would be to deny that we know that there is meaning in life. But this point takes us back to the kind of case examined above, that of a young child who is suffering. Surely, relieving the pain of a toddler's burn has conferred some meaning on a parent's life, as have more 'grand' projects such as dismantling apartheid, caring for lepers, discovering that space-time is curved, proving that human beings are a product of natural selection, writing *The Brothers Karamazov*, and painting *Guernica*. Just as Cottingham is pretty sure that cruelty is wrong, and in all possible worlds (5.4–5.5), so most readers are, I presume, confident that these are sources of meaning. It is not merely that those in the field have thought that these would be good examples of meaning, if it were to exist; instead, they tend by and large to hold the stronger claim that meaning exists and is typified by these examples.[4]

8.6 Developing the imperfection thesis

At least if the last two objections that I have made to supernaturalism and its central motivation, the perfection thesis, are sound, then the natural question to ask is: "Which less than perfect values, to be found in the natural world, comprise meaning in life?"

This question, in turn, should be broken down into two distinct ones, depending on whether one is interested in knowing which naturalist conditions can add *some* degree of meaning to a life or which of them can make a life meaningful *on balance*. It is my task in part three to answer the former question, about those conditions in a purely physical universe that are capable of making life somewhat more meaningful. There, I defend an objective form of naturalism, roughly according to which the relevant less-than-perfect values that can enhance meaning in life are those relating to the sophisticated exercise of our rational selves contoured towards fundamental objects (12.3).

In the rest of this chapter, I address the other facet of the question, namely, what amount of value relevant to meaning would be sufficient to consider a life to be meaningful *on balance*? Under what conditions would one sensibly say of a person that she has had a meaningful life, analogous to the way one would say of a person that she has had a happy life? Recall that I hold a mixed view, according to which both parts of a life and the life as a whole can exhibit independent dimensions of meaning. Therefore, if I want to know whether a life is on balance meaningful or not, in principle I must assign a weight to the part-life facets that have been meaningful, subtract any anti-matter exhibited in the parts, and then assign a weight to the whole-life facets

---

[3] For theists who are adamant about having knowledge of God's existence, perhaps the argument could be reformulated to grant that point, but still invoke a large discrepancy in the degree of justification for believing in God relative to that for believing in meaning.

[4] Of course, this is not true of all interlocutors, and I address such nihilists (pessimists) briefly at the end of the book (Chapter 13) and in other work (Metz 2011a).

that have been meaningful, subtract any anti-matter exhibited there, and finally add the two sums to form an overall assessment of the life. The question I address in the rest of this chapter is how great those combined values, of parts and whole, must be in order for the life not to be aptly described as 'meaningless'.

I begin by criticizing naturalist claims that the amount of value needed for one's life to be meaningful on balance is determined in an individualistic way (8.6.1). According to this view, the standards for judging the meaningfulness of lives vary from individual to individual. Then, I examine and reject existing social conceptions of the value needed for meaning (8.6.2). This kind of view is that whether an individual's life is significant logically depends on facts about a certain group she belongs to, which is the species in the most powerful versions. I propose a new version of the social view, setting out the basic instance of the imperfection thesis that a naturalist ought to put forth as a counter to the perfection thesis, and I also indicate some ways that naturalists could reasonably question and refine it (8.6.3).

8.6.1 An individual conception

In the literature, one finds several accounts of which amount of less-than-perfect value is sufficient for a meaningful life. It is useful to divide them into two different kinds of principles. On the one hand, there are social views according to which a person's life is meaningful on balance depending on how she fares in comparison to some group of which she is a member. On the other, there are individual views according to which a person's life can be properly judged to be meaningful on balance, regardless of the group she belongs to and solely by virtue of certain facts about her. In this section, I consider and ultimately reject the most powerful version of the latter view.

Some who have explicitly rejected the perfection thesis (or something close to it) have asserted individualism but rejected subjectivism with regard to appraising whether a life *in toto* counts as meaningful. In particular, E. M. Adams (2002) has contended that whether a life is meaningful on balance is a function solely of facts about the individual considered apart from his membership in any group, but maintains that what is valuable is not simply a matter of what individuals happen to value. In a way, this is an odd combination of views, for the natural explanation of why meaningfulness is relative to individuals would be that it is logically dependent on their desires or other variable attitudes. However, it is coherent to hold that there are objectively good conditions—i.e., good not merely because they are the object of an individual's preferences—and that the only ones pertaining to whether a person's life is meaningful are those that are in some sense 'available to her'.

Addressing the nature of these objective goods is not pertinent here; they could be part-life factors such as achievements and relationships, as well as whole-life patterns. The point is that there are some things that are good not merely because of our attitudes towards them, and there is also a fact of the matter about how much of these conditions a given individual can realize. People vary biologically, psychologically, and socially, with some being able to realize more objective good than others. Those who are healthy are *prima facie* more able to exhibit or promote value than those who are sick; those who have self-esteem on the face of it have a greater capacity to be and do good than those who are depressed; and those who live in societies with minimal repression and substantial resources are in some respects in a better position to achieve worthy ends than otherwise. However, the proper way to judge a person's life is not by judging how much value she brought about in comparison with others or with an idealized condition such as living forever in a certain way. According to Adams:

> The most important question for us is not the length of one's life but how well one used one's time in developing one's powers, defining one's identity and life plan, and working for an order of goodness....One's life is meaningful to the extent that it is the life reality requires of one, given one's abilities and circumstances. There are normative requirements impinging on one from within one's own constitution and from within one's time and place (2002: 80).[5]

I read Adams' remarks as pointing towards the following version of the imperfection thesis:

> (IT_1) A human person's life is meaningful on balance if and only if she has achieved a high proportion of the maximum amount of *pro tanto* meaning that she could have brought about.

This view entails that if one did, say, 90 per cent of what one could have done to promote the 'order of goodness', then one's life would be meaningful on balance, whereas if one did only 35 per cent, it would not be. It follows that perfection is not necessary for meaning in life, supposing it is not within our reach. Instead, meaning is simply a function of how well one does at promoting imperfect value in light of what is within one's particular and limited capacities.

Note that the concept of being able to realize a condition is not the same idea as a condition being within one's control, which I have already discussed and rejected above in the context of some of Baier's and Trisel's ideas (8.2). I cannot always control whether others love me, but, if they do, then it is something within my reach. I also cannot control the remote consequences of my actions, or at least not nearly as much as the action itself or its proximate consequences, but if my action brings about a certain state of affairs in the long run and far away, then I was indeed able to achieve it.

Notwithstanding the differences between Adams and Baier/Trisel, one of the objections I made to the latter applies equally to the former. Recall the case in which your child and spouse have died before you do. Even though it was beyond your capacity to determine when (or that) they died, their death would surely have had some effect on the meaning of your life. The principle inspired by Adams entails that if the prevention of their death were not within your power, then it would be just as irrelevant to the meaning in your life as is, for him, the fact that you cannot prevent your own death. However, it is counterintuitive to think that meaningfulness is not lessened by the loss of loving relationships when others die.

[5] For similar views, see Fabry (1968: 40–1); Kushner (1987: 148–9).

In addition to the claim that a state of affairs must be achievable for an individual for it to count with regard to the meaning in his life, Adams claims that meaning is a function solely of the degree to which the good states of affairs that are achievable have in fact been achieved. Here are two counterexamples to the latter claim. First, consider an innocent, young person who has been kidnapped and held in a cage, with absolutely no means of escape. If Adams were correct that meaningfulness is a function of the percentage of one's capabilities that one has used, then whether the person's life is meaningful on balance and how much would not be negatively affected by the kidnap. Once the person's opportunities were unavoidably reduced, according to the view as expressed in principle (IT_1), meaningfulness would be a function of the extent to which those opportunities that remain would be used. Indeed, this view entails that great meaning would be possible in this scenario, more than in the course of a normal life, if the kidnapped person's opportunities were few in number and easily 'maxed out'. However, this is absurd.

Consider, too, the suggestion that people who are severely handicapped and impoverished as a result of natural forces on balance might not have meaningful lives because of these conditions. Suppose, for example, that people were born onto an isolated island with meagre resources, so that their time were nearly entirely taken up by searching for food. Adams's view entails that they could have terrifically meaningful lives, if they fully realized whatever few opportunities to promote 'the order of goodness' were available to them. However, my intuition suggests that the number and kind of opportunities available to people affects whether their lives can count as meaningful.

Some readers might baulk at this judgement for sounding harsh. In particular, some judge that any plausible account of meaning must entail that everyone is at least capable of living a meaningful life (e.g. Cottingham 2003: 69-70). I, however, believe that the present cases are more powerful than this view; they show that it is more plausible to think that meaning can be undercut by dramatic losses at the social and natural lottery than that meaning can always survive any such losses.[6]

Furthermore, readers should reflect on why they presumably do not want more ignorance, worse health and less wealth. Why would you prefer not to become an Alzheimer's patient, or, less dramatically, to lose 30 IQ points? Part of the best explanation, I submit, is that one senses that a life with much fewer capabilities would lack the meaning available to a life with more of them. If it is true that being (born) very dumb, sick, and poor can affect the degree of meaning in one's life, then it is false that meaning is a function merely of how much one has exercised whatever capabilities one has.

The individualist account of meaningfulness implies that whether a person's life is meaningful or not is a function solely of facts about that person, considered apart

---

[6] Remember that judging some people's lives not to count as meaningful on balance is consistent with believing that everyone's life is equal from a moral point of view (1.2). Indeed, the fact that someone has little meaning in her life could ground the moral judgement that she is entitled to more help and resources than those with more meaning.

from membership in any group. The extent to which one has realized the objective goodness of which one is capable involves no essential reference to group membership. However, the counterexamples I have presented suggest that a different, social standard for assessing meaningfulness is correct. When one judges that being severely impoverished or permanently kidnapped reduces the available meaning in a life, one is probably comparing such lives with others that do not suffer from these afflictions. Judging these lives not to be very meaningful requires appealing to conditions beyond what the given individual is capable of achieving.

8.6.2 Social conceptions

Whereas the above, individualist version of the imperfection thesis is that meaningfulness is a function of whether a person realized a certain proportion of the value available to her, social versions imply that it is a matter of whether she has realized value available to her that is a certain proportion of some group-based standard. They entail that one cannot ascertain whether someone's life is meaningful on balance in isolation from society or humanity. I critically explore social conceptions of meaningfulness to be found in the recent literature, and argue that none is completely attractive.

Baier posits a broad social standard for specifying the amount of *pro tanto* meaning that one must obtain in order to have a life that is on balance not meaningless. Over a span of 40 years, Baier (1957, 1997) has maintained that one appropriate way to judge the meaning (or, more generally, worth) of a human's life is in comparison with the typical lives of other humans. When comparing human lives, the absence of immortality or some other perfect condition does not matter. As Baier remarks, '(O)ur question would be more difficult if some human beings were immortal or at least survived the death of their body. That all must die appears to make the finality of death irrelevant to assessing the comparative worth of human lives' (1997: 66). This comparative standard is vague as it stands, and I now construct and then criticize more specific versions of it.

In one text, Baier develops a concrete way to judge lives by drawing analogies with the proper way to judge other things, such as tallness and students:

Ordinarily, the standard we employ is the average of the kind. We call a man and a tree tall if they are well above the average of their kind.... The same principles must apply to judging lives.... A good and worthwhile life is one that is well above average. A bad one is one well below.... The Christian evaluation of earthly lives is misguided because it adopts a quite unjustifiably high standard.... We do not fail every candidate who is not an Einstein (1957: 127).

These remarks suggest the following standard for when one's existence is all things considered significant:

(IT_2) A human person's life is meaningful on balance if and only if it has (much) more *pro tanto* meaning than the average human's life.

One obvious problem with this principle is that it entails that meaning is 'zero-sum', i.e., that it is *impossible* for everyone's life to be meaningful (Joske 1974: 286; Belliotti 2001: 73). Of course, it is implausible to think that everyone's life is in fact meaningful, but that is not the point. So long as there is a possible world in which everyone's life is meaningful, Baier's suggestion about which lives count as 'meaningful' is incorrect, since it entails that those people with less than the average amount of meaning must have meaningless lives on balance. And if we can conceive of a world in which everyone develops herself through close relationships, intricate skills and sophisticated projects, then there is a possible world in which the lives of all are significant.

There is another way to read Baier so that this counterexample is avoided. Instead of deeming a human being's life to be meaningful *in toto* if and only if it has more objective goodness than the average human's, one might deem it to be one that has at least as much as the typical human.

(IT_3) A human person's life is meaningful on balance if and only if it has at least as much *pro tanto* meaning as most human beings.

So, instead of *better than average* as per (IT_2), according to (IT_3) one must be (at least) *equal to most*, for one's life to count as meaningful. This latter principle would have us ascertain the range of objective goodness that a large majority of human beings realize and then see whether a given individual at least falls within that range. This principle leaves open the possibility that everyone falls within the same range, and hence does not entail that some must have meaningless lives if some have meaningful ones. And the principle continues to serve the function of denying that an immortal soul in a certain ideal state, or any other instance of perfection, is required to have a meaningful life.

However, (IT_3) still has the counterintuitive implication that it would be impossible for a person whose life has somewhat less *pro tanto* meaning than most to have enough for her life to be meaningful on balance. Imagine a world in which the lives of most human beings have quite a lot of the relevant final value in them and that those who have somewhat less still have enough to give them what appear to be significant lives on the whole. Return to Baier's student analogy; one can readily think of a group of students who are generally excellent, with a minority not performing as well but nonetheless well enough to count as satisfactory.

Furthermore, (IT_3), along with (IT_2), entails that a promising way to make one's life meaningful on balance would be to kill off most human beings and then subject the few remaining ones to torture (Joske 1974: 286; Belliotti 2001: 73). That way, despite the great amount of anti-matter in it, one's own life would be not only better than average, but also at least equal to most.

Where has Baier gone wrong? Perhaps he is led astray by the analogies he draws, which are, upon reflection, not as strong as he believes. Tallness is different from meaningfulness, on the face of it. Tallness seems inherently comparative in the sense that something small must exist in order for there to be something tall. However, meaningfulness is not obviously comparative in this way; one can imagine a world in which human lives are all meaningful or all meaningless. In addition, while being a good student is not comparative in the way that tallness is, it seems plausible to think that whether a student is on balance good is not merely a function of whether she is better than average or equal to most. For, again, the middling and the typical might be terribly good—or terribly bad. So, tallness is perhaps a function of an average but is disanalogous to meaningfulness, whereas being a good student seems more analogous to meaningfulness but is not merely a function of usual performance.

Let me suggest a way to flesh out Baier's hunch that a human norm fixes when a life counts as meaningful that not only differs from the better than average and equal to most principles, but also avoids the problems I have raised for them. Consider this way to ground meaning on normality:

(IT$_4$) A human person's life is meaningful on balance if and only if it has at least as much *pro tanto* meaning as is characteristic for human life.

What is characteristic of human life is, I think, well understood as a dispositional feature of it, rather than an actual one. Whereas the criteria of better than average and equal to most peg the relevant final value sufficient for meaningfulness to an amount that human beings have in fact exhibited, the present criterion does not. Instead, it invokes an amount that human beings are biologically, psychologically, and socially disposed to realize. It could be that the amount of *pro tanto* meaning that human beings have in fact realized is greater or smaller than what is characteristic for them to realize, just as most human beings in fact have fewer teeth than the 32 that are characteristic for them, as Elizabeth Anscombe famously noted long ago.

So construed, (IT$_4$) avoids the counterexamples that plague Baier's (IT$_2$) and (IT$_3$). It entails that everyone's life could be meaningful, since everyone could realize the amount of final value that is characteristic of human life, avoiding the problem with (IT$_2$). It entails that a person's life can be meaningful on balance, even if she has realized less *pro tanto* meaning than most have, so long as she has realized what is characteristic of human life, thereby evading the objection to (IT$_3$). And it entails that a life that is better than the lives of all others who are tortured is not necessarily meaningful, for that life could still fail to realize the amount of meaning-conferring final value that is characteristic of human life, hence bypassing the objection that applies to both of the prior principles.

So far, so good. However, (IT$_4$) faces the criticism that it oddly entails that it is impossible for everyone's life to be meaningless, a criticism that also applies to (IT$_2$) and (IT$_3$). Someone is always going to be better than average,[7] someone is always going to fit into the category of most, and someone is presumably always going to meet what is characteristic of the species. However, it seems possible for there to be a world in which human lives are universally meaningless; simply imagine we were all thrown into hell.

---

[7] Except in the rare case in which everyone has exactly the same amount of value—in which case no one's life would be meaningful, since no one would be above average!

In reply, the friend of (IT_4) might suggest that it in fact does entail that it is possible for all lives to be meaningless. Even if we were all thrown into hell, that condition might not be characteristic of human life; we could be disposed for something better, even if a contingent force unluckily ended up damning everyone.

However, I am not sure what it would mean to speak of what is 'characteristic' of human life, or what such a life is 'disposed' to be like, if in fact all human beings were in hell. It seems that a necessary condition of something's being characteristic of a species is that a certain percentage of the members of that species meet this norm, or at least approximate it.

Another problem with (IT_4) is that one often appraises what is characteristic of human life in light of a more absolute scale. To see the point, consider an analogy. Childbirth is characteristically extraordinarily painful for human females, and so if the appeal to what is characteristic of human life exhausted our evaluation of its disvalue, we would lack the resources to claim that a typical childbirth is 'bad'; the only bad labour would be one that is worse than what is characteristic, which is counterintuitive. Just as we should not restrict our evaluation of childbirth to what is characteristic of it, so we should not restrict our evaluation of the meaningfulness of human life to what is characteristic of it. The amount of final value that grounds meaning in a characteristic human life could intuitively be low.[8] The same is true of the amount of final value that is average or typical, meaning that the present objection also applies to (IT_2) and (IT_3).

In the rest of this chapter, I suppose that judging whether a life is meaningful on the whole requires a standard independent of the human norm. Since the average, typical, or even characteristic human life could have an intuitively low amount of meaning, and since appeal to a human norm entails that it is impossible for all human lives to be meaningless, I seek a more absolute scale, albeit one still fixed by reference to facts about the human species.

Baier, in fact, has proposed such a scale. In his last work to address meaning in life, Baier notes that a comparative standard is not the only one we sensibly use to appraise the worth or meaning of a life. While being better than the norm would make one life more meaningful than another, Baier acknowledges that it would not necessarily be enough to make it count as meaningful on balance when he says (1997: 66):

How much more valuable, significant, and rewarding than the average life must a life be to be a good one? Must it be vastly, perhaps infinitely more so than the best possible earthly lives, as those would seem to imply who claim that if human existence ends in death, our earthly

---

[8] As Nagel has pointed out in the context of death: 'Normality seems to have nothing to do with it, for the fact that we will all inevitably die in a few score years cannot by itself imply that it would not be good to live longer. Suppose that we were all inevitably going to die in agony—physical agony lasting six months. Would inevitability make that prospect any less unpleasant?' (1970: 10).

lives cannot be meaningful, significant, or good ones? Is not this an exorbitantly high standard?....Might not the appropriate standard be one that relates to the best possible lives here on earth?

Although this passage suggests that a life could be meaningful only if it were better than average to a certain extent, I ignore that claim, since it invites the kinds of counterexamples considered above. What I find more promising in Baier's remarks is this principle:

> (IT_5) A human person's life is meaningful on balance if and only if it is not very far below the maximum amount of *pro tanto* meaning that a human could have on earth.

By virtue of being more absolute, this account of how much less than perfect value is constitutive of a life that is not meaningless avoids the counterexamples raised to the social accounts considered so far. Unlike the better-than-average, equal-to-most and characteristic views, it leaves open the possibilities that all human beings could live meaningless lives, if they all were quite far below the maximum possible, and also that what is normal for human life is too low. Summing up, (IT_5) is independent of norm and is for that reason compelling, in my view.

However, as attractive as (IT_5) is, it is in some respects arbitrary and poorly motivated, as currently expressed. It seems unprincipled to suggest that whether a human being's life avoids being meaningless should be a function of any location such as the earth.[9] Why choose the earth? Why not the earth plus the moon and an orbiting space station? Or why not something less than the earth, say, its habitable continents? Any suggestion about where to draw the line in terms of location will be capricious, I submit. After all, suppose human beings will colonize Mars and age on that planet much more slowly than us. Imagine they will live twice as long and without mental or physical impairment. When judging whether the amount of meaning in a human's life is sufficient to make it meaningful on balance, such a possibility seems relevant and should not be automatically excluded. Another problem with (IT_5), as it stands, is that it is vague. Speaking of the value that a human 'could have' leaves unsatisfyingly open the sort of possibility involved.

8.6.3 A new social conception

(IT_5) is close to the mark and just needs to be cleaned up a bit. I replace geographical talk of the 'earth' with 'human existence in the natural world'. And with regard to the sort of possibility involved in speaking of what a human 'could have' there, it strikes me as not only in line with Baier's intentions, but also plausible from a naturalist viewpoint, to suggest that the relevant sort of possibility is nomological, as opposed to, say, logical or metaphysical. Most naturalists believe that meaning in life is a function

---

[9] Perhaps by 'earth' Baier means not a planet but instead an 'earthly or this-wordly existence', viz., as what I label '(IT_6)' below.

of certain conditions of a world as known by science, with remotely possible worlds being irrelevant. I therefore revise (IT_5), as follows:

> (IT_6) A human person's life is meaningful on balance if and only if it is not very far below the maximum amount of *pro tanto* meaning that a human can achieve in the physical universe, given the laws of nature.

I am not aware of anyone in the literature who has explicitly advocated (IT_6), though it is not far from (IT_5) and might be what Baier had in mind. I think (IT_6) is at least close to the sort of imperfection thesis that a naturalist ought to find attractive. If one believes that perfection is not necessary for a life to be meaningful on balance and that a meaningful life is available in the world as known by science, then, by virtue of the dialectical progression in this section, one should find (IT_6) compelling, as it not only avoids all the objections to previous principles, but also provides a plausible explanation of them: (IT_1) is incorrect because it is individualist (as opposed to social); (IT_2), (IT_3), and (IT_4) are incorrect because they focus on a human norm (as opposed to a human maximum); and (IT_5) is incorrect because it is arbitrary (as opposed to principled).

In the rest of this discussion, I present objections to (IT_6). I find the objections to be more speculative and less forceful than ones made to previous principles. I raise them tentatively as issues on which naturalists and other friends of the imperfection thesis could (more) reasonably disagree.

One apparent problem with (IT_6) is that it might counterintuitively make meaning too far out of our reach. Although our lives are not as meaningful as 'freaks' such as artistic and intellectual geniuses who have put their talents to effective use, most of our lives are probably 'not very far' below theirs, and so are still able to count as meaningful. Suppose, however, that a 'superfreak' were realized, a once in ten billion occurrence of a human being whose powers are so great that everyone else's lives turn out to be 'very far' from his and hence all meaningless.

Perhaps a superfreak is nomologically impossible, and if it were, then no revision to (IT_6) would be necessary. However, if a superfreak were possible in the light of the laws of nature, then to avoid the counterexample, I suggest this principle:

> (IT_7) A human person's life is meaningful on balance if and only if it is not very far below the maximum amount of *pro tanto* meaning that a human is likely to achieve in the physical universe, given the laws of nature.

When I speak of the 'maximum that a human is likely to achieve', the reader should think of the range of the most meaningful lives that humans could lead, and then ascertain which lives in this range are the least improbable to occur. These lives are the relevant maximum to use when judging the meaningfulness of everyone's life.

I am not sure whether (IT_7) resolves the problem, for it might still entail that our lives, or at least the lives of our ancestors, have been meaningless. What humanity has produced so far could turn out to be quite far removed from the kinds of superlative lives that are most probable.

However, if that were true, one might be prepared to bite the bullet. After all, we now judge our ancestors' lives to have been nasty, brutish, and short; we live on average to about 80 with anaesthesia, whereas they made it to 30 or so without it. Perhaps one day we will discover so much about what humanity is capable of with regard to meaning in life that it will seem plausible to judge what we have done up to now not to be worthy of the title of 'meaningful on balance', even if we have had some meaning.

Another, more fantastic counterexample to both (IT_6) and (IT_7) is the case of a 'transhuman superfreak', something like Nietzsche's *Übermensch*. Imagine, for instance, that knowledge of the human genome enabled scientists to manipulate the genes of a human being so that he would become a great individual but no longer human. Suppose that such genetic manipulation resulted in much longer life, greater intelligence, and better health. Suppose that after the process the person's DNA were no longer human and that he could not mate with humans. And suppose that the choice of whether to undergo the process were widely available. It might be that, under these conditions, the relevant maximum to consider when appraising human lives would not be based on facts about our species.

Note that the transhumanist thought experiment should not be dismissed out of hand, at least if one finds the Mars counterexample to (IT_5) compelling. The Mars case had us imagine that human beings could substantially enhance the powers they have, and hence the meaning available to them, simply by being in a location other than the earth. If the reader is inclined to think that changing the environment could in principle affect the relevant maximum available to invoke when appraising meaningfulness, then, analogously, she should think that an altered constitution could do the same.

However, some might try to drive a wedge between the external and the internal; that is, try to argue that biological species, or at least objective kind, is alone relevant to meaningfulness, marking a difference between the Mars case and the case of the transhuman superfreak. For instance, Martha Nussbaum has argued that it is inappropriate to judge the value of human lives with reference to any non-human lives, saying this of an eternal life:

> (I)ts constitutive conditions would be so entirely different from ours that we cannot really imagine what they would be. Nor, if we could, would they be of any immediate importance for us....We are not attempting to show that an immortal existence could not have value, beauty, and meaning internal to itself....What we are attempting to show is the extent to which our values would be absent in that life; and that is the thing most relevant to a therapeutic treatment of our always impossible wish to have that life in place of our own. It is, as well, the only perspective on value from which we can coherently proceed, in asking a question for ourselves: for in asking about ourselves there is not much point in asking whether a certain life seems good from the point of view of creatures that we have no chance of ever being, or rather creatures becoming identical to which we would no longer be ourselves (1989: 340).[10]

Nussbaum here offers three reasons for thinking that the case of the transhuman superfreak should not lead us to revise (IT_6) or (IT_7). One is that we could never become

---

[10] For similar claims, see Lenman (1995) and Kass (2001: 20–1).

a transhuman superfreak, but the thought experiment, of course, invites us to suppose that we could and then to see whether we would be inclined to judge the quality of our lives in the light of the possibility.

Another is that we cannot know enough about the life of any non-human creature to know whether it would be desirable, but that seems false. I really would not want to be a polar bear in a typical zoo in the summertime. Just because we invariably know ourselves best does not mean we cannot justifiably claim to know what it would be like to have a quite different sort of life.

Nussbaum's third and most interesting suggestion is that even if we could become another being and could know what it would be like, the possibility would not be relevant to appraising our lives. This point is more asserted than defended in the quote above, but the most straightforward motivation for it is the idea that a meaningful life cannot involve suicide, losing one's identity. Put that way, the point sounds irrefutable.

Upon reflection, however, it does not obviously tell against the idea that our lives should be judged relative to the transhuman superfreak, for two reasons. First, one can grant the point that for a condition to confer meaning on one's life, the condition must not alter what one essentially is, but deny that being human is what one essentially is. One plausible way to deny that[11] is to suggest that we are essentially a non-branching stream of consciousness intimately connected causally and intentionally over time (Parfit 1984); if so, then one could remain who one essentially is even if one were to change species and become a transhuman superfreak.

Second, one could reasonably reject the principle that one must remain essentially who one is in order to obtain meaning from a state of affairs. After all, it is typical to think that one's life can obtain meaning by virtue of sacrificing it in order to help others. Suppose one dies as a result of donating a kidney to a family member or that one jumps on a grenade to protect one's fellow soldiers and to promote a just cause. Meaning arguably can accrue to one's life upon its end. If killing oneself for the sake of others can confer meaning on one's life, then becoming a transhuman superfreak who is not oneself, but who is closely related thereto, can probably do the same.

If one holds the judgement that whether our lives are meaningful on balance could be a function of the extent to which we approximate the meaning available to the transhuman superfreak (and I am not sure that one should), then one should find the following appealing:

> (IT_8) A human person's life is meaningful on balance if and only if it is not very far below the maximum amount of *pro tanto* meaning that a being that was born human is likely to achieve in the physical universe, given the laws of nature.

One might wonder, now, whether (IT_8)'s requirement of initially being human is truly necessary. Imagine that an inhuman superfreak like Superman lived among us, but that we had no opportunity to become one of his species. Should being in the presence of

---

[11] For another, see LaPorte (1997).

another, greater species, or perhaps merely being able to imagine one, be considered relevant to judging whether our lives count as meaningful or not, if one is a naturalist friend of the imperfection thesis?

My answer is a firm 'no'.[12] After all, some naturalists are theists; that is, some who believe that a meaningful life is possible in a world lacking a perfect, supernatural realm believe that such a realm does exist, for example, that we are in God's presence. Awareness of being in God's presence is not enough for a naturalist to hold the perfection thesis or supernaturalism, i.e., is not sufficient to think that a maximally conceivable condition possible only in a transcendent realm is the relevant standard to use when judging whether a human life has avoided being meaningless on balance. And so, even if a naturalist were acquainted with (or could merely imagine) beings lesser than God but greater than us, she should not think that these beings set the relevant standard for whether our lives are meaningful, if the amount of meaning they can acquire were in no way available to us. Hence, if one is inclined to depart from (IT_6), one should probably not go beyond (IT_7) or (IT_8).

8.7 Implications of the imperfection thesis

This chapter has sought to identify conclusive reasons for most readers to reject supernaturalism and its motivating perfection thesis and to begin to articulate the sort of naturalist, imperfect alternative that is most defensible. It began by considering extant arguments against supernaturalism that, I argued, misfire against it, and then considered three objections to it that are more promising. In particular, and perhaps most forcefully, I maintained that just as it would be incoherent to believe in divine command theory, because we do not know that God exists but do know that wrongness exists, so it would be incoherent to believe in supernaturalism, because we do not know whether anything spiritual exists but do know that meaningfulness exists. Then, I began to spell out the sort of imperfection thesis that a naturalist ought to hold with regard to how to judge whether lives on balance avoid being meaningless or not. I presented several rival principles in the literature, objected to them, and then sketched a new principle that avoids and explains the objections. According to this view, a life is on balance meaningful if and only if it is not very far below the most degree of meaning (in terms of both part and whole) that a human can physically achieve, given the laws of nature.

Before beginning part three, in which I critically discuss the naturalist factors that might make life *pro tanto* more meaningful, I bring out two important implications of the criticisms of supernaturalism I have made. First, the incoherence argument against supernaturalism also tells against non-naturalism. Non-naturalism is the view that meaning in life is constituted by properties that are neither physical

---

[12] For the opposite answer to this sort of question, see Benatar (2006: 82–6).

nor spiritual. A non-natural property is one that neither is composed of atoms in space-time, nor inheres in a spiritual substance. It is difficult to point to any uncontroversial example of a non-natural property, but here are three possibilities relevant to the field of life's meaning. First, if one finds *Euthyphro* concerns about the divine command theory of normativity (or my objection to it in 5.5) compelling, if one doubts that good reasons for action are physical objects or relations, and if one thinks that meaningfulness provides a good reason for action, then one should be inclined to find non-naturalism attractive (Audi 2005: 337-41). Second, those who find libertarian freedom the only sort worth wanting with regard to meaning in life, as per Immanuel Kant (discussed in Williams 1999), might be drawn to postulate the centrality of a non-natural realm. Third, if one believes that something can obtain meaning only from something apart from it, then human life, which is discursable, must obtain its meaning from 'what lies beyond the human', or what is not discursable, i.e., a realm of reality as it is exists in itself, apart from our spatio-temporal apprehension of it (Cooper 2005).

The incoherence objection to divine command theory and supernaturalism applies with equal force to such non-naturalist positions. A very large majority of readers know that meaningfulness exists, but they do not know whether non-natural properties such as libertarian free will truly exist. If so, then they cannot consistently believe that meaningfulness is a function of such properties. The incoherence objection pushes enquirers to base a theory of meaning on an ontology of which we have good reason to claim knowledge. Virtually no one disputes that there is a material world, and so that is the most promising metaphysical view on which to ground a conception of meaning in life.

A second implication of my criticisms of supernaturalism, however, is that an attractive conception of life's meaning need not be grounded *solely* on the natural. Recall that the two most powerful objections were the idea that meaning would exist even if there were no spiritual realm and that it would be incoherent to think that meaning is exhausted by the spiritual, since we know that meaning exists but do not know whether the spiritual does. Both arguments leave open the possibility that while meaning is *constituted* by natural properties, it should not be *identified* with them. For X and Y to be identical is for them to be one and the same thing in all possible worlds. For all that has been said, it would be hasty to think that meaningfulness is identical to combinations of physical properties; it appears that what matters in life *could be* something in addition to mere matter, and if a spiritual (or non-natural) realm were to exist, then it probably would be.

The point I am making is similar to 'multiple realizability' arguments in the philosophy of mind. Since a mind could be present in a silicon-based biology, it is implausible to think that a mind is one and the same thing as a carbon-based brain, even though it may well be constituted alone in such a brain, viz., we have no reason to think that any silicon-based organism exists. Similarly, if there *were* a spiritual realm populated by persons, or if there were a non-natural realm that included libertarian free will, then it would plausibly be the case that meaning in one's life could be *enhanced* by engaging with these things in certain ways. My overall conclusion, then, is that while meaningfulness is constituted by natural properties, such that neither God nor a soul is necessary for meaning in life, meaningfulness could be constituted by other properties if they were to exist. That is something that those who reject supernaturalism (and non-naturalism) have tended not to admit,[13] but that there is no good reason to deny.

---

[13] Hooker (2008: 199) is an exception.

III

Naturalist Theories of Meaning in Life

9
Subjectivism

'It is precisely objective reasons that the will craves in determining what is worth pursuit: often it could not go forward unless it thought it had them.'

David Wiggins ('Truth, Invention, and the Meaning of Life')

9.1 Overview

This, the initial chapter of part three, begins the search for an attractive naturalist theory of meaning in life. The previous part established that the underlying motivation for any plausible supernaturalist view is the perfection thesis that meaning requires engagement with a maximally conceivable value, and it also provided reasons to doubt both the perfection thesis and supernaturalism. The two strongest objections both appealed to the claim that we know that some aspects of our lives, and some of our lives on balance, are meaningful, and then contended, first, that we would deem them to be meaningful even assuming there were nothing perfect or spiritual (8.4), and, second, that we would be incoherent to think that they were meaningful because of something perfect or spiritual, since we do not know the latter exist (8.5). These objections have led me to seek a firmer metaphysical basis for meaningfulness than supernaturalism, and naturalism, the view that meaning is constituted by certain ways of living in the physical universe, is the most sensible alternative (8.6–8.7).

At the end of the previous chapter, I began to sketch what a naturalist ought to say about when a life is meaningful on balance. I argued that the number of *pro tanto* meaningful conditions one must have in order for one's life to count as meaningful on balance is not very far below the maximum degree that a human could achieve in the material world, given the laws of nature. While that is a clear and robust principle, one that I have argued is superior to several rivals in the literature, it is abstract, and, of course, incomplete as a theory of meaning in life. It leaves open, after all, what rightly count as *pro tanto* meaningful conditions. Specifying those, especially (but not solely) as they concern part-life factors, is the task of part three. Which ways of living in a physical universe confer some meaning on a person's life, such that she should aim to realize no less than what is not very far below the most a human could achieve there, holding the laws of nature constant?

In the following section, I remind the reader of some of the essentials of a general naturalist perspective, and distinguish between two major forms of it, subjectivism and objectivism (9.2). Then, I spell out the subjectivist view in some detail, taking care to draw distinctions between types of subjectivism, some of which are more defensible than others (9.3). I next consider arguments for subjectivism, and maintain that they are unpersuasive, often because the considerations invoked are most plausible on an objectivist analysis; for instance, subjectivists appeal to the values of love and authenticity, but, as David Wiggins suggests, prizing these values includes an awareness of objective considerations (9.4). Finally, I argue that subjectivism has counterintuitive implications about what constitutes meaning in life (9.5). This criticism is common to make in the field, and so I seek to advance the debate by considering in some detail how the subjectivist might plausibly try to reply to it, especially by appealing to less well-known forms of her view. Ultimately, I conclude that no version of subjectivism can avoid the problem (9.6).

9.2 Naturalism: subjectivism and objectivism

Naturalism about meaning is the broad perspective that it is constituted by physical properties, ones that inhere upon substances located in space and time, composed of atoms and are best known by scientific methods. This view implies that life could be meaningful even if there were neither God nor a soul, although, as I have pointed out, it is open to the naturalist to hold that a supernatural realm could enhance meaning in life. His defining point is that the natural is *sufficient* both for *pro tanto* meaningfulness and for a meaningful life on balance, which leaves open the possibility that the supernatural, if it existed, would contribute to meaning.

Among those who believe that a significant existence can be had in a purely physical world, the most salient division concerns the extent to which, and the respects in which, the human mind constitutes it. A subjective theory maintains that what makes a life meaningful depends on the subject, where this subject need not be a spiritual substance. More carefully, subjectivism is the view that lives are meaningful solely by virtue of obtaining what the field calls objects of 'propositional attitudes', mental states such as wants, emotions, goals, and the like that are *about* states of affairs. Propositional attitudes are characteristically capable of 'that clauses'; for example, one has a desire that something were the case, or one is proud that something is true of one. Propositional attitudes are typically held to differ from experiences such as pleasure and from moods such as irritability, which normally lack objects. Self-described 'subjectivists' do not tend to appeal to the latter mental states, and for good reason, in light of my contention that they are logically irrelevant to constitutive facets of meaning in life (2.4).

The subjectivist in general maintains that there is no standard independent of people's propositional attitudes to determine which states of affairs are meaningful. Instead, for him, states of affairs are meaningful just insofar as they are objects of propositional attitudes that have been obtained. For example, what makes it the case that being a chess grandmaster is meaningful or not, for the subjectivist, is entirely determined by the attitude that someone (or some group) has taken (or would take) towards it. If I wanted to be a chess grandmaster, and if I succeeded in becoming one, then typical subjectivists would claim that (some) meaning would be conferred on my life. If I wanted to be a chess grandmaster, and if I failed to become one, then standard subjectivists would claim that no meaning would have been conferred on my life. And if I did not want to be a chess grandmaster, but somehow became one despite that, then, again for exemplary subjectivists, no meaning would be conferred on my life.

The objective naturalist, as construed here, need deny neither that obtaining the object of a propositional attitude would enhance meaning in life, nor even that it would be necessary for it. The objectivist's defining claim is that subjective conditions such as a person getting what he wants or realizing his aims are *not sufficient* for his life to be either *pro tanto* or on balance meaningful. Instead, for the objectivist, certain states of affairs in the physical world are meaningful 'in themselves', apart from being the object of propositional attitudes. Some conditions are such that they *ought to be* wanted, chosen, valued, and so on, even if people have not done so. Returning to the chess grandmaster case, if an objectivist were to deem this project to be meaningful, she would appeal not *merely* to the fact that a person has adopted it, but instead, or at least also, to facts independent of his propositional attitudes, such as that taking up chess would: be intrinsically good because it is complex; develop facets of his rational nature; or improve others' quality of life by teaching them or entertaining them. If someone had a talent for chess but did not like it, or were not pursuing it, an objectivist might recommend that he change his mind, that he cultivate an interest in it, so as to bring more meaning to his life. Such advice would make little sense from a subjectivist perspective, since, roughly, a condition would confer meaning on someone's life only if she is already taking an interest in it (or should in light of other, higher-order interests that she already has).

9.3 Types of subjectivism

A variety of subjective theories are possible, and it is worth distinguishing between them, since some versions are more vulnerable to criticism than others. The most central distinctions are these: whether the propositional attitudes (simply 'attitudes', from now on) are negative or positive (9.3.1), what the nature of the attitudes are (9.3.2), whether the attitudes are those of an individual or a group (9.3.3), whether the attitudes are actually exhibited or hypothetically would be (9.3.4), and, finally, whether the attitudes would be invariant among human persons or would instead vary among them (9.3.5). Along the way, I point out which forms of subjectivism are dominant and which are minority positions, and I also mention some additional distinctions that have played much less of a role in the literature, but that are probably essential for a full characterization of subjective factors insofar as they plausibly contribute to meaning in life.

9.3.1 Valence

With regard to most of the attitudes relevant to meaning, it is standard to maintain, at least implicitly, that only positive ones confer meaning on a life, as opposed to negative ones. A positive or 'pro-' attitude is, roughly, one that includes a wish that a state of affairs obtain, whereas a negative one includes a wish that it not.[1] Positive attitudes include wanting, preferring, appreciating, liking, being proud, loving, and seeking out, whereas negative ones would include opposites such as disliking, detesting, being ashamed of, hating, and avoiding.

I recall having encountered only one or two people in the field, specifically, Milton Munitz (1993: 83) and Charles Starkey (2006: 94), who have explicitly maintained that negative or 'con-' attitudes could confer meaning on life. The best example, in my view, is that anger can be a source of meaning when it is directed towards injustice. Perhaps some would suggest that the negative emotions could have merely an instrumental value with regard to meaning, so that anger would be desirable only insofar as it caused positive attitudes or reactions, which would in fact constitute the meaning. However, insofar as many do have the intuition that meaning inheres in fighting undesirable conditions such as poverty, injustice, and ignorance, then it becomes reasonable to think not merely that actions aiming to combat these conditions confer meaning, but also that additional aspects of one's self such as desires, emotions, and intentions negatively directed towards them can do so too. The relevance of negative attitudes should be taken seriously by anyone who believes that there is some kind of subjective dimension to meaning in life, even if she is not a strict subjectivist.

9.3.2 Nature

This distinction concerns the genus and species of the attitude involved. For example, many subjectivists focus on conation as the central genus, with some then maintaining that desire in particular is the relevant conative attitude (Taylor 1970; Griffin 1981), and others appealing to preference (Darwall 1983), which seems different, perhaps for implicitly involving a comparison of at least two objects.

Second, there are those who focus on our capacity for intentional action, holding that meaning in life is constituted by making choices (Sartre 1946), pursuing 'ground projects' (Williams 1973) or adopting purposes and acting to realize them (Ayer 1947: 226-7; Nielsen 1981; Smart 1999: 16).

A third salient genus is emotion, with some claiming that the relevant instances are being satisfied with something (Ayer 1990: 189-96; Martin 1993: 593-5), and additional

---

[1] Gordon (1987) distinguishes between the valence of emotions in this way, but it strikes me that the distinction applies to a much wider array of attitudes.

suggestions including the ideas that taking joy in things is what matters (Klemke 2000: 195-7), that care is key (Frankfurt 1982), or that there even exists an irreducible 'emotional sense of meaningfulness' (Starkey 2006: 96-9).

Finally, less often, one encounters a fourth category of cognition, with some sympathetic to the idea that meaning in life is simply a matter of being keenly aware of the details of a particular object (Murdoch 1970), and others inclined to find one's beliefs and judgements to be central (Markus 2003).

Subjectivists have not much debated the issue of which attitudes do the work. Invariably, they have been concerned merely to articulate a position and to weigh it against supernaturalism or objective naturalism. They have not systematically addressed the issue of whether it is conation, emotion, intentional action, cognition, or some combination that fundamentally matters.[2] Too often subjectivists unwittingly shift between different attitudes and rest content with vague formulations such as being 'attracted to' or 'engaged by' an object.

It is worth being sure that more than one attitude is necessary before suggesting a plural account. William James (1899) is a classic theorist who has done so, arguing with care that meaning consists of, roughly, the passionate pursuit of an ideal. Arjan Markus (2003; see also Boylan 2008: 22-4) is a contemporary philosopher who has been fairly careful not merely to specify attitudes, but also to argue that more than one genus is necessary, appealing most explicitly to both cognition and volition. He contends, for example, that a fully meaningful life is one in which a person's beliefs about what is meaningful avoid outright logical contradiction, cohere with one another, fit with one's other, non-evaluative beliefs about the nature of reality, could realistically be applied to one's life over time, and in fact are acted upon. Finally, below I also strive to provide good reason to think that more than one attitude is central to subjective matters, arguing that, in fact, all propositional attitudes are relevant, to the extent that they are parts of one's rational nature (12.3).

9.3.3 Bearer

The present distinction concerns which persons in the natural world exhibit the relevant attitudes. On the one hand, what makes a person's life meaningful could be a function of her own attitudes, a view that I call 'individual subjectivism' but that many in the field simply refer to as 'subjectivism' (using a more narrow sense of that word than I do). On the other hand, a person's life could be meaningful by virtue of obtaining the objects of the attitudes of some group of which she is a part but whose attitudes she may or may not share. 'Intersubjectivism' nicely characterizes this view. The individualist variant is much more popular than the intersubjective one, with nearly all theorists referred to above holding the former, and only a handful in the literature advocating the latter (Darwall 1983: 164-6; Brogaard and Smith 2005; Wong 2008).

---

[2] Though Starkey (2006) is a salient exception.

9.3.4 Status

By the 'status' of an attitude, I mean whether it is actual or hypothetical. A large majority of subjectivists maintain that the relevant attitudes are actual ones, whether held by the individual or a group to which she belongs. For example, most believe something to the effect that an individual's life is meaningful insofar as she gets whatever she in fact wants, or that a person's existence matters insofar as she conforms to standards that her social group happens to accept.

There are some in the recent literature, however, who appeal to hypothetical attitudes. On the intersubjective side, Stephen Darwall (1983: 164-6) maintains that a state of affairs confers meaning on a life roughly insofar as all human agents *would prefer* it to obtain, when reflecting on it from an impersonal standpoint. On the individualist side, James Griffin (1981: 54-8) suggests the view that what confers meaning on a person's life is whatever she *would desire* if she were informed about the options available.[3] There is more leeway on hypothetical views to claim that individuals are misguided in their present pursuits; pointing out that they would have different inclinations upon engaging a certain reasoning process should, according to hypothetical theorists, give them reason to change their mind about how to live.

9.3.5 Variation

The last major distinction I draw among types of subjectivism concerns the degree to which meaningful conditions vary among human persons. The degree of variation entailed by a subjective theory is largely a function of the bearer and status of the attitudes involved, that is, whose attitudes count and whether they are idealized in any way.

A subjective theory likely entails that meaningful conditions vary extremely widely, if it holds that the relevant attitudes are the actual ones of individuals. Since people's desires, projects, loves, and other dispositions vary tremendously, this *actual-individual* form of subjectivism entails that what is meaningful varies substantially from person to person. The subjective theory most likely to underwrite a fairly strong invariance in the conditions that confer meaning on people's lives is Darwall's *hypothetical-intersubjective* theory, according to which someone's life is meaningful just insofar as she pursues objects that all human agents would prefer when abstracting from the way they would bear on their own lives. In between these two theories will be *hypothetical-individual* theories and *actual-intersubjective* ones. What individuals would want under certain conditions, and what a group such as a society or species actually wants, are both likely to be more uniform than what individuals in fact want and less uniform than what a suitably large group would want under idealized conditions.

In this section, I have pointed out that subjective theories can vary in at least five major ways, making literally dozens of different kinds of them in principle possible.

---

[3] And perhaps if she then suitably changed her desires.

In the following, I focus in the first instance on the dominant form of subjectivism, which appeals to pro-attitudes such as wants and aims that are actually held by individuals and probably vary widely among them.

(S_1) A human person's life is more meaningful, the more that she obtains the objects of her actual pro-attitudes such as desires and goals.

I do, however, draw on forms of subjectivism less influential than (S_1), when doing so would be useful to help subjectivists avoid criticism of the standard version.

9.4 Arguments for subjectivism

In arguing for (S_1), some will find it tempting to suggest that it must be true because, given the relevant attitudes, the individual cannot but find a condition 'meaningful to' her. If one believes that a state of affairs would be meaningful, wants it to exist, has willingly brought it about, or is taking pride in it, then surely it is 'meaningful to' her. 'What else could be meaningful apart from that?', so the rhetorical question is posed.

However, objective naturalists and supernaturalists reasonably doubt that something's being 'meaningful to' an individual is sufficient for her life to in fact be meaningful. Recall that meaning-talk is a matter of enquiring into those purposes beyond one's pleasure that one ought to rank highly, or those conditions worthy of great admiration, or those facets of one's life that transcend one's animal self in the right way (2.5). It is fair to ask whether merely *deeming* something to be meaningful is the best conception of these concepts.

(S_1) was popular for much of the twentieth century, but is now largely out of favour among philosophers who address the question of what makes a life meaningful. It used to be that theorists with quite different philosophical commitments, in particular pragmatists (James 1899), positivists (Ayer 1947), existentialists (Barnes 1967), and Humeans (Williams 1976) found common cause in a subjective account of life's meaning. However, as many of these philosophical sources have dried up, there is much less of a confluence towards subjectivism these days. Rather than examine the particulars of these philosophical worldviews, I address arguments that were common to many of them and that continue to be influential in the recent literature.

9.4.1 Meta-ethics

First, then, meta-ethical considerations drove many of those philosophical perspectives, and they continue to motivate contemporary adherence to (S_1) and subjectivism in general. Specifically, many subjectivists accept some version of Jean-Paul Sartre's (1946) famous argument that since there is no God, since only God could ground objective values, and since there are values, all values are subjective. Although there are 'value nihilists' who would dispute the claim that there is anything good or bad (i.e., who would claim that all assertions about value are false, in the way that those about witches are), most philosophical debate centres on the nature of good and bad. Now, suppose that traditional *Euthyphro* concerns about divine command theory (or my criticism of it in 5.5) have led one to reject the view that final goodness is grounded in God. And suppose that one finds queer the idea that nature independently of us could be a source of values; after all, one apprehends nature principally through the five sense organs, and one cannot see or smell meaningfulness, let alone any other sort of final goodness. What else is left? If one thinks that some lives have meaning in them, one will be led to think that values such as meaning in life must be constructed by us, that is, must be a function of the responses of valuers, whether that be in the form of their desires, emotions, or choices.

One influential subjectivist who buys this kind of argument is Harry Frankfurt (1982, 2002; see also Trisel 2002: 73, 79, 2004: 379). According to Frankfurt, what is significant for a person's life is a function of what she cares about or loves, a particular sort of caring. For Frankfurt, caring about or loving something is sufficient to confer importance on one's life. One does not need to appeal to the objectivist idea that there are certain things that one ought to care about, apart from the perspective of something else that one already cares about. Frankfurt says, 'Devoting oneself to what one loves suffices to make one's life meaningful, regardless of the inherent or objective character of the objects that are loved', with his major argument for this view being that 'efforts to make sense of "objective value" tend to turn out badly' (2002: 250).

At this point, it is tempting to engage in some meta-ethical debate regarding the metaphysics, epistemology, and language of evaluative judgements. However, just as I earlier avoided taking up the sundry arguments for the existence of God, when doing so would have provided the strongest defence of my view, I abjure a full critical discussion of the reasons for and against thinking that objective value exists. This is a book devoted to the topic of meaning in life, not meta-ethics, where the former is under-explored relative to the latter. Therefore, I rest content with making two points. First, I argue in the rest of this section that many of what initially appear to be motives to adopt subjectivism lead us, upon reflection, to think that there are things that are good not solely because they are the object of a propositional attitude. Second, the main reason that theorists would resist the temptation to accept the reality of objective value is that they cannot see how it is possible. However, the account of a mind-independent morality that I articulated earlier clarifies how it is (5.5).

Let me elaborate on these two points in the context of Frankfurt's particular version of subjectivism, according to which one's life has meaning just insofar as one loves something. Against this view, Susan Wolf (2002) has pointed out that there intuitively are mind-independent standards governing what one ought to love. She notes that it is fairly uncontroversial to think that more meaning would come from loving people than, say, loving torture of them, which seems best explained by the idea that some things are objectively worth loving more than others. Hence, upon a fuller consideration of when one is most inclined to think that love confers meaning on life, one is drawn to the theory that the more one loves what is objectively worth loving, the more one's life is meaningful.[4]

Now, despite finding plausible Wolf's suggestion that objective value plays some essential role in the way love makes our lives matter, some readers, like Frankfurt, will be reluctant to believe it, since they cannot understand from where objective value could come. They likely have one of these concerns in mind: 'It is implausible to think that values would exist in the physical world if we did not exist'; 'We cannot apprehend goodness through any of the five senses, our primary vehicles for detecting mind-independent physical properties'; 'There is enormous variation in people's value judgements, which we would not expect to encounter if there were mind-independent values with which all minds are acquainted.' Such considerations drive naturalists who believe that values exist to think that they are the kinds of things that we must create with our minds, rather than discover with them; they are not aware of any plausible alternative.

However, the account of moral realism that I sketched above (5.5) indicates how objective value is possible. Recall the standard realist explanation of how objective facts about water, for example, are possible. Roughly, we have dubbed a clear, odourless liquid found in the ocean and in the rain with the term 'water', such that there is a mind-independent fact of the matter with regard to the existence of any deeper properties of the stuff we have called 'water' that might account for the more surface properties that occasioned our calling it that. We have learned over time that the essence of the stuff we call 'water' is H_2O, viz., that this chemical composition is what is ultimately responsible for what we call 'water' being a clear, odourless liquid found in oceans, etc., such that anyone who disagreed with that claim would be mistaken.

The moral or, more broadly, evaluative realist suggests a parallel way to understand how mistakes about the nature of value could be possible, i.e., how there could be goods that exist independent of our minds that we could either apprehend or fail to apprehend. The evaluative realist can hold that there are certain conditions of the world that we have dubbed 'good for their own sake' (or, more specifically, 'meaningful'), with relevant surface properties including: there is always some good reason to protect these conditions and to exhibit pro-attitudes towards them; people are drawn towards them under normal conditions; exemplary instances of them include moral supererogation, intellectual achievement, and aesthetic creation. Just as in the case of water, the evaluative realist points out that having dubbed these conditions 'good for their own sake' leaves open the possibility that there are deeper properties of these conditions that might account for the more surface properties that occasioned our calling them 'good for their own sake'. In short, there might be an underlying pattern

---

[4] In one place, Frankfurt suggests that while loving anything is sufficient for meaning, loving some things rather than others could confer more meaning in virtue of the worthiness of the object of love (2002: 246). However, in the same text, he suggests that there is reason to care about morality as opposed to immorality only on the basis of other things one actually cares about (2002: 248).

to what we have called 'good for their own sake', at least in certain, specific domains such as meaningfulness, that really exists, waiting to be discovered.

Let me now briefly return to the concerns that probably prevent a reader from accepting objective value, adumbrated earlier. First, on the realist picture, at least of the sort I find attractive (5.5), it is true that no essence of value would have obtained had human beings (or other agents) not existed, in two respects. First, supposing we know that human beings and their lives are the quintessential forms of value, it is obviously true from a realist perspective that a world without human beings would be devoid of value, or at least have much less than it does now. Second, recall that on the realist picture a dubbing process, undertaken by a language speaker, fixes the nature of value with respect to his language. Just as the nature of water is the underlying structure of the surface properties that have occasioned us to use the term 'water', so the nature of final goodness is fixed by us upon having referred to something as 'good for its own sake'. However, calling something 'good for its own sake' is compatible with what we have called by that phrase having a mind-independent essence that we can learn about over time.

Second, it is true that one does not literally see or smell values such as love, democracy, or knowledge, but the same is true about scientific properties such as causation, time, and gravity, which are respectably deemed to be objective. Granting that the five senses are the primary vehicle for apprehending natural facts, we need not restrict our descriptions of the world to languages germane to our sensory organs (5.5.2). We need not be immediately sensorily aware of objective properties in the natural world to know they exist; we can infer that they exist, particularly insofar as positing them best explains what we do immediately sense.

Finally, just as those curious about nature for a long time did not know that water is H_2O, even though that is a mind-independent fact, so it could be that those curious about value have not yet acquired a firm grasp of its objective composition, resulting in widespread disagreement about it. As Parfit has pointed out (1984: 453-4), it has been only a few decades that many people have spent their lifetimes thinking about goods and shoulds, and have done so without appealing at bottom to supernatural considerations. Perhaps it will not be long before widely accepted naturalist theories of value take hold, similar to the way that certain scientific principles are now virtually unquestioned among the educated.

Of course, to show that objective value is possible (in the way objective facts are) is not to show that it is actual, something I do not have the space in this book to undertake. If I am correct, though, that considerations driving philosophers to hold subjectivism, such as love and authenticity, are in fact better support for objectivism, and if what often prevents them from accepting objectivism is failing to grasp how it is possible, then this analysis will have done something real to advance the objectivist cause, while remaining squarely focused on the issue of meaning. In the following two sections, I again demonstrate that certain motives for subjectivism backfire and instead, upon reflection, provide no less, if not more, support for objectivism.

9.4.2 Authenticity

Accepting that objective value is possible or even actual need not commit one to objectivism in the particular realm of meaningfulness. One could think, say, that objectivism is true about morality, but not about meaning. One reason for thinking that meaning might not be objective has been best articulated by Charles Taylor (1992; see also Frankfurt 1982), even though he ultimately rejects it. Taylor contends that the appeal of authenticity is a driving force behind (S_1). He suggests that what leads people to think that meaning in life is utterly constructed by the individual subject's pro-attitudes is the idea that life's meaning consists in being true to oneself, from living according to what one finds within.

Taylor ultimately objects to this argument, for an interesting reason. He contends that the idea of being true to oneself is incoherent unless one accepts that it, and the meaning in life it underwrites, is at least partially independent of one's attitudes. 'Which issues are significant, I do not determine. If I did, no issue would be significant' (Taylor 1992: 39). Why? Because, according to Taylor, the very idea of significance includes not being arbitrary, and mind-independence is the only way to avoid arbitrariness (for similar claims, see Darwall 1983: 164-6; Wiggins 1988: 135-41).

I doubt this objection is successful, since there are forms of subjectivism that are more invariant than others, e.g., intersubjectivism (9.3.5). At best, Taylor's objection applies to authenticity only insofar as it would ground the standard form of subjectivism, according to which meaning varies from person to person, depending on the individual's contingent desires and aims.

An objection that promises to apply to authenticity conceived as a rationale for any sort of subjectivism is to point out that subjectivism accounts for the ideal of authenticity no better than certain objective theories, particularly those that include a subjective component. As I discuss later, the most widely held account of meaning in life as I write is this view: 'Meaning arises when subjective attraction meets objective attractiveness' (Wolf 1997a: 211), the principle that life's meaning consists of the positive orientation of an individual's attitudes towards certain kinds of mind-independent final goodness (10.2). Even granting the point that meaning is a function of realizing one's deepest nature or true self, the latter could plausibly be done in the context of objectively worthwhile projects, for example, those involving moral relationships with other people (Railton 1984: 164-71). By virtue of making certain attitudes necessary (but not sufficient) for a meaningful life, this objective theory accommodates the intuition about authenticity well.

Of course, some subjectivists would respond that, say, a serial killer could not be true to himself only in the context of moral or otherwise objectively desirable projects. But, then, it becomes implausible to think that authenticity is something worth all that much, or has the kind of importance that could ground considerations of meaning in life. When the idea of living in a way that is true to oneself has the most force, as something worth desiring strongly, then taking it to be a matter of acting according to one's pro-attitudes is no less plausible than deeming it to be a matter of directing one's pro-attitudes towards objective value.

9.4.3 Absorption

A third common argument for (S_1) appeals to the idea that meaning is intuitively a function of losing oneself, i.e., of getting absorbed in, or taken over by, an awareness or activity (James 1899; Schlick 1927: 65; Taylor 1970; Frankfurt 1982: 89–91; Ayer 1990; cf. Murdoch 1970). That idea well explains why religious communion with God or other people, intense creativity, and passionate love are what many people take to be central to meaning in life, and it would also account for the widely held intuition that boredom is usually incompatible with meaning, and might even count as anti-matter (cf. 7.5.1).

However, the view that subjectivism follows from the premise that meaning comes from losing oneself is open to the criticism that there are better and worse ways to do so, and that the better ways appear to involve objective value. To see the point, consider a thought experiment that Richard Taylor constructed 40 years ago (1970), one that has been widely discussed in the field. Taylor reflects on the myth of Sisyphus, the figure imagined by the ancient Greeks to have been condemned by the gods to push a rock up a hill forever. Taylor poses the question of precisely what it is about the condition of Sisyphus that leads us to think that it is meaningless, and, conversely, what would have to change about his condition for us to find it meaningful. According to Taylor, the trouble with Sisyphus is that he hates and is bored by what he is doing; once we imagine a situation in which the gods have implanted in Sisyphus 'a keen and unappeasable desire to be doing just what he found himself doing' (1970: 330), then, Taylor suggests, his life would be meaningful.

However, virtually no one in the field, so far as I am aware, has shared Taylor's appraisal of this case. And, in fact, Taylor himself changed his mind about it, later objecting to it in two ways (1981: 6–8, 19–24, 1987: 679–82). First, Taylor argues that Sisyphus's life would not be meaningful because of the way that his desire originated (see also Kekes 1986: 80–4; Thomson 2003: 57–8). Since his urge to roll the stone is the product of the gods' manipulation, it is not truly his, and hence its satisfaction confers no meaning on his life. That criticism, however, is one that a subjectivist should not be unhappy to accommodate. One could easily amend subjectivism to require that the relevant attitudes be autonomously formulated.

Taylor's second objection to his earlier judgement cuts deeper. Even supposing Sisyphus of his own accord strongly desired to spend his life pushing a rock up a hill, it is simply counterintuitive to think that such a life would be meaningful by virtue of this desire being fulfilled. Most readers will be inclined, with the later Taylor, to find meaning only when a person is absorbed by a condition that intuitively *merits* it. Or, at the very least, they will tend to find *much more* meaning when a person is absorbed by an objectively good condition than when he is not. Hence, the consideration of absorption, like those of authenticity and love, seems most powerful as a source of meaning in contexts that take one beyond subjectivism.

9.5 The argument against subjectivism

There is only one standard argument against (S_1) in the literature, that it has seriously counterintuitive implications about which lives count as meaningful. Not only would it entail that Sisyphus's life could be meaningful merely for having fulfilled a desire to roll a stone, it would also entail that a person's existence could become significant by merely: staying alive (Kekes 1986: 81); harming others (Dahl 1987: 12); growing more corn to feed more hogs to buy more land to grow more corn to feed more hogs to buy more land to grow more corn, and so on *ad infinitum* (Wiggins 1988: 137); orienting her life around a single colour (Morris 1992: 58); maintaining 3,732 hairs on her head (Taylor 1992: 36); engaging in conspicuous consumption and being self-absorbed (Singer 1995: 200-13); collecting bottle tops (Singer 1996: 113); memorizing the dictionary, or recounting the number of tiles on the bathroom floor (Wolf 1997a: 210, 218); watching reruns of television series such as *Buffy, The Vampire Slayer* (Belliotti 2001: 75, 86-7); lining up balls of torn newspaper in neat rows (Cottingham 2003: 21); trying to make flowers sing or becoming addicted to drugs (Belshaw 2005: 115-22); or (best of all!) ingesting her own excrement (Wielenberg 2005: 18-23). Such cases have led a majority of those currently writing on life's meaning to reject subjectivism.

Here are two ways that the friend of (S_1) can reply.[5] First, she might bite the bullet, by maintaining that these conditions would indeed make people's lives meaningful if they were truly desired, chosen or what have you. In a recent essay, Brad Hooker (2008: 184-92) adopts this strategy to defend the idea that subjective conditions can be sufficient for meaning in life. He carefully advances the case of a 'sybarite', a person whose foremost aim is to develop 'an exquisite sense of the luxurious'. Hooker's intuition is that realizing the aim of cultivating a refined ability to detect and appreciate opulence could confer some meaning on one's life, the best explanation of which is that she has achieved the contingent object of her pro-attitude.

In response to Hooker, some naturally espouse a contrary intuition, as Cottingham (2008: 260-2) explicitly has, in effect suggesting that Hooker's willingness to bite the bullet has cracked his teeth. My response to Hooker is different, and is instead that the appeal of his case implicitly smuggles in some objectively valuable conditions. He points out, for instance, that developing an exquisite sense of the luxurious would require 'the development and exercise of capacities beyond that of sub-human intelligence' (2008: 186), including paying close attention, exhibiting determination,

---

[5] In correspondence, Alexander Alexis has suggested that focusing on the right sort of propositional attitude can help the subjectivist avoid such counterexamples. For example, Moritz Schlick (1927) contends that meaningfulness, and specifically that of the good, the true, and the beautiful, is exhausted by the orientation of being playful, where presumably one would not be playful when rolling a rock for eternity, watching reruns, counting tiles, ingesting excrement, and so on. However, one problem with Schlick's view is that play does not appear to be necessary for meaning in life, for it is hard to suppose that Mother Teresa was being playful when changing bedpans. Another problem is that play does not appear to be sufficient for meaning; playing draughts (or *Halo* when stoned, as the kids these days prefer) does not count. Such counterexamples will plague any form of actual-individual subjectivism, I submit.

grasping complexity, learning to discriminate between aesthetically similar properties, and so on. To the extent that these are indeed valuable facets of our rational nature, those sympathetic to objectivism can sign on to the idea that the sybarite's life would not be utterly devoid of meaning. In short, the case of the sybarite differs in important ways from the case of Sisyphus and many of the other counterexamples above.

A second reply is to let go of the standard form of subjectivism, which is most vulnerable to the counterexamples, and instead to opt for a less popular version that would be less vulnerable to them. Setting aside (S_1), then, I consider social versions of subjectivism, and conclude that none of them successfully resolves the problem.

Berit Brogaard and Barry Smith (2005) advance a powerful form of actual-intersubjectivism. They maintain that one's life is meaningful roughly to the degree that one has engaged in projects in a way that ranks highly on public criteria for success. Their basic idea is that a life counts as meaningful when one's activities have largely measured up to the community's standards. Roughly, then:

> (S_2) A human person's life is more meaningful, the more that she engages in activities that are the object of her community's pro-attitudes (such as judgements that her actions are successful or were worth doing).

This principle includes three necessary and sufficient conditions for a life to have engaged in non-trivial projects by virtue of satisfying communal norms. First, for an activity to confer meaning on life, one's community must have a way to judge it as better or worse, or as successful or not. To use Brogaard's and Smith's example, daydreaming is not a meaningful activity, since there are no public criteria for judging it to have been done well or poorly. Second, one's community must be aware of the activity, or at least have criteria that 'can be easily applied in the public light of day' (Brogaard and Smith 2005: 446) to the activity. This is the reason Brogaard and Smith suggest for thinking that mass murder is not meaningful, for it needs 'to some extent to be practiced in the dark' (2005: 447). Third, the activity must in fact be successful in the light of the community's criteria, where, say, making bad art is not.

(S_2) is still vulnerable to counterexamples, and I start with cases where it cannot entail that certain lives are meaningful that intuitively are, differing from the previous ones where subjectivism entailed that certain lives were meaningful that intuitively are not. For one, Brogaard's and Smith's theory oddly entails that intimate love-making, which might not be practicable if it had to be subject to public scrutiny, cannot be a meaning-conferring activity. For another, many of Europe's early scientists had to practice their work 'in the dark', viz., could not share their findings without risking ostracism, punishment, or even death.

One might, in reply, revise Brogaard's and Smith's account, dropping the requirement that the community must be aware of the activity, which would not be a major modification. Even so revised, though, their view would have counterintuitive implications, depending on the community at stake. Which community matters when evaluating an individual's life? Suppose that the relevant community is the one with which a given person identifies most strongly. In that event, he might identify most with the Nazi Party or the Flat Earth Society and do well by their standards at, say, purifying the race or spreading the word, respectively. However, surely little or no meaning would come from these activities.

To avoid the problem of being too permissive as to what counts as meaningful, then, one might go to the opposite extreme and suggest that the relevant community is the human race as a whole, or humanity at a given time. However, this view is too restrictive, implying that majorities are *necessarily* correct when it comes to what counts as a non-trivial pursuit. Consider, again, the first scientists in the modern era. The dominant institutions of the time, as well as the general populace, failed to appreciate the grand project of empirically testing hypotheses, positing physical unobservables as the best explanation of observables, appealing to Ockham's Razor, explaining events by appeal to mechanism rather than *telos*, and so on. Brogaard's and Smith's view still cannot account for the meaningfulness of their work, even dropping the publicity condition, as mentioned earlier.

The natural reply to make on their behalf is a claim that Wai-hung Wong, another actual-intersubjectivist, has advanced (2008: 134-44). Wong considers a case akin to mine of the first scientists, focusing on Mahler, a composer whose works were not appreciated in his lifetime (2008: 143).[6] According to Wong, even if Mahler's life was not meaningful when he lived, it *became meaningful posthumously*, once there arose a community that appreciated his music.

Here is the problem with that manoeuvre: it implies that the individual cannot do something meaningful until his fellows deem it to be, which appears false. Whereas the actual-intersubjectivist can, at best, claim that Mahler's life *became* meaningful only after his death because a community at that point *appreciated* his music,[7] it is much more natural to say that the community at the time Mahler was alive failed to recognize the meaningfulness of his life that was *already present* because his music *merited* appreciation.

I submit that any community or public the actual-intersubjectivist posits as determining whether an activity is meaningful must suffer from being either too permissive, for enabling an individual to obtain meaning from virtually anything (Nazi Party, Flat Earth Society), or too restrictive, for entailing that minorities are always wrong about what confers meaning (early scientists, Mahler). It is plausible to appeal to actual-intersubjectivity as a reliable *marker* of significance. If one wants to know whether a condition is meaningful, (*some*) *evidence* that it is meaningful is that a majority of one's community has judged it to be. However, this epistemic relationship between community and meaning does not entail a robust metaphysical one; that is, the fact that one can be *justified in thinking* that a life is meaningful by appealing to the

---

[6] I first discussed such a case in Metz (2005b: 326).

[7] Or, more carefully (having dropped the publicity requirement), developed standards that entailed that the community should appreciate his music, upon becoming aware of it.

judgement of one's fellows does not mean that conforming to the judgement of one's fellows is what alone *makes* a life meaningful.[8]

I now turn to the hypothetical-intersubjective form of subjectivism, which has been articulated as an account of meaning in life by Stephen Darwall (1983: 164-6). According to Darwall, a state of affairs confers meaning on a life roughly insofar as all human agents would prefer it to obtain, when coolly and calmly reflecting on it from an impersonal standpoint. More specifically, he suggests:

> (S_3) A human person's life is more meaningful, the more that she lives in ways all human agents would prefer upon the dispassionate consideration of their properties while abstracting from the way in which they would bear on their own lives.

This theory is subjectivist in that preference is ultimately doing the work of determining what makes a life meaningful. However, it differs from typical subjective theories in that the relevant preferences are those that *everyone would have* when he avoids thinking about how its object would affect anyone, such as himself, in particular.

These alterations most promise to rescue subjectivism from the counterexamples considered so far. This particular appeal to unanimity might well avoid the two problems of permissiveness and restrictiveness. With regard to permissiveness, the attitudes of idiosyncratic individuals and communities cannot determine what is meaningful, on Darwall's view, for the preferences of all human agents are alone what count. Presumably *not* all human agents would prefer advancing the cause of the Flat Earth Society, let alone living like Sisyphus or eating their own excrement, upon coolly and calmly considering the nature of such activity apart from its bearing on their particular lives. And with respect to restrictiveness, perhaps it is the case that *all* human agents would prefer the activities of scientists and composers, if they were under the idealized circumstances of carefully abstracting from their own viewpoints when reflecting on these activities.

Darwall's theory of meaningfulness has formal characteristics that would reasonably lead one to find it the most promising way for the subjectivist to avoid some powerful counterexamples. Despite its *prima facie* appeal, however, I reject Darwall's theory in favour of an objective naturalist theory because it appears to have the serious implication that no one's life is meaningful. My concern is that there is in fact no state of affairs that everyone would prefer when reflecting on it from an impersonal standpoint. Darwall's theory can entail that some lives are meaningful only if there is, dubiously, a striking uniformity in the motivational makeup of human beings. For anything to be meaningful, all human agents must prefer the same things, and it is

---

[8] I admit that, *ceteris paribus*, a project would be more significant, the more one's community appreciated it. We normally think that meaning comes from positively affecting others, where the more others whom one affects, and the more positively they are affected, the more meaning that accrues to one's life. Note, however, that this point most plausibly indicates an effect that would *enhance* the meaning in one's life, and does not entail the view that community recognition is *necessary* for it, let alone *solely constitutive* of it.

unlikely that everyone would, even setting aside any thought of how they might bear on themselves.[9]

Furthermore, for anything to be meaningful to a specifiable extent, all agents must prefer them to the same degree. But, again, it is extremely unlikely that all human agents, even if they preferred the same things, would prefer them to the same degree, even upon abstracting away as much as they can from their own idiosyncrasies.

The most straightforward way for Darwall to deal with these indeterminacy problems would be to focus on a community of agents much smaller than the class of all human persons. However, since idiosyncratic communities of Flat Earth Society fans and excrement eaters are possible, the original objection to subjectivism presumably resurfaces.

9.6 The 'whac-a-mole' problem

In this chapter, I distinguished four major types of subjectivism, which differ with regard to whose attitudes constitute meaning in life, those of individuals or of groups to which they belong, and the status of the attitudes, viz., whether the relevant agents actually exhibit the attitudes or would under certain, idealized conditions. I pointed out that the standard version of subjectivism is the actual-individual form, according to which what is meaningful for a human person is based solely on the objects of the desires, emotions, goals, and the like that she in fact has. I spelled out three rationales for subjectivism that are influential among contemporary theorists, namely, meta-ethical doubts about the existence of objective value and the ideas that subjectivism best explains the intuitions that meaning is a function of an authentic existence and of being absorbed by something. Against the latter arguments, I objected that objectivism can account no worse, and indeed often better, for the respects in which authenticity and absorption are relevant to meaning, and against the former, meta-ethical argument, I pointed out how the realist conception of morality I articulated in the context of purpose theory indicates how values in general could be objective.

Then I turned to the major problem with subjectivism, that it has counterintuitive implications about which lives are meaningful. The problem is particularly glaring for the standard version of subjectivism, and I considered the extent to which other forms could resolve it. However, for each version that I addressed, the objection resurfaced in 'Whac-a-mole' fashion. Subjectivism would become a bigger player in today's field if it could be shown that some version of it does much better at entailing that intuitively meaningful conditions are meaningful and that intuitively meaningless conditions are meaningless. For now, I conclude that there is enough reason to seek out another theory that has fewer counterintuitive implications.

---

[9] A similar point applies to a related version of hypothetical-intersubjectivism, according to which meaning is a function of what all ideal enquirers would agree is meaningful, which Alexander Alexis has suggested in correspondence.

10

Objectivism I: Being Attracted, Meriting Attraction, and Promoting Consequences

'To give life meaning cannot be just a matter of pursuing worthy projects, for that account fails to cope with phenomena like Tolstoi's arrest of life, or John Stuart Mill's during his mental crisis of 1826....What one often finds, on reading the reflective autobiographies... is that in asking questions about the meaning of his life, the author is asking how he can relate the pursuit of various valuable ends to the realising of a certain kind or form of life, the thought of which evokes in him the response: "The pursuit of these goals really concerns me, matters to me!"'

R.W. Hepburn ('Questions about the Meaning of Life')

10.1 Objective naturalism

To summarize my appraisal of the theories of life's meaning so far, I have argued that supernaturalism cannot be correct, since we know that some lives have meaning in them but do not know whether any spiritual substance exists, and I have argued that subjective naturalism cannot be correct since, although we surely have propositional attitudes, no configuration of them alone can determine what confers meaning in a way that matches our intuitions. The major theoretical category that remains is objective naturalism, the view that there are material properties that confer meaning on our lives to some degree apart from being the object of anyone's propositional attitudes. According to this view, a condition is meaningful at least in part because of its inherent physical nature, independent of whether it is believed to be meaningful, desired, liked, or sought out.

Engaging in beneficent projects, making scientific discoveries, and creating beautiful artworks are widely held instances of actions that confer meaning on life, while rolling a stone up a hill forever, urinating in snow, and eating ice cream (let alone excrement) are not, with objectivism of some kind, widely thought to be the best explanation for these respective kinds of judgements. The former are actions that are meaningful not merely because some agent or group believes them to be, wants to do them, or seeks to engage in them, and they are rather conditions towards which people ought to exhibit pro-attitudes. The latter are actions that cannot become meaningful even if some agent or group were to exhibit such attitudes towards them, which they ought not do (or at least not for the sake of meaning).

A majority of philosophers currently writing on the meaning of life are objectivists of some kind, aiming to detect a common denominator among the various ways of living in the physical world that confer meaning at least partially in virtue of mind-independent factors. As I bring out in this chapter and the next, some of the extant objective theories to be found in, or at least suggested by, the literature have the problem of being too permissive, entailing that a condition is meaningful when our intuition indicates otherwise. Many more have the opposite problem, of being too restrictive, entailing that certain conditions counterintuitively lack meaning. My goal in the rest of this part of the book is to articulate an objective theory that does better than existing rivals at accounting for our comparatively uncontroversial judgements of all and only those conditions that confer meaning on life.

To appraise objective conceptions, I focus predominantly on the three exemplars that have been recurrent themes in the field and in this book, namely, the good, the true, and the beautiful. I consider it to be a major plus for an objective theory if it can entail and well explain the fairly uncontested judgements that moral supererogation, intellectual achievement, and aesthetic creativity are central sources of meaning in life, and it is a serious strike against one if it cannot.

Looking at the existing forms of objectivism from a bird's-eye perspective, I distinguish three major kinds, two of which I address in this chapter. By far the most popular version is the theory that meaning in life is a matter of being attracted to what merits attraction, so far as I can tell first articulated in the analytic literature by R. W. Hepburn (1966), but recently developed with care by Susan Wolf (1997a, 1997b, 2002, 2010). Against it, I argue that attraction is not a necessary condition of meaningfulness, and also that no defender of this theory has indicated what makes something merit attraction, which would require constructing an additional theory altogether (10.2). A second form of objectivism is consequentialist, the view that the more final value one promotes, the more meaningful one's life (10.3). I devote the most space to arguing against the utilitarian version of this account, according to which meaning is roughly a function of making those in the world better off, and I reject it mainly on the ground that there are other plausible ways to obtain meaning besides promoting well-being. I also consider non-utilitarian forms of consequentialist objectivism, but contend that, because of their inherent teleological structure, they cannot account for some deontological or 'agent-relative' respects in which the good, the true, and the beautiful clearly confer meaning on life (10.4). I conclude the chapter by indicating what kernels of truth there are in the two rejected theories, and what mistakes in them must be avoided by the most defensible instance of objectivism (section 10.5), which I aim to develop in the rest of this part of the book.

10.2 Subjective attraction to objective attractiveness

I have indicated that an objectivist is someone who maintains, roughly, that realizing an object of propositional attitudes is not sufficient for meaning in life. Instead, a necessary condition for one's life to obtain some meaning in it is to live in the material world in ways that have a certain final value not merely by virtue of being the object of propositional attitudes. Now, there is nothing in this analysis of objectivism entailing that mental states play no role in constituting meaning in life, and it is important to see that this is so, in two respects.

First, it could be that certain mental states are themselves objectively valuable. It is typical, for example, for objectivists to think that exercising one's rational capacities would be meaningful, and not merely insofar as it is the object of propositional attitudes such as, say, believing that doing so would confer meaning, or wanting to do so. Second, it is consistent with objectivism to hold that realizing the object of propositional attitudes would *partially* confer meaning on life. Indeed, almost all objectivists hold such a view. Few objectivists are 'pure' in the sense of maintaining that being the object of conation, intentional action, emotion, and cognition play absolutely no constitutive role in making life meaningful (although there are consequentialists, examined below, who are). Instead, a large majority of objectivists believe that a life is more meaningful not merely because of objective factors, but also in part because of subjective ones.

More specifically, the dominant form of objectivism is the hybrid view that meaning is constituted by subjective attraction to objective attractiveness. Susan Wolf nicely sums up this conception:

(A) meaningful life must satisfy two criteria, suitably linked. First, there must be active engagement, and second, it must be engagement in (or with) projects of worth. A life is meaningless if it lacks active engagement with anything. A person who is bored or alienated from most of what she spends her life doing is one whose life can be said to lack meaning. Note that she may in fact be performing functions of worth.... At the same time, someone who is actively engaged may also live a meaningless life, if the objects of her involvement are utterly worthless (1997a: 211).

Wolf's theory is not merely the most commonly held version of objectivism, but probably the most commonly held view period among contemporary philosophers who have thought about meaning in life.[1] Although others have independently advanced what Wolf does in the quote above, I focus on Wolf's terminology and concepts, since she has articulated her theory with particular care and thoughtfulness and since it has been particularly influential of late.

---

[1] Hepburn (1966); Joske (1974: 286–9); Bennett (1984); Perrett (1985: 243–4); Kekes (1986, 2000); Wiggins (1988); Graham (1990: 181–4); Munitz (1993: 84–5); Dworkin (2000: 237–84); Kurtz (2000: 56); Raz (2001: 10–40); Schmidtz (2001); Martin (2002: 206–7); Wolf (2002, 2010); Belshaw (2005: 120–2); Starkey (2006). For related views, see Brogaard and Smith (2005); James (2010).

It is natural to ask precisely what counts as subjective attraction or active engagement, on the one hand, and as objective attractiveness or projects of worth, on the other. Different theorists cash these out in different ways, and it will be useful to look at one concrete illustration of how this might be done, from Wolf (2002, 2010), who has held that meaning in life is nothing but loving what is worth loving. If one loves nothing, then one's life lacks meaning, and if one loves something but it is objectively unworthy of love, then, again, no meaning accrues to one's life.

There are specific questions one might ask about Wolf's view, particularly about what counts as love, which objects are worthy of love, and how worthiness to be loved should affect the love one expresses. However, I set those concerns aside, instead presenting two objections to her theory that promise to tell against the attraction to attractiveness view more broadly.

10.2.1 Subjective attraction is not necessary

One should initially question whether one must love, or otherwise be attracted to, what one is doing in order to obtain meaning from it. I do not have in mind the idea that some *other* pro-attitude, such as desire, might rather be involved; instead, my first point is that *negative* attitudes towards undesirable conditions such as injustice, sickness, and poverty might be factors relevant to the subjective aspect of meaning (9.3.1). In the present context, one could fairly suggest that con-attitudes would be pertinent with regard to the subjective facet. For instance, even if one did not love or otherwise exhibit a pro-attitude towards fighting injustice, meaning might plausibly accrue so long as one hated the injustice one is fighting.

Second, there are counterexamples suggesting that meaning is possible despite not having any of the purportedly relevant attitudes. Return to some cases considered above in the context of boredom (7.5.1). Recall, for instance, the case of someone who volunteers to be bored so that others do not suffer boredom. Imagine that he exhibits no positive attitude whatsoever to his condition, and rather hates being bored. Even so, *some* meaning would plausibly accrue to him for having made the decision he did.[2]

Consider as well the case of a Mother Teresa who is, in stereotypical fashion, doing all she can to alleviate serious pain and heal grave injuries and illnesses. It is unlikely that Mother Teresa would have been terribly cheerful emptying bedpans and putting bandages on lepers. In any event, imagine that she lacked any potentially relevant attitude. Suppose that she loved neither the people she helped nor the activity of helping them, that she was not inspired by her work, but instead did it out of fear that she would face eternal damnation for not doing it, that for large periods she wondered whether human beings were really worth all the trouble, etc. Even so, my intuition

---

[2] One might be tempted to point out that the decision to take on boredom is itself a pro-attitude. However, it is not in the relevant sense; such a decision is far from what Wolf and others have in mind by phrases such as 'attracted to', identified with', 'actively engaged with', and so on.

is that she would have acquired some meaning in her life simply by virtue of having substantially helped so many needy people.

I think these cases provide reason to doubt that *any* propositional attitude, positive or negative, is *necessarily* constitutive of one's life being *somewhat* more meaningful. That criticism against Wolf's and related views is consistent with related claims that I accept, for instance, that certain attitudes would enhance meaning in one's life (10.4-10.5, 12.2-12.3), and that they might even be necessary for one's life to count as meaningful on balance (12.4), as the quote from Hepburn plausibly suggests.

10.2.2 'Objective attractiveness' is vague

My second major objection to Wolf's theory and related views is that her and others' talk of what is 'worth loving' or, even more broadly, 'objectively attractive' or a 'project of worth' is vague, too vague to make the theory one that could clearly entail and plausibly explain why the good, the true, and the beautiful lie at the heart of meaning in life. It will not suffice to say that the classic triad are 'objectively attractive' or are 'projects of worth'; that the reader already knows. What is needed at this stage of debate is some sense of what the good, the true, and the beautiful (among other conditions) have in common in virtue of which they have a mind-independent final value. Unfortunately, no advocate of the attraction to attractiveness theory has offered a precise answer.

Some prominent proponents of this theory have not done so because they believe that there is a wide plurality of answers that cannot be reduced to a single principle or handful of them (Dworkin 2000: 237-84; Kekes 2000; Schmidtz 2001; cf. Wolf 1997b: 12-13). However, as I have argued above (1.3), to have substantial justification for believing that there is no underlying unity among a myriad of conditions, one must have searched diligently and come up empty-handed. Given how nascent the field is, the search has arguably not gone on long enough. The discussion in the rest of this book can be viewed as contributing to this search, for the objective theories of life's meaning that remain include specific, unifying accounts of which projects in effect count as objectively attractive or worth pursuing.

10.3 Making people in the world better off

The second major form of objectivism is the basic view that the more final goodness one produces, and the more badness one reduces, wherever and however one can in the long-term, the more meaningful one's life. This view is a 'consequentialist' theory; it is the idea that meaning in one's life is solely a function of the net amount of dis/valuable results one has brought about in the universe.

Consequentialism is, on the face of it, an improvement over the attraction to attractiveness theory, in that it avoids the objections facing the latter. Consequentialism does not imply that subjective attraction is necessary for meaning in life; for all that matters is the extent to which one has promoted certain objectively attractive states of affairs. And, as I demonstrate below, consequentialists have been fairly specific about the relevant sort of objective value.

In addition, consequentialism provides a *prima facie* reasonable account of what would make a life meaningful. I have argued that to enquire into the meaning of life is to ask about three overlapping ideas (2.5), and the consequentialist fills them out with plausibility. First, to enquire into meaningfulness is, in part, to ask about which purposes people ought to pursue besides their own pleasures as such, where the consequentialist construes a purposeful life in terms of one that is useful, i.e., that serves the function of producing good states of affairs and reducing bad ones. Second, to enquire into meaningfulness is, in part, to ask about conditions in which one identifies with something greater than oneself, or transcends one's animal nature in particular. Consequentialism says that the relevant way to go beyond one's limits is to act in accordance with the 'point of view of the universe', as Henry Sidgwick has famously expressed it, that is, from an impartial standpoint of aiming to maximize benefits and minimize costs, whomever may be affected (regardless of their relationship to oneself), or wherever they may be in the physical world (regardless of their proximity to oneself).[3] Third, to enquire into meaningfulness is, in part, to ask about conditions that merit admiration on the part of others, and we tend to admire most those who have made a substantial, positive impact on humankind.

Another maxim often associated with meaning in life is the thought that one's existence is significant insofar as it leaves the world a better place than it was when one entered it, which readily admits of a consequentialist interpretation. Now, what counts as 'better' differs among consequentialists. I start by analysing and criticizing the most salient version of consequentialism, which is welfarist. According to this utilitarian theory, one leaves the world a better place *by having made beings within it better off*. I first spell out this theory (10.3.1), after which I present objections to it. My strategy is to present an objection to the most basic sort of utilitarianism (10.3.2), after which I refine the theory so that it avoids the objection, but then present a new objection to the refined theory, after which I again modify the view, but again present a new objection and so on. I conclude that the strongest version of utilitarianism cannot avoid criticism, which will lead me to explore a non-utilitarian form of consequentialism in the next section.

10.3.1 The utilitarian theory of meaning

According to the present conception of meaning, the more one benefits people (and animals), and the more one prevents harm from coming to them, the more meaningful one's life, regardless of whether one enjoys benefiting them, believes they should be aided or works particularly hard to help them. Although not expressed precisely this

---

[3] For classic articulations of this standpoint, see Adam Smith (1759: 235–7) and William Godwin (1793: 377–82).

way, such a theory is naturally read into the remarks of Peter Railton (1984: 164-71), Peter Singer (1993: 314-35, 1995: 171-235) and Irving Singer (1996: 101-40). So stated, this is a 'pure' objective theory, one according to which certain actions or other conditions can confer meaning on your life regardless of your propositional attitudes towards them.

One might wonder what counts as an improved or worsened quality of life. Some utilitarians would say that hedonism is true, so that a life is *ceteris paribus* better for having more pleasure and less pain, whereas others would broaden their conception of a quality of life to include other types of experiences, states, or relationships, such as contemplating an artwork or maintaining friendships, where these need not involve, or at least be exhausted by, pleasure. In the following, I work with the broader account of well-being,[4] which is intuitively more promising as a way to ground the meaning of the good, the true, and the beautiful.

Because readers are no doubt more familiar with utilitarianism about morality than utilitarianism about meaning, it is worth comparing and contrasting the two. Any attractive version of the utilitarian theory of meaning will differ from contemporary moral utilitarianism in at least two respects. First, the utilitarian theory of meaning will not require agents to produce the maximum amount of welfare available. Apparently unlike moral rightness,[5] meaningfulness comes in degrees. Hence, it is not plausible to suggest that an action is meaningful only if it maximizes (or even satisfices) available interest satisfaction. It is more sensible to maintain that an action is meaningful to the extent that it satisfies interests.

The second way that utilitarian meaning will differ from utilitarian morality concerns the act/rule distinction. Rule-utilitarianism is an appealing moral theory to the extent that it fleshes out a certain impartial perspective associated with the question, 'What if everyone did that?'. Impartiality is an essential (even if not exhaustive) aspect of the moral point of view. While various utilitarian moral theories are different conceptions of even-handed behaviour, theories of life's meaning are not. Hence, there is no reason to think that the forms of impartiality relevant to utilitarian morality are relevant to utilitarian meaning. The standpoint from which one asks 'What if everyone did that?' has comparatively little purchase with regard to living a life that realizes worthy purposes beyond one's own pleasure or that merits great admiration. At first glance, the actual effects of one's life on the world are more relevant to its significance than are counterfactual considerations about what the world would have been like had everyone acted as one did.

The utilitarian theory of meaning should appeal to an audience broader than just act-utilitarians about morality. Of course, the latter should find attractive the idea that an independent reason for people to act in accordance with a utilitarian conception

---

[4] More carefully, well-being insofar as it excludes the interest of a person in meaning (cf. the last note in Chapter 4).

[5] Of course, some classic utilitarians about morality, e.g., Jeremy Bentham and John Stuart Mill, do think in terms of degrees of rightness.

of right action is that it would confer more meaning on their lives, but a much wider array of theorists should also find the present conception of meaning worth taking seriously. First, nearly anyone who believes that helping others is the quintessential source of meaning[6] should be drawn to a theory that systematizes that judgement. Second, as Ronald Dworkin acknowledges (2000: 251-2), such a theory, which he calls 'the model of impact', is one plausible way to unify widely held intuitions to the effect that the larger the difference one makes to the world, the more meaningful one's life. Third, Dworkin also admits that the utilitarian theory on the face of it does well at explaining why the good, the true, and the beautiful are key sources of meaning. He notes,

We admire the lives of Alexander Fleming and Mozart and Martin Luther King Jr., and we explain why we do by pointing to penicillin and *The Marriage of Figaro* and what King did for his race and country. The model of impact generalizes from these examples; it holds that the....value of a life....is parasitic on and measured by the value of its consequences for the rest of the world (Dworkin 2000: 251-2).

When moral achievement (King), scientific discovery (Fleming), and artistic creation (Mozart) intuitively confer meaning in life, they appear to do so just insofar as they make many people's lives go much better. Conversely, when conditions intuitively fail to confer meaning in life, actions such as eating ice-cream, maintaining a certain number of hairs on one's head, and rolling a stone up a hill forever, what appears to be missing is anything of use for society. In short, the theory that the more one helps people, the more meaningful one's life, needs to be taken seriously, even by those who are resistant to utilitarianism about morality.

10.3.2 Objections to the utilitarian theory

I start by evaluating the most straightforward version of the utilitarian theory of meaning:

(UTM_1) A human person's life is more meaningful, the more that she makes those in the world better off than they would have been without her influence.

(UTM_1) says that meaningfulness is constituted by, and proportional to, the net amount of well-being contributed, taking into consideration all places where the agent could produce benefit and reduce harm. Degrees of the significance of a person's existence track overall degrees of welfare she has produced in the universe. (UTM_1) is inspired by some remarks of Peter Singer on the meaning of life (1993: 334):

(E)thics comes into the problem of living a meaningful life. If we are looking for a purpose broader than our own interests, something that will allow us to see our lives as possessing significance beyond the narrow confines of our own conscious states, one obvious solution is....the

---

[6] Representative examples include Baier (1957: 128–9); Dahl (1987: 5–10); Rescher (1990: 161–5); Teichman (1993: 159–60).

standpoint of the impartial spectator. Thus looking at things ethically is a way of transcending our inward-looking concerns and identifying ourselves with the most objective point of view possible—with, as Sidgwick put it, 'the point of view of the universe'.

From the standpoint of the universe, according to Singer, everyone's interests have equal weight. The more one acts from what would be recommended by this standpoint, working to promote the highest amount of interest satisfaction wherever and however one can, the more meaningful one's life.

A serious problem facing (UTM_1) is that it allows the agent's own welfare to count as enhancing the meaning of his life, but that seems counterintuitive.[7] Singer emphasises that action from the perspective of the impartial spectator transcends narrow self-interest, but it does not do so enough; for there are times when one can contribute more to the well-being of oneself than to that of others, and under such conditions (UTM_1) entails that one's life would be most meaningful for benefiting oneself. However, this is absurd.

For one illustration of the problem, consider Bear, a case in which you and a friend are unexpectedly exposed to a hungry grizzly. Suppose that you and your friend's influences upon other people's well-being would be equal, that you are capable of marginally more long-term well-being than your friend, that the bear will catch at least one of you, and, to be complete, that the bear would find no difference in taste or nutrition between you and your friend. The present utilitarian theory would say that, of the actions available, escaping would confer the most meaning on your life, even if this required throwing your friend to the bear. However, I submit that your life would not be made any more meaningful for so acting. The case shows that the promotion of your own welfare, even when this constitutes promoting the most welfare available, cannot enhance the significance of your life.

For an additional example, think about Mum. In this case, a mother gives up food for her young son, where we suppose that the food would do her somewhat more good than it would him. (UTM_1) entails that more meaning would accrue to the mother by eating the food herself (and staving off guilt) than by giving it to her son, since she would thereby promote more well-being. But this seems implausible. The mother's life would be no more meaningful for enhancing her own welfare. If there is a meaningful action here, it would be sacrificing her own good for the sake of her son's, even when this would result in less welfare overall in the universe.

An interesting way for the defender of (UTM_1) to respond to Bear and Mum would be to note cases that apparently point to the opposite conclusion. Consider, for example, Auschwitz. It seems that one's life would be meaningful for fighting to stay alive and to obtain as much comfort as possible in a Nazi concentration camp (see Frankl 1984). Auschwitz suggests that promoting one's own welfare can in fact make one's life matter.

---

[7] It might not even squarely count as a theory of life's meaning, insofar as welfare is a matter of pleasure and meaning-talk, by definition, is about final goods beyond one's own pleasure in itself (2.4).

One could question whether Auschwitz is truly a case where meaning is conferred on a life. Perhaps surviving under such conditions is a matter of preserving some necessary conditions for meaningful activity, but is not itself meaningful. Suppose, though, that it would be meaningful to uphold a tolerable existence in the face of such challenges. I question whether it is *qua* benefiting oneself that the meaning arises. In Auschwitz the meaning is arguably instead a function of demonstrating courage and maintaining self-respect. Here it takes substantial strength of will to maintain a decent state, and staying alive is needed to stand up for one's dignity. I suggest that it is such excellences, and not any welfare produced, that are guiding intuitions about Auschwitz (cf. 11.3). This hypothesis would also explain why one would have different intuitions about Bear and Mum; it neither takes courage nor maintains self-respect to outrun a bear and to eat food that could have gone to one's hungry child. I am therefore inclined to think that the Auschwitz case backfires against the friend of (UTM_1).

I now propose an alternative utilitarian theory, one that does not imply that making oneself well off confers meaning on one's life:

> (UTM_2) A human person's life is more meaningful, the more that she makes others in the world better off than they would have been without her influence.

Whereas an egoistic theory gives the agent's interests alone any weight, and (UTM_1) gives everyone's interests equal weight, (UTM_2) gives the agent's interests no weight. If I were constructing a moral theory, the strict impartiality of (UTM_1) might be preferable, but I am instead searching in the utilitarian tradition for a theory of what makes a life meaningful that coheres with considered judgements about the subject. Such intuitions have driven me to (UTM_2), which obviously solves the problem facing (UTM_1), for it denies that benefiting oneself ever makes one's life matter. Instead, benefiting others is alone what counts. Irving Singer approximates (UTM_2) when he says:

> A significant life.... requires dedication to ends that we choose *because* they exceed the goal of personal well-being. We attain and feel our significance in the world when we create, and act for, ideals that may originate in self-interest but ultimately benefit others.... The greater the benefit to the greater number of lives, the greater the significance of our own (1996: 115, 117).

Singer speaks of performing actions 'because' they will help others, but one's life could plausibly be more meaningful for unintentionally benefiting others. Imagine that a scientist makes a discovery that unexpectedly turns out to help cure a disease. This case suggests that a person's life might be significant for being part of a beneficial causal chain that she did not have in mind (or at least did not aim for) when acting, at least if the probabilities are fairly strong that one link will follow another. (UTM_2) may be understood to accommodate such a case.

The major problem facing (UTM_2) is that it counterintuitively entails that *anytime* one confers a benefit on another person, one's existence is more significant. For a counterexample, consider Humiliation, a case in which a person lets others denigrate him for the fun of it. He associates with people who make him the butt of racist or otherwise derogatory jokes, where there is no urgent reason to be in their company. Here, a fellow makes others happier than they would have been without him, which (UTM_2) says confers some meaning on his life, but his life seems no more meaningful for this. For another example, consider Prostitution. Selling sex for money could bring pleasure to many people, but, *contra* (UTM_2), would arguably fail to confer any significance on a life, at least when done merely to support an expensive cocaine habit.

Exactly what fundamental problem with (UTM_2) is suggested by Humiliation and Prostitution? One possibility is that immorality, conceived in non-utilitarian terms, pollutes the benefits produced. Immoral types of welfare such as sadistic or racist pleasures do not seem to count. And immoral ways of producing (what might be the right type of) welfare—for example, by treating oneself like a commodity—fail to confer meaning on a life. So, we might try:

> (UTM_3) A human person's life is more meaningful, the more that she, without being immoral, makes others in the world better off than they would have been without her influence.

(UTM_3) introduces a 'side-constraint' into the way an agent may promote well-being and thereby acquire meaning. Considerations of meaningfulness restrict agents from promoting welfare, no matter how large, by performing immoral actions, no matter how small.

Once the view is stated this way, the problem with it becomes obvious. There are clear cases, which appeal not merely to utilitarians, in which the beneficial results of an immoral action can make one's life meaningful. For example, consider the classic case of Broken Promise, where instead of keeping one's promise to a dying man to bequeath his wealth to his undeserving son, one uses it to start up a shelter for homeless youth. Here it appears that meaning is indeed possible, despite immorality.

What, then, is the difference between Broken Promise, on the one hand, and Prostitution and Humiliation, on the other? My answer is: although in the former case there is an *infringement* of moral rules, the infringement is not a *violation*, since it is justified on moral grounds. All things considered moral reasons prescribe breaking a promise to help the poor, whereas (supposing non-emergency circumstances) they do not recommend letting others humiliate oneself or hiring out one's sexual organs.

Hence, one might be inclined to revise the utilitarian theory of meaning as follows:

> (UTM_4) A human person's life is more meaningful, the more that she, without violating moral constraints, makes others in the world better off than they would have been without her influence.

However, this version, too, is vulnerable to powerful counterexample. There are cases in which a life appears to be meaningful for having performed an all things considered immoral action. Consider the widely discussed case of Gauguin, who is reputed to have abandoned his family in order to develop his painting. Usually this case is raised with a view to ascertaining whether Gauguin had sufficient reason to do what he did, i.e., whether non-moral considerations can outweigh moral ones. I use this case to make a different point, namely, that meaningful action seems possible despite moral violation. It appears as though Gauguin's life was meaningful, at least for having stimulated people's aesthetic sensibilities, even though he did so by failing to live up to responsibilities he had assumed.[8]

Now, why is there meaning in Gauguin, but not in Prostitution or Humiliation, when in all these cases there is welfare produced by an-all-things considered immoral means? One plausible suggestion is that the kind of moral violation involved differs. In Prostitution and Humiliation, welfare results from a particularly degrading sacrifice, whereas the same is not true in Gauguin. Let us accept this admittedly rough account of the interplay between meaning and morality and reformulate the theory accordingly:

> (UTM_5) A human person's life is more meaningful, the more that she, without violating certain moral constraints against degrading sacrifice, makes others in the world better off than they would have been without her influence.

Of the versions considered so far, this particular form of 'restricted' or 'constrained' utilitarianism fits best with common-sensical ideas about what makes a life matter.

However, there is an important objection that applies even to (UTM_5) and that now cuts to the heart of the utilitarian perspective. I doubt that meaning in life comes merely from making people better off. Besides satisfying other people's interests, developing virtues or excellences in oneself and in others intuitively confers meaning on one's life, and is part of the best explanation of why the good, the true, and the beautiful make life a meaningful. First, with regard to morality, a person who exhibits integrity or courage might thereby acquire significance, even if being overly deferential or wimpy would make everyone happier. Second, with respect to intellectual enquiry, making certain substantial scientific advances plausibly confers significance on a life, even if these discoveries have no technical application and cannot be understood by laypeople. Finally, making great works of art makes a life meaningful, even if they are not widely appreciated (recall the discussion of Mahler in 9.5).

Utilitarians will, of course, suggest that welfare can entail the intuitively right answers in these cases. Specifically, some will wonder whether a 'two-level' hypothetical theory might not do the work here. If everyone routinely did things such as create artworks and maintain integrity, then welfare would be promoted more in the long run than if people did not commonly do such things. Hence, for an individual to do them now is meaningful.

Rather than construct a counterexample to this proposal, I note two things. First, this version of utilitarianism is poorly motivated relative to (UTM_5), for recall that rule-utilitarianism is better understood as a moral theory than as a theory of life's meaning. Second, even if welfare would be maximized if people routinely did things such as create artworks and maintain integrity, it would not be these hypothetical

---

[8] For a different judgement about the case, see Cottingham (2003: 25–8).

consequences that best *explain* the meaningfulness of these activities. Creating artworks and maintaining integrity, I submit, can confer *pro tanto* significance at least partially in themselves, apart from *any* long-term results, hypothetical or actual. There is something about the state of the individual agent, apart from his influence on others' quality of life, that at least partially accounts for the meaningfulness of the good, the true, and the beautiful.

10.4 Making better people in the world

Defenders of a consequentialist approach to life's meaning should at this point suggest that the problem with (UTM$_5$), the most defensible version of the utilitarian theory, is its focus on well-being and woe as the relevant objective good and bad. Perhaps the problem has been interpreting an instruction to leave the world a better place to mean that one ought merely to make those in the world better off. Suppose that one instead were to understand 'better' not (solely) in terms of happiness or welfare, but (also) excellence. Thomas Hurka (1993) and Quentin Smith (1997: 179–221) have presented intricate and powerful consequentialist theories of this form. I argue that while they, or some version close to them, do better than (UTM$_5$) at accounting for the good, the true, and the beautiful, there are enough problems with them—ultimately because of their consequentialist structure—for us to consider some other objective naturalist theory of meaning in life.

These forms of consequentialism invoke the distinction between well-being, what is *good for a person*, and perfection, what a *good person* is. A person's self-interest, or what makes her well off, is one thing, while a person's self-realization or what she does well is another. Living a happy or pleasant life is different from living a virtuous one or one that develops its nature. Hurka and Smith both suggest that it is only the perfection of one's nature, or excellence, that can make life matter, but they differ over what counts as the relevant sort of excellence. Hurka focuses on human persons alone, and maintains that what is excellent for them is solely the development of their rationality. With regard to meaning, then, for Hurka:[9]

(PT$_1$) A human person's life is more meaningful, the more that she promotes rational behaviour anywhere that she can.

Hurka has a complex and fascinating account of what is involved in the exercise of rational nature, but I need not take that up in order to give this perfectionist theory its due, relative to competitors.

There are two *prima facie* problems with Hurka's account, which Smith's theory resolves with some plausibility. First, Hurka's view excludes the idea that one's life could be somewhat more meaningful for developing the natures of any non-human

---

[9] Hurka (1993) is professedly putting forth what he takes to be an attractive *moral* theory, but one could fairly, and I think even more plausibly, read his view as an account of meaning in life.

beings. For instance, it is intuitive that preventing the torture of many animals would confer some meaning on one's life. As Smith points out, Peter Singer's life is reasonably deemed to have some meaning in it by virtue of his helping to reduce the extent to which painful tests are conducted on animals for trivial reasons, such as the development of 'two-in-one' shampoo-and-conditioner.

Second, Hurka's view is overly narrow in an additional way, in that it restricts the human essence to rational nature, conceived as a certain kind of thought and action (cf. 11.3.7), thereby failing to capture the respect in which promoting other people's well-being can, under certain conditions, also be a source of meaning in one's life. This appears to be the kernel of truth in utilitarianism. I have encountered no reason to doubt that enhancing others' welfare can make one's life meaningful; surely relieving a child's pain could do so to some degree (8.4), as could, say, pleasing an audience by playing music to it. Hurka's theory must say, implausibly, that such activities are meaningful only insofar as they are exercises of one's rational beliefs or goals or lead others to exercise theirs.

Smith's form of consequentialism improves on Hurka's in these two respects. For Smith, the relevant natures to be developed are, in principle, those of all beings in the universe. For Smith, even rocks, weeds, and bacteria have natures there is always *some* (very small) reason to foster. He says, 'In determining the best goal, I do not take into account just humans, but all humans, all other animals, all plants and other organisms, and all physical things' (1997: 213). However, Smith also contends that there is a ranking of natures, such that higher natures are those that include other natures. So, a plant's nature is higher than a rock's, since, in addition to including spatio-temporal and atomistic elements, the plant also has the properties of life, growth, and reproduction. An animal's nature is higher than a plant's, since it includes everything the plant's does and then also additional capacities such as perception, goal-directed activity, and the like. And then, of course, a human person's nature is higher than an animal's, for the usual consideration of including not only everything the animal has, but also the property of rationality.

In addition, Smith is explicit that he believes that beings have basic rights (1997: 206-10), which, based on the discussion of (UTM_3) and (UTM_4), is appropriate. Therefore, consider this statement of a non-welfarist, and, specifically, perfectionist consequentialism:

(PT_2) A human person's life is more meaningful, the more that she, without violating certain moral constraints against degrading sacrifice, develops the nature of beings anywhere, with more meaning accruing to her life, the more beings whose natures she develops and the higher their nature.

(PT_2) appears able to avoid the counterexamples to (UTM_5), since the distinctively human excellence of rationality typically has a greater value than human happiness, a capacity shared with animals. Without going into the details of Smith's account of reason (see 11.3.7), it is plausible to think that a person who exhibits integrity or courage when others would be happier if she did not would manifest the virtue of strength of will. A person who made a scientific advance that admits of no technical spin-offs would exhibit the virtue of theoretical intelligence. And a person who made great works of art that went unappreciated would have displayed the excellence of creativity.

In short, the friend of perfectionist consequentialism à la (PT_2) can suggest that when the lives of the upright, scientists, and artists seem to be meaningful not for promoting welfare (either in themselves or in others), they are realizing in themselves (and in others) the superior exercise of their rational capacities in the forms of will, intellect, and imagination. As Smith remarks, 'people admire or approve of Einstein's theoretical efforts, Beethoven's musical efforts, Sylvia Plath's literary efforts, and these efforts maximize (or at least are believed to maximize) the development of human nature, which includes the creator's nature as well as the audiences' natures' (1997: 206).

So far, so good, for the friend of perfectionist consequentialism. Here, however, are three objections that apply not only to it, but also to (UTM_3) and consequentialism as such,[10] which lead me to search for a different kind of objective theory altogether. None of them invokes the most common objection made to consequentialism about morality, namely, that it disregards the existence of side-constraints or restrictions, i.e., prohibitions on intuitively immoral ways to produce objective value. (PT_2), as a form of constrained consequentialism, is not vulnerable to that objection. Instead, the root problem with consequentialism as an account of meaning is its 'teleology' or 'agent-neutrality', the fact that it deems meaningfulness to be a function solely of outcomes that in principle have no special relation to the agent who promotes them.

10.4.1 How to promote the consequences

For a first objection to (PT_2) and to any other consequentialism, consider a thought experiment from Nozick (1974: 42-5), who imagines that a 'result machine' existed, such that it could be programmed to bring about pretty much any outcome whatsoever. If consequentialism were true, then a maximally meaningful life would be one that programmed the machine, or pushed the relevant buttons, so as to bring about as much objective value as it could.

However, that judgement is counterintuitive. Even if it were the case that one *ought* to get the machine running, one's life would not be *maximally* meaningful for having done so. Instead, a fully meaningful life, insofar as it involves the promotion

---

[10] I mention here a fourth objection, namely, a version of the incoherence argument that I have invoked in the contexts of divine command morality (5.5), supernaturalism (8.5) and non-naturalism (8.7). A typical consequentialist is committed to the following claims: (1#) An act confers meaning on a life just insofar as it produces objective goodness and reduces objective badness in the long run. (2#) At least some lives have meaning in them. (3#) No one knows whether any particular action has produced objective goodness and reduced objective badness in the long run.

(3#) is the claim that 'we do not have a clue about the overall consequences of many of our actions. Or rather—for let us be precise—a clue is precisely what we do have, but it is a clue of bewildering insignificance bordering on uselessness—like a detective's discovery of a fragment of evidence pointing inconclusively to the murderer's having been under seven feet tall' (Lenman 2000: 349–50). (3#) is controversial, and not unreasonable to reject; but if it turned out to be justified, then, on pain of inconsistency, one would have to reject (1#).

of objective value, requires *effortful* or *hands on* activity. Less fantastically, merely cutting a check to Oxfam, donating funds that one was lucky to inherit, would not be as meaningful as labouring in the poor community that Oxfam seeks to benefit, supposing the latter produced only marginally fewer objectively good long-term outcomes than the former. However, (PT_2) cannot account for that intuition, since the nature of the means by which the objective value is brought about does not matter (so long as it does not violate constraints against degradation); by that principle, the more that is brought about by whatever (permissible) mechanism, the more meaningful one's life.

10.4.2 Where to promote the consequences

The second objection to consequentialism focuses less on *how* objective value should be produced, and more on *where*. The instruction to promote as much objective value wherever one can is too crude, and does not capture the intuition that, even if one were robust in producing objective value, there would often be most reason of meaning to realize it *in oneself*, even if that meant somewhat less objective value in the world overall.

For a drastic sort of case, consider Organs. You could, at this very moment, go to the hospital, announce that you are donating your organs to people who would otherwise die without them, and then shoot yourself. Assuming that would save four lives, where those lives would have a comparable degree of excellence as yours would, you would be maximizing the amount of excellence in the universe if you were to commit suicide in this fashion. While some meaning might well accrue to your life if you were to sacrifice it for the sake of others, it is implausible to think that this would be the way to make one's life maximally meaningful.

In reply, one might suggest that such self-sacrifice would count as a violation of a moral constraint against (self-)degradation. It would perhaps be to treat oneself merely as a means to a worse degree than prostitution. Consider, therefore, another case. Think of a Married Couple, both of whom are talented, indeed so talented that they have precisely calculated that the most excellence would be produced in the long run if the wife stayed home and supported the husband in his professional career, more than if he instead took care of the household or if they both worked and shared the domestic labour. Suppose that the amount of extra excellence realized by the husband through his work would be marginal relative to the other options. Now, on a consequentialist view, the higher the overall amount of excellence one has produced, the more meaningful one's life. This theory entails that, to maximize meaning in her life, the wife ought to stay home and support the husband.

But that is counterintuitive. Even if the wife had worked hard at home (avoiding the first problem with consequentialism above) to enable her husband to perfect his human nature, to a slightly higher degree than that of which she were capable, she would have had more meaning in her life if she had instead exhibited quite a lot of excellence in herself.

In short, insofar as the production of excellence confers meaning on life, it particularly does so when the excellence is realized in oneself, something that consequentialism denies with its requirement to act in accordance with the 'point of view of the universe'. With respect to at least some objective goods, such as excellence, more meaning can come from exhibiting them oneself, rather than from enabling others to exhibit them, even somewhat more of them.

10.4.3 Not merely promoting consequences: the relevance of subjective attraction

The third objection to consequentialism springs from a case of someone who has to choose between maximizing good outcomes in the long run for the world, but not 'being into it', on the one hand, and producing somewhat fewer desirable results, but satisfying deep drives within himself, on the other.[11] I have the intuition that, despite the production of a somewhat larger amount of objective value (however conceived) in the former case, this person would have had a more meaningful life in the latter. If that is right, then *producing* good and *reducing* bad are not the only responses to dis/value that would make one's life significant. What can also partially affect the degree to which a condition confers meaning is whether it is the object of certain of one's propositional attitudes.

Earlier, I objected to Wolf's subjective attraction to objective attractiveness view on the ground that it implies that the former is necessary for a condition to be *pro tanto* meaningful (10.2). And I appealed to cases that a consequentialist would readily find attractive, such as a person volunteering to be bored so that others would not be, and a Mother Teresa who helps others enormously but is alienated from her work. Now I argue, against the consequentialist, that although subjective attraction is not *necessary* for a condition to be *pro tanto* meaningful, it would *increase* its meaning.

According to the standard form of consequentialism, exhibiting a propositional attitude is never partially constitutive of meaning, and can have only instrumental value for causing us to promote desirable results. For example, if one strongly wanted to help others, one would be more likely to do so than if one did not want to. That, however, does not exhaust the respect in which strongly wanting to do something is relevant to meaning. I advocate a position that stands between consequentialism, which deems subjective attraction to be incapable of constituting meaning in life (and instead, at best, to be instrumental for it), and the subjective attraction to objective attractiveness theory, which deems subjective attraction to be (in part) necessary for it. Midway between pure consequentialism and the Wolfian hybrid theory is the view that having certain propositional attitudes towards objectively good activities is not necessary to obtain meaning from them, but would *constitutively enhance* it.[12] According to this view, a stereotypical Mother Teresa, bored by her work, would have

---

[11] Such cases are explicitly discussed in Metz (2003: 62–8, 2007c) and Affolter (2007), but are implicit in discussions by Wolf (1997a) and other friends of the attraction to attractiveness theory.

[12] A view I first proposed in Metz (2003); see also Audi (2005: 344); Hooker (2008); James (2010).

had *some* significance in her existence because of the work, but would have had an even *more* significant existence insofar as she had exhibited attitudes such as love of people, a sense of pride in what she did, belief in its intrinsic importance, and perhaps hatred of undeserved suffering.

How might the friend of consequentialism reply to this objection? Here is one revealing way. She must admit that there is something to be preferred about a Mother Teresa who is not alienated from her work, as opposed to one who is. However, the consequentialist may question the claim that what is preferable about the former life is that it would be *more meaningful* than the latter. It would clearly be a *happier* life. Now, recall the claim that one's own pleasure as such logically cannot enhance the meaning of one's life (2.4), as well as the cases of Bear and Mum, which grounded the objection to (UTM_1) that promoting an agent's own well-being substantively cannot make her life more meaningful. Accepting these points, the defender of consequentialism might respond that Mother Teresa's subjective attraction, viz., her happiness, would not make her life any more meaningful.

I respond by contending that being subjectively attracted to promoting objective value is not exhausted by being happy doing so. It is not a matter of *welfare* to exhibit attitudes such as identifying closely with a project, or concentrating intently on it, or setting an end and realizing it. And even if it were, I submit that these subjective conditions have an additional, non-welfarist property that is the factor conferring meaning on the agent's life. Roughly, it is not the happiness that makes a non-alienated Mother Teresa's life more meaningful, but rather, as I maintain below (12.3), the fact that more aspects of her rational self are positively engaged.

10.5 Agent-relativity as intrinsic to meaning

In this chapter, I have argued against two of the three major forms of objectivism, the general view that meaning in life is constituted by ways of living and being in a purely physical world that are desirable not merely because they are the object of anyone's propositional attitudes. I noted that the most popular version of objectivism is the view that meaning is constituted by subjective attraction to objective attractiveness. Against Hepburn, Wolf, and many others, I provided reason to doubt that subjective attraction is necessary for a condition to confer (*pro tanto*) meaning on a life; a meaningful condition in an individual's life need not be 'meaningful to' her. I also pointed out that no friend of this theory has articulated what all objectively attractive conditions have in common, requiring one to construct a separate principle, such as the next major version of objectivism.

The second objective theory that I addressed is consequentialism, the view that meaning in life is a matter of producing objective goodness, such as well-being or excellence, wherever and however it can be realized in the long run, as well as similarly reducing objective badness. I began by considering and rejecting utilitarian versions of consequentialism. I argued not only that *well-being* is not the only final value the promotion of which can confer meaning on life, but also that not every *promotion* of well-being matters, as it must be done in a way that avoids violating certain moral constraints against the degrading sacrifice of oneself or others.

Then, I considered a non-utilitarian consequentialism, and rejected it and other forms of consequentialism, for three reasons. My first criticism was that bringing about final value *with any (permissible) mechanism whatsoever* does not exhaust the respect in which realizing it can confer meaning on life; *ceteris paribus*, promoting goodness for its own sake in a robust, active, or intense way would confer more meaning than, say, merely pressing a button on a result machine.

My second objection was that bringing about final value *anywhere one can* does not account for the intuition that certain goods would confer the most meaning on one's life if one were to exhibit them oneself. The case of Married Couple indicated that there is more meaning to displaying virtue oneself than to producing the same amount of virtue, or even a bit more, in someone else.

My third objection was that *bringing about* what is non-instrumentally desirable is not the only way to relate to it so as to accrue meaning in life. Reflection on Mother Teresa ultimately suggested that an agent's life would be more meaningful to some degree insofar as she exhibited subjective attraction to objective value, to the point where considerations of meaning can counsel producing less objective value overall if doing so would involve more of one's propositional attitudes such as conation, emotion, and cognition being attuned to it.

I sum up most of these considerations by saying that there are 'agent-relative' or 'deontological' facets of part-life meaning.[13] Meaning depends, in part, on whether the agent: promotes well-being in others in morally permissible ways; promotes well-being in others in ways that robustly involve her agency and effort; realizes excellence in relation to herself; and is subjectively attracted to what she is doing. Acting 'from the point of view of the universe' might be one way to confer meaning on one's life, but it is not the only way; for the agent's non-teleological relationships to good states of affairs matter as well. In the next two chapters, I seek a theory that accommodates these points.

---

[13] Cf. 3.5.2 for discussion of non-teleological facets of whole-life meaning.

11

Objectivism II: Non-consequentialism

'The true, the beautiful, the good: through all the ages of man's conscious evolution these words have expressed three great ideals: ideals which have instinctively been recognized as representing the sublime nature and lofty goal of all human endeavour.'

Rudolf Steiner ('The True, the Beautiful, the Good')

11.1 Honing in on that sublime nature and lofty goal

In the previous chapter, I presented and criticized two of the three major objective naturalist theories of meaning in life, the attraction to attractiveness theory and consequentialism. Against the former, I argued that another theory altogether would be required to provide content to the concept of attractiveness, and that being attracted is not necessary for meaningfulness. Against consequentialism, I argued that there are several respects in which meaning is 'agent-relative'. In this chapter, I critically address the third major form of objectivism, which avoids the problems facing the other two forms, and which comes the closest, in my view, to capturing all and only those conditions that confer meaning on life. This third branch of objectivism is naturally called 'non-consequentialism', for it directs one to realize objective value in ways other than by merely promoting it wherever and however one can in the universe as much as possible over the long-term.

I begin by explaining what is inherent to non-consequentialism, indicating how it differs from the other two major forms of objective naturalism (11.2). Then, I take up the most prominent and promising non-consequentialist theories that one finds in the literature. Sometimes I argue they are too narrow, for being unable to capture salient respects in which the good, the true, and the beautiful confer meaning on life, while at others times I argue that they are too broad, for deeming meaning to inhere in conditions that it intuitively does not (11.3). I conclude by summarizing the lessons learned from the discussion, that is, both the kernels of truth in the existing theories that must be retained, and the problems in them that must be avoided, if I am to develop a theory that better captures, in Rudolf Steiner's words, 'the sublime nature and lofty goal' of moral action, intellectual reflection, and artistic creation (11.4).

11.2 Non-consequentialist objective naturalism

Non-consequentialist objective naturalism in general is the view that meaning is possible in the physical universe alone, by virtue of living in certain ways that are meaningful neither merely because they are the object of anyone's propositional attitudes, nor merely because they bring about desirable long-term consequences somewhere in the universe. There is *something* about, say, making a substantial ethical accomplishment, obtaining deep insight into the workings of nature, and creating a masterpiece that are significant 'in themselves', apart from any relation to a spiritual realm, to our attitudes, and to the state of the universe in the long run.

This is the last major theoretical perspective on what makes a life meaningful that I address in this book. I have argued, roughly, that naturalism is to be favoured over supernaturalism (Chapter 8), that objective naturalism is to be favoured over subjective naturalism (Chapter 9), and that non-consequentialist objective naturalism is to be favoured over consequentialism (Chapter 10). What remains is for me to specify the sort of non-consequentialism that I find most defensible.

In this chapter, I evaluate existing theories primarily with regard to the nature of the relevant objective value and the proper ways for the agent to relate to it.[1] And I do so by focusing particularly on whether they adequately capture intuitions about the meaningfulness of the good, the true, and the beautiful. By the end of the chapter it will be clear what needs to be jettisoned from existing non-consequentialist theories, what needs to be retained in them, and what needs to be added to them, so as to entail and plausibly explain the significance of morality, enquiry, and creativity.

11.3 Existing non-consequentialist theories

In this section, I organize and criticize several distinct theories according to which what makes a life meaningful are ways of living in the universe that are intrinsically valuable, not only in the sense that they are good for their own sake, apart from their usefulness, but also in the sense that they are good in themselves, at least to some degree apart from any relationship to God, to our attitudes or to the long-term future of the universe. Although I associate the theories with the remarks of certain philosophers, my aim is not to provide a thorough analysis of anyone's views. I am most concerned to present coherent and promising theories of meaning in life, and hence use a given person's words to illustrate a principle, not to capture the intricacies of her

---

[1] I set aside, for now, issues such as whether the principles can account for the ideas that supernatural conditions could enhance meaning in life, even if they are not necessary for it (8.7) and that long-term desirable consequences of one's life, even posthumous ones, could do the same (3.4-3.5, 4.4).

Weltanschauung. So, when I criticize a theory, the reader should consider whether my criticism is fair to it, not necessarily to the thinker whose remarks I associate with it.

11.3.1 Love

The non-consequentialist theory that I begin with is one that has been most clearly advocated by Terry Eagleton (2007: 158-9, 164-73) in a light, lively, and widely read book on life's meaning. Eagleton firmly rejects the subjectivist, or what he calls 'postmodern', approach to meaning, and instead sums up the nature of meaningfulness by suggesting that it inheres solely in loving relationships with people.[2] Consider, then:

(NCT$_1$) A human person's life is more meaningful, the more that she loves other beings worth loving.

(NCT$_1$) is a broader principle than what Eagleton's remarks suggest, as it both includes the idea that loving animals such as pets could confer meaning on a life and also makes room for the idea that, if some people are not worthy of love, no meaning would accrue from loving them.

(NCT$_1$) is attractive in that failures to confer meaning on life, or forms of anti-matter, typically involve some kind of isolation from others, for example, being put into solitary confinement, watching sit-coms, engaging in prostitution to feed a drug addiction, throwing rubbish out of a moving vehicle, interacting with people on the basis of a deep neurosis, living in an experience machine akin to *The Matrix*, rolling a rock up a hill for eternity à la Sisyphus. (NCT$_1$) well captures the intuition that, as Will Durant has put it, 'Solitude is worse than war' (1932: 105). It can also account for the meaningfulness of the good with some plausibility, insofar as one is willing to characterize the moral point of view as including at its core an ideal of impartial love, the prizing of harmonious relationships or an ethic of care.

However, the glaring counterexamples to (NCT$_1$) are the true and the beautiful. Of course, scientists and artists often produce a given work for an audience whom they expect will be interested in it. However, that is much too thin a basis on which to hold that their activities are meaningful merely for involving loving relationships. To see this, recall the two properties that philosophers most often take to characterize the essence of love (7.2.2), and note how meaningful science and art can be in their absence.

First, it is typical to think of love as inherently involving togetherness between people. Those who love each other think of themselves as a 'we' (and not merely as an 'I') and engage in joint projects. In short, they share a way of life. However, many scientists and artists are isolated, and sometimes downright self-centred, and yet can have meaningful lives all the same.[3]

---

[2] For additional views that take meaning in life to be largely a matter of love for persons, see Sampson (1969: 496-9); Van Schaik (1977); Wolf (2002, 2010); Thomson (2003: 128-31); Baggini (2004: 181-4).

[3] For interesting discussion of some real life figures, see Storr (1988). Note that this point also tells against Velleman's (1999) interesting account of love, as an arresting apprehension of another's value that tends to break down emotional barriers, construed as an account of meaningfulness. Scientists and artists clearly can find meaning in their activities without exhibiting this kind of emotional state or producing it in others.

The second property commonly deemed to be part of the essence of love is mutual aid, a condition in which people help one another for each other's sake and often do so consequent to sympathizing with one another. However, often the motivation of scientists and artists involves no awareness of others, whether sympathetic or not; in the first instance, many simply want to discover something about the world or to express an interpretation of something in a way that pleases their own eye or ear, regardless of how others may receive the findings.

Furthermore, it is implausible to think that the findings must be benefit-conferring on an audience in order for them to be meaning-conferring on the scientist or artist. Recall the objection to (UTM$_3$) made above, that discovering abstract natural laws could plausibly confer meaning on a scientist's life, even if doing so would have no technical application and even if they were not widely appreciated by laypeople, let alone the generally educated. If love 'means creating for another the space in which he might flourish, at the same time as he does this for you' (Eagleton 2007: 168), then, even if that is what many musicians in bands do, it is not what is alone meaningful about aesthetic creation or scientific discovery.

11.3.2 Culture

It is, of course, reasonable to think that loving others is part of a meaningful life, and even a large part, to be placed under the heading of 'the good', but it is implausible to think that it exhausts it. Indeed, there has lately been a strong movement in the field away from prizing loving and other personal relationships and towards deeming work of various kinds to be the most important source of meaning. One theory that I think goes too far in this direction is Joseph Mintoff's intricate view (2008). According to Mintoff, meaning, or at least great meaning, is a function of being attached to certain kinds of 'transcendent' ends. Transcendent ends, for him, are states of affairs that are not arbitrary in that they exhibit mind-independent goodness, are long-lasting in duration, and are broad in scope, roughly involve lots of people and not merely the agent. In light of this perspective, Mintoff's theory can be expressed this way:

> (NCT$_2$) A human person's life is more meaningful, the more that she is attached to transcendent things, where things are more transcendent, the more they are objectively valuable, last a long time, and affect many people.

According to Mintoff, this theory ultimately entails that '(t)he life of cultured leisure...one primarily spent in intimate appreciation of the *so-called* "higher" things in life, philosophical and scientific theories, but also literature and poetry, art and nature, etc...is the best way to make our lives more meaningful' (2008: 77). He presents three major reasons for thinking so.

First, helping people, or more generally acting morally with regard to others, is principally a matter of giving people what they need in order to do other things. Freedom, money, health, and an absence of pain are morally significant mainly because they are instrumentally valuable for being able to do much of anything with one's life. From this, Mintoff concludes that the mind-independent value of moral action is not as transcendent or high as that of other things, such as appreciation of beauty and truth, which moral action would enable.

Second, Mintoff argues that the objects of cultural appreciation are capable of both a kind of breadth and durability that the objects of moral action cannot be.

> Though we can influence only a small part of the world, we have the ability (through science) to understand the universe as a whole and how it works.... Further, the principles which occupy our scientific thoughts, and perhaps even the truths about human nature we may discern from literature, are of greater stability and permanence than anything we are likely to achieve (Mintoff 2008: 79).

In short, preventing people's woe is for nearly all of us (except for, perhaps, Bill Gates or the United Nations) a very local and temporary success.

Third, distinguishing between two ways one could relate to objective value, namely, 'doing good' and 'knowing good', Mintoff argues that knowing is generally a more intense way of relating than is doing. He points out that because craftspeople, politicians and readers of novels have an awareness of what is doable that is much deeper and greater than what they can in fact do, interpretation of a transcendent thing is likely to be a more intense relationship to it than is acting on it.

Based on these considerations, Mintoff concludes that while doing good, in the form of acting morally, can confer some meaning on life, knowing good would be best. One finds some similar ideas in the work of Hannah Arendt. She can be read as saying that the greatest meaning has come from putting valuable objects into the public realm that have remained there for many generations (Arendt 1961: 197-226). For her, culture is the height of human activity, and 'the only nonsocial and authentic criterion for judging these specifically cultural things is their relative permanence....Only what will last through the centuries can ultimately claim to be a cultural object' (1961: 202).[4]

(NCT_2) seems to entail and explain why the good, the true, and the beautiful are often sources of meaning in life. However, one serious problem with Mintoff's theory concerns the good; it excludes loving relationships altogether from the domain of meaning, an objection of which he is aware (2008: 80-4). Mintoff admits that '(s)eeking transcendence implies that one's path in life will climb from the warm lowlands of the personal to the chilly heights of the impersonal' (2008: 81), but he is willing to bite the bullet. To buttress his revisionist position, Mintoff points out that personal relationships such as friendships will invariably *accompany* meaningful projects, even though they will not constitute them, and he suggests that we should think of their value in terms of *worthwhileness*, rather than meaningfulness.

However, I find Mintoff's position too counterintuitive, here. It is worth taking seriously the way I read Eagleton and others, as suggesting the view that love exhausts

---

[4] For additional sympathy towards the idea of cashing out meaning in terms of what endures, see Ostwald (1906); Durant (1932: 105-6); Nozick (1981: 582-5); Partridge (1981); Kushner (1987: 15-26, 159-73); Baggini (2004: 33). For thoughtful criticism, see Trisel (2004).

meaning in life, only because love must constitute meaning to some noticeable extent. As Will Durant says in a characteristically bald statement, 'You will see that what you need is not philosophy, but a wife and a child, and hard work' (1932: 109). Durant himself finds the project of rearing children to be meaningful because they will outlive oneself, something Mintoff could conceivably also sign on to as something transcendent (viz., objectively valuable, permanent), but that rationale would fail to account for the meaningfulness of marriage, which is common-sensically preferred to a string of one-night stands because of its meaningfulness.

A second major objection to Mintoff's theory is that it fails to capture the respects in which the truth and beauty that he prizes are particularly meaningful. Recall the core of Mintoff's view: '*knowing* good is a better way to make our lives more meaningful' (2008: 76) than is doing good. That apparently entails that *creating* a work of art or a true scientific theory is no more meaningful than merely *apprehending* one, but surely part of the reason we find Einstein's, Darwin's, Picasso's, and Dostoyevsky's lives to be particularly meaningful is that they came up with their works, whereas we can merely take them in. It will not do for Mintoff to suggest that creating is a particularly intense form of knowledge; after all, his reason for prizing knowledge is that it is a more intense relationship with the transcendent than is any kind of non-cognitive activity (2008: 78). Mintoff notes that the 'life of cultured leisure...is more like the intimate and delightful contemplation of some grand temple than the tedious labours of removing graffiti from one of its walls' (2008: 79). However, he neglects to consider that more meaningful than both would be *constructing the temple in the first place.*[5]

My third worry about Mintoff's theory is that it has counterintuitive implications about the comparative meaningfulness of different kinds of knowledge. Much of the intellectual reflection that intuitively confers meaning on life is not a matter of knowing *good*, one of the central elements of a transcendent condition. When Mintoff explains the respect in which knowledge is a way to be attached to transcendent things, he discusses the transcendence of the object of knowledge, what it is that a person knows (2008: 77-9). And an object of knowledge is more transcendent, for Mintoff, the more it is objectively valuable rather than arbitrarily a function of one's dispositions, the more it is broad in scope rather than concerning just oneself or one's intimates, and the more lasting it is as opposed to fleeting. On that reading, however, the following objects of knowledge do not count as 'transcendent', so construed: knowing that action at a distance is real; knowing that $E = MC^2$; knowing that the Rutherford ('solar system') model of the atom is inaccurate. None of these objects of knowledge is itself objectively valuable.

In reply, Mintoff might relax the idea of value being essential for a transcendent condition, at least when it comes to knowledge. In that case, though, although knowledge of physics could be considered meaningful, it might be given too much weight.

---

[5] As Einstein is reported to have said, 'All meaningful and lasting change starts first in your imagination and then works its way out. Imagination is more important than knowledge.'

Knowledge of the universe is *extraordinarily* transcendent, as Mintoff himself points out (2008: 79). For just about *any* construction of a rich theory about the spatio-temporal world of sub-atomic particles, it will be *much* more meaningful than one about human nature, by virtue of the *massive* difference in extension and permanence. And I presume that is counterintuitive; the insights of Darwin, Marx, Freud, Weber, and the like are comparatively important.

11.3.3 Open-ended activities

I turn now to another theory that, like Mintoff's, plays up the work side of meaning and plays down the relationship side, while not entailing that love confers no meaning on life and more generally avoiding the objections to (NCT_2). Neil Levy (2005) argues that the greatest significance possible is a function of productive activity that involves striving to realize extremely desirable ends that cannot be fully achieved, and cannot be fully achieved for the particular reason that we can have no stable conception of what it would be like to do so. He explains:

> The practice of artistic creativity, when it is carried out at the very highest level, is paradigmatic of such an open-ended activity. We have only to think of how the avant-garde movements of the Twentieth century would have been perceived by earlier generations of artists to see at once how the ends of art themselves evolve along with the activities which aim to achieve them. Like the pursuit of the good and right, and the pursuit of truth, it is an inherently open-ended activity insofar as its ends are at stake within the activity itself. The ends of superlatively meaningful activities cannot be achieved, because as the activities evolve, so the ends at which they aim alter and are refined (Levy 2005: 186).

So, in the realm of the true, Levy suggests that we have no clear and distinct conception of what it would be to have complete knowledge of nature. Levy's claim is not merely that such a conception is 'too big', but that any apprehension we have of it changes as we obtain more knowledge. And when it comes to the good, Levy maintains that we cannot achieve a perfectly just world because we have no detailed conception of what it would be to live in one. As humanity makes progress towards a just world, for example, by transcending tribalism and accepting universal conceptions of moral status, our interpretation of the ethically ideal goalpost shifts. These remarks ground the following theoretical statement:

> (NCT_3) A human person's life is more meaningful, the more that she actively makes progress towards highly worthwhile states of affairs that cannot conceivably be realized because our knowledge of them changes as we strive to meet them.

One worry about (NCT_3) is that it, like (NCT_2), appears to exclude love altogether. To be fair, Levy purports to be putting forth a theory of what provides the *most* meaning in life, and so he is not intending to capture the meaningfulness of love, which he thinks is meaningful (*contra* Mintoff), but not superlatively so. However, I submit that Levy's ideas can plausibly be put to a broader use. One might reasonably suggest that the project of love has an open-ended structure similar to the other conditions that Levy has discussed. It is worth considering whether love is meaningful insofar as it is a relationship with another person that cannot be brought to full fruition because our knowledge of it changes as the relationship progresses. One has no stable view of what it would be like to have a *completely* loving relationship with another individual, because as one learns more about the other, oneself, and the nature of the relationship, and as the lovers adjust themselves to accommodate and please one another, one's idea of what to strive for to make the relationship better alters.

So, in contrast to Mintoff, Levy could include love in the realm of (superlatively) meaningful projects, as a plausible interpretation of his own theory apparently entails that love is (highly) meaning conferring. Furthermore, there is nothing in (NCT_3), unlike (NCT_2), to suggest that knowledge of the universe is always much more meaningful than knowledge of local conditions such as human nature; both could well be open-ended. Finally, Levy can account for the intuition that creating a theory or artwork is more meaningful than merely apprehending one that has already been created in that

> though ordinary meaning is available to almost all of us, through participation in the goods of family and the appreciation of art, through friendship and interaction with the natural world, *superlative* meaning requires much more: active engagement with projects. But engagement in a project, at a level which can secure sufficient achievements to confer superlative meaning, is available only to a few. For instance, only a very few of us can participate (as opposed to being interested spectators) in the project of the pursuit of knowledge (Levy 2005: 188).

On the face of it, then, Levy's theory obtains the advantages of (NCT_1) and (NCT_2), roughly, that both love and work are meaningful, but without their respective disadvantages.

One worry about (NCT_3) is whether moral achievement, intellectual reflection, and aesthetic creation, which Levy, with the field more broadly, takes to be quintessential sources of great meaning in life, in fact share the feature of open-ended activity. I have some doubt about that, particularly in light of the theorization of justice that has been undertaken in the past 40 years. However, I grant it here, so as to make a more incisive point.

Suppose, for the sake of argument, that Levy is correct that whenever the good, the true, and the beautiful confer superlative meaning on a person's life, the person has striven towards an end that we cannot conceivably realize because our understanding of it changes as we progress towards it. My claim is that it does not follow that this common feature is what *best explains why* they confer superlative meaning. A theory of meaning aims to provide not merely its 'supervening' conditions—that is, the properties that *co-vary* with meaning—but rather the factors that *constitute* it, which are of much greater theoretical interest. And my first real objection to Levy is that he has at best indicated certain features that *accompany* (great) meaning; he has not given an account of *by virtue of* what (great) meaning obtains.

To see this, consider why Mandela's life is important for having ended apartheid and done so in a way that tended to minimize the use of unjust means. The natural explanation of why this accomplishment has conferred great meaning on Mandela's life does not appeal to the idea that he made progress towards an end that we cannot conceivably realize, since our understanding of it alters as we advance towards its realization. Mandela himself was unlikely to be thinking that what he was doing was important for the reason that he was progressing towards a perfectly just world that he could neither conceive nor achieve, and we need not think of him as doing so in order to apprehend the great significance of what he did. One fails to capture Mandela's greatness by saying, 'He brought South Africa closer to perfect justice, although we have no clue as to what that might be, and in principle can never reach it.' Instead, achieving the 'closed-ended' goal of ending apartheid seems to carry within it the ground of superlative meaning.

Similar remarks apply to the true and the beautiful, for example, Darwin's account of the origin of the human species and Picasso's painting of *Guernica*. Their significance is not best explained by the fact (supposing it is one) that they progressed towards the realization of ends that we cannot conceive of reaching because, as we approach them, our conception of them changes; something else is doing the work.

There is a second problem with (NCT_3), namely, that a crucial part of it begs the question, or at least does not provide as much information as one might reasonably like. The question I am seeking to answer in this book is, 'Is there a single property, or small handful of properties, by virtue of which the variety of meaningful conditions are all meaningful, and, if so, what is it?' Answering that meaningful conditions all involve progress towards 'valuable' or even 'supremely valuable' goals (Levy 2005: 182, 184, 185, 186) does not adequately answer the question, and instead naturally invites asking, 'What is the basic thing by virtue of which meaningful conditions are (supremely) valuable?' I presume that most readers are like me in wanting to know more about the nature of the value that constitutes meaningful conditions. Is there any common denominator as to their value? It would be revealing if we could find a fairly specific answer to this question.[6]

11.3.4 Organic unity

Consider, now, a fourth theory, one that accounts for both love and work, that is fairly specific about what counts as objectively valuable, and that presents a plausible candidate for what is constitutive of meaning, as opposed to what merely accompanies it. This is the view articulated with most care by Nozick (1981: 574-619, 1989: 162-9). Recall that he can be read as claiming that the concept of meaning is captured by

---

[6] One might be tempted to suggest on Levy's behalf that an end is highly valuable just insofar as it is open-ended in the way he describes. However, that will clearly not do. The promotion of injustice, for example, has exactly the same open-ended structure as the promotion of justice (supposing the latter indeed has it).

the idea of transcending limits (2.4.2), and, so, naturally his conception of meaning is an account of the relevant limits and of how to transcend them. His basic idea is that meaning consists of positively relating to things apart from oneself that have intrinsic value, conceived in a certain way.

The particular things or causes people find make their life feel meaningful all take them beyond their own narrow limits and connect them up with something else. Children, relationships with other persons, helping others, advancing justice, continuing and transmitting a tradition, pursuing truth, beauty, world betterment—these and the rest link you to something wider than yourself....[M]eaning is a transcending of the limits of your own value, a transcending of your own limited value. Meaning is a connection with an external value.... [where] intrinsic value is degree of organic unity (Nozick 1981: 595, 610, 611).

When Nozick says that intrinsic value is 'organic unity', he is claiming that a thing's inherent properties are good for their own sake by virtue of forming a whole that brings together a diverse array of parts, i.e. being complex. All together, then, we get the following theory:

> (NCT_4) A human person's life is more meaningful, the more that she positively connects to intrinsic value, namely, a whole that integrates a high degree of differentiated elements, that is beyond herself.

This principle on the face of it well captures the meaningfulness of the good, the true, and beautiful. Nozick points out that a person is valuable insofar as it is a unification of a wide array of different experiences, beliefs, desires, emotions, and other mental states into a single self (1981: 417, 445, 460-9, 517-22, 594), which would account well for the intuitive meaning of having children, developing intimate relationships, helping others, promoting justice, and otherwise treating others in beneficent or moral ways. Furthermore, Nozick reminds us that important works of art are often construed as unifications of form, content, technique, tone, and so on into a single object (1981: 415-16), which would mean that his view successfully entails that the production and apprehension of artworks can be meaningful. And with regard to the true, or knowledge, Nozick could argue that it is valuable insofar as it is about organic unities such as galaxies, solar systems, ecosystems, people, animals, and plants.

There are three important counterexamples that drive me away from this powerful theory in search of another. Beginning with the beautiful, a focus on organic unity obviously has difficulty capturing important forms of art from the twentieth century. Certain forms of surrealist painting intentionally forgo the search for unity, instead highlighting the random and the otherwise illogical. Minimalism, too, eschews the attempt to unify a wide array of divergent factors, often using repetition in music and plain colours and simple lines in painting. Perhaps there is a way to view these movements in terms of organic unity—maybe their worth resides in being parts of a larger, historical narrative about art. I find it hard to believe that the value of these movements can be reduced to these ideas, but I leave further reflection for those better trained in aesthetics than I.

Moving from the beautiful to the true, the view that meaning comes from positive relationships with organic unities beyond oneself can account well for the importance of making certain contributions to the social sciences, which are about people, as well as to natural scientific fields such as biology, ecology, and astrophysics, supposing, as is plausible, that animals, plants, ecosystems, solar systems, and galaxies are indeed also organic unities. However, a stumbling block is the most abstract thought characteristic of metaphysics and the most basic natural science of physics. Knowledge of quarks, time, and action at a distance is important, but not because it is about intrinsic values *qua* organic unities. Nor are certain metaphysical theories about, say, causation, reference, or necessity, significant for this reason. The objects of these forms of knowledge are not organic unities, but are clearly relevant to their ability to confer meaning on life.

The most promising way to reply on behalf of Nozick,[7] I think, would be to grant that metaphysical and natural scientific enquiry is often not about any organic unity, but to put forth a revisionist position, namely, that such enquiry can constitute an organic unity *itself*. Nozick maintains that a theoretical belief just is an organic unity in that it synthesizes a diverse array of data into a single principle (1981: 417-18, 619-27). If so, then developing a *theory* about quarks and causation could confer great meaning on a person's life.

An apparent problem with this interesting suggestion is that it follows that a *theory about anything at all* would confer great meaning on life. Developing the theory of quantum mechanics confers more importance on a person's life than a theory of, say, which personal ads generate the most responses (which, by the way, has been done), but since they are both *theories*, they both confer meaning, by the present hypothesis.

Nozick may fairly respond that the theory of quantum mechanics unifies much more data than any theory of personal ads could, making the former a good candidate for great meaning, unlike the latter. Indeed, the search for a grand unified theory in physics appears to be very important, since it would be a theory of all physical phenomena and therefore the most comprehensive organic unity possible for a theory, supposing that only the physical exists.

The suggestion is worth considering, but does not work, in the final analysis; for this rejoinder oddly entails that just about any knowledge of physics and also chemistry is always more important than any knowledge of human beings, since the former sort will invariably unify more data than the latter, which will obviously be limited to the planet earth, one of the same problems with Mintoff's view. Surely the linguistic

---

[7] At one point, Nozick suggests that reality as a whole is also an organic unity (1981: 524–7). Setting aside my inclination to question that claim, I doubt that it is its purported organic unity that makes reality as a whole worth investigating. Surely it is true that, even if the natural universe did *not* form an organic unity, it would be important to discover certain facts about it.

theories of Saul Kripke and Hilary Putnam were important, for example, despite the limited amount of data they unify compared to those in the basic natural sciences or metaphysics. Overturning the theory of reference that had dominated philosophy for over 2,000 years was no mean feat (Putnam 1988: 19–41).

For the third counterexample to Nozick's theory, consider the good. Consider virtues or excellences of the sort discussed in the previous chapter (10.4). Connecting with complexity beyond one's person does not entail the meaningfulness of maintaining integrity in the face of difficult circumstances à la Nelson Mandela, Rosa Parks, and Steve Biko. Overcoming a tendency to exhibit weakness of will, and then exhibiting courage and determination, too, are aspects of good character that intuitively enhance the significance of one's existence, but that need not involve connecting with an external organic unity. Similarly, overcoming a neurosis such as narcissistic personality disorder would be a meaningful achievement.

Notice that the problem of there being internal goods such as virtue and excellence applies not merely to (NCT_4), but the other previous non-consequentialist theories as well. (NCT_1) focuses on love of others, (NCT_2) appeals to transcendent ends,[8] and (NCT_3) invokes activity directed towards achieving states of affairs the realization of which requires the input of many generations of human beings. In contrast, the remaining non-consequentialist theories promise to capture the internal dimension of meaning in life, and hence are not exclusively directed *outward*.

11.3.5 *One's rational capacities*

That said, one ought not swing entirely *inward* either, as Alan Gewirth, for one, does. Gewirth makes remarks suggesting that meaning is a function solely of realizing what is best within oneself, where this is a matter of the superlative exercise of one's rationality. Indeed, the standard way that philosophers account for the nature of internal goods such as virtues or excellences is to appeal to the idea that they are constituted by the exercise of one's rational nature. Gewirth does so as follows:

(S)piritual values consist in ideals of moral, intellectual and aesthetic excellence. They involve that one goes beyond one's narrow personal concerns of self-aggrandizement, that one in effect surrenders oneself to the pursuit of goodness, truth and beauty....The ground for calling such pursuits 'spiritual' is precisely that one goes beyond oneself, i.e., beyond concerns focused solely on oneself; one recognizes the demands of a broader moral, intellectual, and aesthetic culture. These demands are experienced as objective because they embody criteria of excellence that one does not make or invent but rather discovers....It may be thought that such bringing to fruition of one's worthiest capacities can serve as an adequate and even final answer to the question of the ultimate values of human life (1998: 177, 183).

---

[8] Responding to an earlier statement of this concern (Metz 2002a: 799), Mintoff suggests that something can confer meaning on one's life only if it is a transcendent end to be achieved (2008: 71). However, in the face of the above cases, this appears to be mere assertion. Standing up for what one reflectively believes to be right, or doing the hard work of ridding oneself of neurotic tendencies, are surely conditions worthy of great pride, purposes beyond one's pleasure well worth pursuing, and desirable ways of going beyond one's animal nature (even if they fail to involve lots of other people or something that will outlive oneself).

For Gewirth, the good, the true, and the beautiful are a matter of transcending one's animal nature by developing one's rational self, or going beyond an 'unimproved self to a self whose best capacities are developed' (Gewirth 1998: 182). These remarks suggest the following principle:

> (NCT$_5$) A human person's life is more meaningful, the more that she actualizes her best capacities, which capacities are abilities to reason in ways that meet objective criteria of distinction.

The invocation of objective criteria, or a broader culture, might suggest that Gewirth is not strictly an 'internalist' about meaning. However, Gewirth is naturally read as saying that although other-regarding conditions are necessary for meaning in life, what *ultimately explains* why a person's life is significant is the self-regarding condition that she has exercised *her* rationality.[9]

And that is too cramped an explanation of the good, the true, and the beautiful. I have encountered no reason to doubt that effects on or relationships with others are at least part of the basic explanation of why beneficent or more generally moral actions, for instance, confer meaning on life. So, for example, consider the meaningfulness of being a coach or a philanthropist. Although Gewirth's view can entail that such people have meaning in their lives by virtue of what they do, its explanation will be narrow, focusing *at bottom* on the way these people exercise their own rational faculties, and not also on the fact that they help others to do so.

The other problem with (NCT$_5$) is that it is of urgent interest to know what these objective criteria of distinction are, or whether they have a common thread to them. The natural question to ask of (NCT$_5$) is: 'What do the objective criteria of excellence have in common?' Retyping *The Brothers Karamazov* would not count as an excellent exercise of one's rationality, but why not? A theory that is more elaborate for enabling one to answer such a question is to be preferred to one that is not.

11.3.6 Creative rationality

The following theory enables one to answer such a question, while capturing the twin values of love and work, as well as avoiding both the Scylla of being strictly outward directed and the Charybdis of being strictly inwardly focused. It more or less suggests that the good and the true can be reduced to the beautiful.

> (NCT$_6$) A human person's life is more meaningful, the more that she develops her and other people's abilities to act rationally in creative ways.

Richard Taylor has been the foremost proponent of something close to this theory (1981, 1985, 1987, 1999),[10] although he is not often recognized as such. Taylor is best known for having advocated a subjective account of meaning in life, such that great

---

[9] For similar views, see Bond (1983: 119–22); Taylor (1985, 1987); and Thomas (2005).
[10] Taylor's view is actually 'internalist' or 'self-regarding' in the way Gewirth's is.

meaning could come merely from the satisfaction of one's strongest desires (9.4.3). However, Taylor eventually rejected this view and opted for this contrasting, objective account:

> A meaningful life is a creative one, and what falls short of this lacks meaning, to whatever extent. What redeems humanity is not its kings, military generals and builders of personal wealth, however much these may be celebrated and envied. It is instead the painters, composers, poets, philosophers, writers—all who, by their creative power alone, bring about things of great value, things which, but for them, would never have existed at all....(1999: 14).

On this view, the reason that retyping *The Brothers Karamazov* would not confer meaning is that it is adaptive, a matter of copying, as opposed to originating. Furthermore, it is natural to understand the meaningfulness of Darwin's and Einstein's lives to be a function of them having created novel theories, or of creatively appealing to data to reach new, insightful conclusions. And even loving relationships, Taylor points out, are naturally understood as being meaningful by virtue of the creativity involved, that is, what distinguishes 'the mere begetting of children' from transforming them 'to well-functioning, happy adults' (Taylor 1985: 119). These are all respects in which individual agents are creative themselves, but I note that (NCT_6) entails that a coach or philanthropist would have meaning in their lives also by virtue of helping others to be creative.

Critics will be tempted to suggest that murder can be creative and yet not confer meaning; serial killing à la the film *Seven* might be a case in point. However, Taylor, like all non-consequentialist theorists, can and should incorporate restrictions on his theory of the sort explored above in the context of (UTM_5) (10.3). He can fairly say that meaning comes from creativity that does not degrade people in particularly serious ways.

Even if Taylor's theory can rightly exclude serial murder as conferring meaning on life, it has real trouble including patent sources of meaning in the moral realm. Consider, for instance, tending to the sick by changing bandages, cleaning bedpans, and alleviating pain in the manner of Mother Teresa. She (or the stereotypical view of her) engaged in no particularly original behaviour (cf. Martin 2002: 207). Or at least it was not *that* facet of her behaviour in virtue of which she obtained meaning in her life; instead, what appears key are the facts of having so much compassion towards others, being devoted to helping them on that basis, and exhibiting the strength of will required to live among revolting conditions and miserable people. So, while I am inclined to retain the idea that the exercise of reason is relevant to making one's life significant, we must broaden the way it could be done.

11.3.7 Theoretical and practical rationality

I now return to the perfectionist theory of Quentin Smith and related views of Thomas Hurka and also Bertrand Russell. In the previous chapter, I rejected Smith's theory because of its consequentialist structure (10.4-10.5), but did not take up its substantive account of what it is to exercise reason in meaningful ways. It is worth considering the latter ideas set in a non-consequentialist theoretical framework.

Recall that Smith holds that central to meaning is the development of the nature of not merely oneself, but also others, where the relevant, highest nature includes the ability to reason. Rationality is the highest nature, since a being with it includes the natures of other beings, though other beings do not include its nature. A human person encompasses the natures of all other kinds of beings in the animal, vegetable, and mineral kingdoms, while also exhibiting the capacity to reason. Human nature is the highest, then, because it is the most inclusive, while also being unique for its rationality.

Smith distinguishes between two basic capacities for reason, theoretical and practical, where theoretical reason is 'the capacity to explain why things are and why things are the way they are. The maximal development of theoretical reason is to know the reason why *everything* exists and why *everything* possesses the properties that are in fact possessed' (1997: 212). And with regard to practical reason, Smith says:

> The phrase 'practical reason,' if it refers to a distinctively human essence, cannot refer just to the ability to know the best means of attaining one's goals, for some nonhuman animals have this ability as well.... Distinctive to humans is the ability to know the most effective means to attain the *morally best goal*.... that is, bringing about the greatest development of things' natures (1997: 213).

Let me bring these ideas together to form this non-consequentialist version of perfectionism:

> (NCT_7) A human person's life is more meaningful, the more that she develops her and other people's natures, with more meaning accruing to her life, the more beings whose natures she develops and the higher their natures, where the highest nature is one that is most rational for being most able to explain features of the universe and to develop higher natures.

The reader must keep in mind the implicit non-consequentialist approach to the realization of selves, here. For instance, the friend of (NCT_7), unlike that of (PT_2), could maintain that exhibiting reason in oneself can be particularly meaningful relative to helping others to exhibit it to a greater extent.

By (NCT_7), a particularly meaningful life would be one that enables lots of human people, including oneself, to realize distinctively human abilities to reason in theoretical and practical ways, which involves them explaining what goes on in the physical world and acting in ways that help people to develop themselves. As Smith points out, such a theory captures many of our intuitions about lives that uncontroversially have great meaning in them. In the realm of science, Smith mentions Galileo, Darwin, Einstein, and Hawking as key examples, and with respect to morality, Smith invokes Florence Nightingale, Gandhi, and Martin Luther King (1997: 212).

I turn now to problems with Smith's view, starting with concerns about the meaningfulness of beneficent action. Smith explicitly deems loving and other personal relationships to be at the animal level, and not distinctively human capacities (1997: 201), which would mean that they are not particularly meaningful.

I suspect, however, that this is merely a misapplication by Smith of his own theory. The abilities to empathize with others, to sympathize with them consequent to empathy, and to act in helpful ways consequent to sympathy require intelligence of a sort that is most clearly exemplified in us, even if there are rudimentary elements in chimpanzees and other 'higher' animals. A broader account of human nature *qua* rational seems able to help Smith, on this one.

Turning to the beautiful, Smith mentions figures in music, literature, and painting, such as Beethoven, Virginia Woolf, and Georgia O'Keefe, but it is not clear that his theory permits him to do so, or at least provides the right explanation of why their lives were meaningful. After all, Smith maintains that there are only two basic kinds of reason—theoretical and practical—and it does not appear that artistic creation is an instance of either one.[11]

Smith would surely reply that creating works of art is a kind of practical reason, in that artists are developing their own natures and those of others insofar as audiences appreciate and interpret their works. Notice, though, that Smith cashes out practical reason in terms of moral behaviour, reducing the beautiful to the good. I submit that the best explanation of why the above figures, as well as my stock examples of Dostoyevsky and Picasso, had great meaning in their lives is not that they *pursued a moral end* better than others. Their creativity, as per Taylor, rather seems to be part of the basic explanation.

Finally, with regard to the true, Smith's theory suffers from the by now familiar problem of worshipping extension. Like Mintoff and Nozick,[12] Smith appears committed to lording the fields of physics and metaphysics over any other areas of enquiry. Note Smith's claim that theoretical rationality aspires to explain 'everything' (1997: 212), with the best theoretician being 'someone who discovers why the universe exists' (Smith 1997: 216). My intuition is that while such knowledge would no doubt be extremely important, certain knowledge of human nature can also be very important, *a not so incredibly distant second*. Darwin's articulation and defence of

---

[11] The beautiful is also a real stumbling block for related views such as Hurka's (1993) and Russell's (1912). For Hurka, developing rational selves is a function of expanding 'extent' and 'dominance', so that, roughly, people's beliefs and goals are more rational, the more objects towards which they are directed and the more they structure other beliefs and goals. Although writing a novel does require working in view of a whole (1993: 124) and subordinating tone, content, narrative, and many other considerations to an overall theme (1993: 125), a work of art can be quite limited and still great. Similarly, for Russell, a meaningful life is basically a function of the 'enlargement of Self', the broadening of 'our interests as to include the whole outer world' (1912: 158), which suggests the view that life's meaning is a matter of one's rationality being oriented towards all beings and all facts in the universe. Again, creating and interpreting important art need not involve that.

[12] And also Hurka and Russell. Hurka most prizes knowledge that is about 'all the objects at all the times in history' (1993: 115), while Russell remarks that 'we start from the not-Self, and through its greatness the boundaries of the Self are enlarged; through the infinity of the universe the mind which contemplates it achieves some share in infinity' (1912: 159).

the origin of the human species was limited in scope, to the planet earth, but was enormously important, as were Kripke's and Putnam's novel and plausible accounts of linguistic reference.

11.3.8 Rationality towards universality

In short, size matters with regard to knowledge, but not to the large extent that non-consequentialists have tended to think up to now. The last version of non-consequentialism that I address most approximates the theory that I develop in the Chapter 12, and it is grounded on some of Gewirth's additional remarks about what makes a life meaningful or, in his words, 'self-transcendent'. Self-transcendence, for Gewirth, is not merely a matter of Eagleton's, Mintoff's, Levy's, and Nozick's basic notion of connecting with something external to oneself (a beloved, a transcendent condition, an open end, an organic unity), which, I have argued, neglects the role of internal goods such as virtue. Like Taylor and Smith, Gewirth accounts for the internal facet of meaningfulness by focusing on the exercise of our rational nature. However, unlike Taylor, Gewirth does not restrict the relevant form of rationality to creativity, and, unlike Smith, Gewirth has a notion of rationality that is not so tied to extension. Consider his remarks:

> (T)here are criteria of control of nature, explanatory and predictive power, and generality that enable us to differentiate degrees of cognitive excellence and thereby to transcend the limits set by the restrictive purview of ordinary sense experience.... The arts provide comparable vehicles of self-transcendence....(T)hey enable us to move outside our narrow sphere of direct experience through compelling modes of artistic disclosure.... An especially eminent area of self-transcendence is found in the universalist moral saints and heroes who risk their lives in times of mortal danger to rescue innocent persons.... (1998: 178).

One way to tie Gewirth's remarks together is to suggest that he is thinking of self-transcendence in terms of a *universal* or *impersonal* orientation. Instead of focusing on one's pleasure or welfare, in the moral realm meaning comes from doing something that treats no relevant being as lacking a dignity, or that takes the interests of all into consideration. Instead of merely perceiving objects from one's own standpoint, or perhaps even the standpoint of a species with particular sensory organs, in science and philosophy an important contribution is a function of discovering laws of (human) nature, principles that are true for all (human) beings. And instead of expressing facets of oneself that are idiosyncratic or interpreting the world in ways that are parochial, when it comes to significant artwork, one creates or interprets objects that address 'universal themes', i.e., themes that could be appreciated by all human persons regardless of their culture, while positively affecting the sensibilities of human beings as such.

As the examples above should make clear, to act from a viewpoint that is universal is not necessarily the same as having an extensive scope, where extension is equated with spatio-temporal size or sheer number of facts. Even so, is exercising rationality from a universal orientation a necessary condition for meaning in life? One might plausibly think not, upon reflecting on love, which seems able to make one's existence significant while being limited in scope. Loving a spouse and children surely confers some meaning on life; one need not love, or even be prepared to love, the human species as a whole, so the objection goes.[13] And, indeed, some would doubt that it is even possible to love all human beings, and, still more, claim that love of the most desirable sort is unavoidably particular.

In reply, Gewirth might, first, remind the reader that love is a form of rationality, as I articulated above (7.2.2, 11.3.7); it takes intelligence in order to become aware of what others are experiencing, to exhibit sympathetic emotions, and to feel a sense of awe that overcomes fear and makes one vulnerable. Then, Gewirth could distinguish between whom it is that one loves and in virtue of what one loves them. It is impossible for the large majority of us to love literally everyone we come into contact with, or at least not intensely in the way we do in romantic relationships. So, love is normally particular with regard to those who receive our (intense) love. However, what it is that we love about intimates, Gewirth may propose, involves a universal orientation. Velleman maintains, for example, that

> loving a person for the way he walks is not a response to the value of his gait; it's rather a response to his gait as an expression or symbol or reminder of his value as a person.... Loving some but not others entails valuing them differently but not attributing different values to them....(1999: 371, 372).

What we love in another person is arguably her *humanity*, as it is expressed in her. Perhaps, then, love is a form of awareness and action done from a universal orientation.

I am not so sure about this response; one might have thought that when one loves another, what one loves is not what she shares with other persons, but rather what sets her apart from them (Fabry 1968: 43, 181). Furthermore, there are other cases in which the exercise of reason appears meaning-conferring and yet not to involve a universal disposition. First, recall Hooker's example of the sybarite (9.5), who might well have some meaning in his life by virtue of his resolute determination, thorough study and acquired ability to draw fine distinctions between aesthetically similar properties. Second, consider the suggestion that chess grandmasters have meaning in their lives (cf. Hurka 1993: 123-6), and, more broadly, that excelling in games and sports can be a source of significance. Third, think about those who grapple with and perhaps overcome mental illness; one should take great pride in having successfully battled against, say, anorexia or dependency. These cases roughly suggest that sophisticated rationality in itself can confer noticeable meaning, with nothing about universality, impersonality, or the like essentially figuring in the explanation.

---

[13] *Contra* Russell, who says, 'The impartiality which, in contemplation, is the unalloyed desire for truth, is the very same quality of mind which, in action, is justice, and in emotion is that universal love which can be given to all' (1912: 161).

I suspect, therefore, that the best way to run with Gewirth's ideas is to suggest the idea that the robust exercise of reason invariably confers some meaning, but that it would do so all the more if it were employed in a way that involves a broadening of horizons. Consider, then, the following principle, inspired by Gewirth's comments:

(NCT$_8$) A human person's life is more meaningful, the more that she employs her reason and in ways that transcend particularity.[14]

This is a view worth taking seriously, especially as transcending one's own viewpoint or idiosyncrasies appears to admit of degrees, whereas acting from a universal orientation probably does not. I think that this theory of what makes a life meaningful is the strongest one to be found in, or rather suggested by, the literature.

I have two major reasons for not resting content with (NCT$_8$) and instead seeking to develop a new theory that improves on it. First, although Gewirth does not give extension a large role to play in his theory (so to speak), the emphasis on universality seems to have counterintuitive implications about which kinds of intellectual reflection are meaningful in comparison to others. So, a chemical theory about the nature of stars would be true of conditions that apply to all agents throughout the universe, and so would appear to be less 'particular', in one straightforward sense, than a theory of how human beings evolved on earth. Or a physical theory that abstracts from our sensory organs and so could be appreciated by non-human interlocutors would appear to be more transcending of particularity than a theory that is informed by facts about colour or sound, senses that other beings might not share. In short, the logic of Gewirth's principle, like that of several others, drives us to prize the most basic natural sciences *significantly* more than the biological and social ones. Even if the former kinds of knowledge are more important, it is doubtful that they are *much* more so, as I have contended above.

My second reason for aiming to do better than (NCT$_8$) is that there are forms of rational behaviour intuitively beyond a particular orientation that intuitively do not confer meaning on life, which indicates that Gewirth's ideas need some sharpening. In the realm of the true, consider a person who seeks out knowledge of non-causal correlations between events, perhaps ones that are present in the lives of all human beings. An enquirer who sought to discover, for example, as many mere co-incidences as she could, would, for all that has been said, be using her reason to apprehend universal truths, facts applicable to all human beings, or even to intelligent species generally.

With regard to the good, think not of a Mandela who vigorously strove to overcome apartheid, but rather to ensure that everyone's toenails were regularly trimmed or that no one had bad breath. Here, he would be acting in cognizance that everyone has a moral status, and would be aiming to satisfy universal interests, ones far beyond his own, but comparatively little meaning would have accrued to his life.

---

[14] This idea can be gleaned from some of Russell's remarks, too (1912: 153–61). There are times when he focuses less on the size of the object with which one is interacting and more on the overcoming of a narrow or private orientation.

Finally, consider the act of writing a novella about losing baby teeth or sweating or excreting, or think about making a painting that represents dust, where these are not the occasion for reflecting on intuitively 'deeper' topics. These are 'universal themes', literally construed. But it would not be great art (or at the very least not by virtue of its object), art that would make one's life particularly more meaningful for having created or apprehended.

11.4 Towards a new non-consequentialism

Although I have provided reason to doubt that we should accept the principle inspired by Gewirth as it stands, it is comparatively attractive for avoiding many of the pitfalls of the others, while retaining much of what is attractive in them. (NCT_8) offers a specific candidate for what constitutes meaning, not resting content with vague appeals to 'highly valuable ends' (NCT_3). It also avoids the idea that meaning is a function solely of something internal (NCT_5), as it is the view that for reason to confer substantial meaning, it must be employed in a way that transcends particularity, is roughly oriented beyond oneself and towards everyone in some respect. It also avoids the other extreme, that meaning is a function solely of connecting with something external to oneself ($NCT_{1,2,3,4}$), and instead, by grounding meaning in the exercise of one's rationality, captures the respects in which internal conditions such as virtues can confer meaning on life. It is, furthermore, not narrow in the way that the other rationality-based theories are; for instance, it does not reduce the relevant form of rationality either to creativity (NCT_6), which fails to capture the good, or to theoretical and practical dimensions (NCT_7), which fail to capture the beautiful.

Summing up, (NCT_8) appears to be on the right track to entailing and plausibly explaining what the good, the true, and the beautiful all have in common and of providing more generally an attractive account of what constitutes meaning in life. In the following chapter, I aim to articulate a theory that retains what is appealing about (NCT_8), while avoiding the two objections I have made to it, particularly the worry that universality is insufficient to differentiate uses of reason that confer great meaning from those that do not.

12

Objectivism III: The Fundamentality Theory

'Must burrowing into the ground
Dig oneself into a hole?
When asked by a frown looking down
"For what goal do you dig there, mole?"
"Depth" was what he sounded out,
The gold one knows as the soul.'

Gabriel Zenk

12.1 Time to get deep

In this chapter, I bring together the discussion in all three parts of the book. Here, I articulate a new theory of what makes a life meaningful that purports to avoid the major disadvantages of all the other theories canvassed, while retaining their key advantages. I start the chapter by reminding the reader of the basic strategy I have employed to assess theories and of the salient pros and cons of the ones I have examined (12.2). After that, I present an initial statement of what I call the 'fundamentality theory', which takes fairly literally the idea that considerations of meaning in life are a matter of deep or profound concerns, which contrast with superficial interests (12.3). Like the other theories of life's meaning in the literature, my first pass at the fundamentality theory addresses *pro tanto* and part-life facets of meaning alone. I begin to demonstrate that an appeal to deep facets of human life grounds the most defensible principle available, granting my previous appraisals of rival theories, before providing a fuller statement of the fundamentality theory in the next section (12.4). There, I bring in additional considerations such as the whole-life facets of meaning, the nature of a life that is meaningful on balance, and, of particular interest, the essence of anti-matter, something that the literature has yet to address at all. I conclude by considering objections, some of which I am not yet sure how to deal with (12.5); I aim to be frank about the limitations of the fundamentality theory, suggesting avenues for the next major stage of systematic analytic enquiry.

12.2 Desiderata for an attractive theory

In Chapters 5 to 11, I have critically explored a variety of theories of meaning in life, attempting to capture in as few properties as possible what all the conditions that confer meaning on life have in common. More specifically, these principles have focused on *pro tanto* and part-life facets of the final good of meaning in life; they have been principles aiming to account at bottom for all those aspects of a life other than its pattern that can make it somewhat more meaningful. In this section, I pull together the main conclusions to be drawn from my critical analyses of these theories; that is, I adumbrate the weaknesses in them that an alternative, more promising theory should avoid, as well as the strengths in them that it should also exhibit.

In the second part of the book, I have considered supernaturalist theories of meaning in life, according to which for one's life to be at all meaningful, there must exist a spiritual realm that one relates to in a certain way. I argued that it would be incoherent for most readers to adopt any supernaturalist theory, because they know that some lives have meaning in them but do not know whether anything spiritual exists. That argument, I pointed out, also applies to non-naturalist theories of meaning in life, and drives enquirers towards a naturalist account, which alone coheres with the metaphysical views of which the field is truly confident. According to naturalism, a life can be meaningful solely in virtue of physical properties, a theoretical perspective that, I highlighted, need not and should not deny that life could be *more* meaningful if God or a soul existed and one related to them in certain ways. An attractive theory of meaning in life ought to account for the respect in which supernatural conditions could add meaning, even if they are not necessary for it (Desideratum #1: Spiritual Realm).

I distinguished two forms of naturalism—subjective and objective—where subjectivism is the view that a person's life has meaning if and only if she obtains the objects of her (or her group's) propositional attitudes. Upon exploring sundry versions of subjectivism, I concluded that no form of it can avoid deeply counterintuitive implications about what can and cannot confer meaning on life. I rejected not only the claim that subjective conditions are sufficient for meaning in life, but also the claim that they are necessary for it; a Mother Teresa who is not subjectively attracted to her projects in any sense can have some meaning in her life simply by virtue of the good she does. Nonetheless, I accepted that exhibiting certain mental states can enhance meaning in one's life. An attractive theory of meaning in life ought to account for that judgement (Desideratum #2: Subjective Conditions).

In contrast to the overwhelming majority of those who believe that subjective factors are at least partially constitutive of meaning in life, I noted that it is plausible to think that con-attitudes such as hating injustice can also be meaning-conferring. I also argued that anti-matter exists, viz., conditions of life that reduce its meaning beyond merely failing to enhance it, and I pointed out that it has yet to receive a theoretical analysis. An attractive theory of meaning in life ought to account for the respects in which judgements of meaning in life involve negative elements such as these (Desideratum #3: Negative Conditions).

Turning to objectivism, I distinguished three forms, and pointed out that the most influential form, according to which meaning arises from subjective attraction to objective attractiveness, is too vague. A suitably rich theory of meaning in life would indicate what counts as objectively attractive, and this would require developing another theory altogether. The second major sort of objectivism that I explored was consequentialism, roughly according to which one's life is more meaningful, the more one makes the world a better place, which, in its utilitarian form, is a matter of making those within the world better off. Although I provided strong reason to doubt that improving people's quality of life is the only sort of final value the promotion of which is meaning-conferring, I could find no reason to doubt that improving people's quality of life or more generally promoting final value, at least in certain ways, could enhance the meaning of one's life. An attractive theory of meaning in life ought to account for that kernel of truth in consequentialism (Desideratum #4: Good Consequences).

I also argued against both welfarist and non-welfarist forms of standard consequentialism by contending that there are constraints on the way that meaning can obtain from promoting desirable results, constraints that have a moral dimension to them. An attractive theory of meaning in life ought to account for the intuition that certain kinds of particularly degrading behaviour undercut the meaning-conferring power of the good consequences that they bring about (Desideratum #5: Moral Constraints).

I then addressed constrained forms of consequentialism according to which meaning in life is a matter of promoting final goodness in the long run as much as one can, however one can and wherever one can, albeit without violating restrictions against doing so in ways that are particularly degrading. Against these views, I pointed out that they fail to accommodate firm intuitions that more meaning comes from exhibiting certain kinds of final value, such as virtue and excellence, in oneself, as opposed to merely producing it in others, and from one's actively bringing it about in others, as opposed to doing so by means of any causal mechanism whatsoever, such as pressing a button or writing a cheque. An attractive theory of meaning in life ought to account for these agent-relative aspects of meaning (Desideratum #6: Agent-Relativity).

Next, I critically appraised the third major form of objectivism, non-consequentialist theories that there are certain ways of living in a purely physical world that can be sufficient for meaning, and not merely by virtue of promoting desirable long-term consequences somewhere in the universe. Among the key conclusions were that meaning is constituted neither by exclusively internal or self-regarding conditions, such as virtue, nor by exclusively external or other-regarding conditions, such as connecting with a beloved, a transcendent end, or an organic unity. An attractive theory of meaning in life ought to account for both the intrinsic and relational facets of how the self can obtain meaning (Desideratum #7: Internal and External).

Finally, I also agreed with other thinkers that the exercise of our rational nature lies at the heart of meaningfulness, but contended that it should not be construed in a narrow manner, as solely a matter of loving others, for example, or exhibiting creativity. I instead suggested that the use of intelligence in general is meaning-conferring, but that it makes one's existence notably significant when directed towards certain objects, perhaps ones that transcend particularity. An attractive theory of meaning in life ought to account for the ideas that the use of one's reason appears to be necessary for meaning in life, but that real meaning comes when the reasoning process is dynamic, viz., sophisticated, creative, and unblocked by temptation, addiction, or emotional weakness (Desideratum #8: Deliberation and Decision) and when it is directed away from particularity and perhaps towards (human) nature as such in some way (Desideratum #9: Object of Rationality).

I have arrived at these nine desiderata by making arguments, that is, by appealing to premises that give one reason to believe conclusions comparatively more controversial than they. Often these premises have been in the form of intuitions—that is, judgements that a particular condition is meaningless or not, or meaningful to a certain degree—and they have appealed to features of what I have been abbreviating 'the good, the true, and the beautiful', which a large majority of those writing on meaning in life take to be exemplary instances of it. In particular, certain kinds of morality, enquiry, and creativity are widely deemed to be clear instances of great significance, and so I have argued that any theory that cannot accommodate those judgements is to be rejected, relative to one that can.

In the rest of this chapter, I articulate and defend the fundamentality theory, a new account of what constitutes meaning in life. I aim to show that the idea of contouring reason towards fundamentality can account for the meaningfulness of the good, the true and the beautiful better than rival theories. By the end of the chapter, I aim to have established that, unlike any other principle in the literature, the fundamentality theory exhibits all nine of the above desiderata.

12.3 The fundamentality theory

I proceed by presenting the fundamentality theory in three major instalments, beginning with the most basic version and then refining it as necessary to deal with additional complexities. In particular, I begin with a version of the view that, like nearly all other theories, addresses solely part-life and *pro tanto* facets of meaning in its positive dimension. Further on in the chapter, I address what the salient whole-life facets of meaning are, when a life counts as meaningful on balance, and how to capture theoretically the negative dimension of anti-matter.

12.3.1 The basic idea

Here is the initial, most basic statement of the fundamentality theory of meaning in life:

> (FT_1) A human person's life is more meaningful, the more that she employs her reason and in ways that positively orient rationality towards fundamental conditions of human existence.

I explain later what I mean by 'fundamental conditions', and start with the other technical aspects of the above principle. By 'reason' and cognate terms such as 'rational nature' I mean something broad, namely, all intuitive facets of intelligence of which human beings are characteristically capable and animals, even higher ones such as chimpanzees, are not. The non-rational self amounts to those facets of a person's existence that are shared with animals, for example, the maintenance of life, the instantiation of a healthy body, the experience of sensory perceptions, and the feeling of pleasure. The friend of (FT_1) can agree that these are final values, but she denies that they are sources of meaning, which requires rationality.

In the first instance, the latter is a matter of certain kinds of cognition and intentional action, as Smith, Hurka and others emphasize (11.3.7). Here, the reader need merely recall the discussion of rationality in the previous chapter, where I considered behaviours such as these: explaining why things exist and have the properties they do; appealing to ideas that account for many other ideas; drawing subtle distinctions; overcoming neurosis; playing games and sports; loving others; advancing justice; and engaging in creative behaviour. There are, without doubt, more (basic) features of rational reflection and volition that philosophers, psychologists, and the like could name.

The way I construe 'rational nature' is, however, even more inclusive than this. I deem such a phrase to signify not merely cognition and intentional action, but also any 'judgement-sensitive attitude', a useful phrase from T. M. Scanlon (1998: 18-22). A variety of propositional attitudes, and not merely intention and action consequent to it, can be responsive to deliberation, and so count as 'rational' in my book. Insofar as conation, emotion, and even affection are under our (perhaps indirect) control and can track cognitive appraisals of value to a greater or lesser degree, I consider them to be part of our rational nature.

So understood, the fundamentality theory is not vulnerable to the initially tempting objection that it is an overly intellectual theory, one that neglects the emotional or more generally sentimental facets of meaning in life. Rationality is not merely deliberation and decision, in my view, or at least that is not the way I use the term. Instead, what some might call 'non-rational' facets of our nature are in fact suffused with, or otherwise dependent on, judgement in a way that is clearly different from animal sentimentality and so counts, for me, as part of our rational self. Desiring not to smoke upon judging it to be harmful, feeling joy upon empathetic awareness of a loved one's success, liking a work of art, and being pleased that one has successfully educated students are all attitudes that either include rational cognition in them or can be affected by it. So, while I still maintain that pleasure *qua* pleasure is beyond the pale as a candidate for meaning in life (2.4), pleasure as something responsive to deliberation and decision is a good one.

For criticism of my use of language here, consider this claim: 'Some would begin by describing romantic love, athletics and music as rational spaces. This is similar to the economist who says that all we do can be translated into a quantified game situation. No. It is not reducible to that' (Boylan 2008: 61). I agree that these sources of meaning are not reducible to the ideas of cognition and intentional action. However, insofar as they confer meaning on life, they do involve these two facets, and, furthermore, their affectionate and emotional facets can be regulated by judgements of worthiness.

As I have discussed, love is a relationship that requires a sophisticated degree of intelligence (7.2, 11.3), and meaning comes from loving what is judged to be worthy of it—say, a person—and not the torture of her (9.4).

Now, (FT$_1$) includes the idea that one's mental states, insofar as they admit of rationality, ought to be 'positively oriented' towards a certain condition. This phrase connotes familiar ideas about pro-attitudes (9.3.1). Recall that propositional attitudes generally admit of a valence—that is, either a positive or negative dimension—depending on the nature of one's wishing in regard to their object. So, a negative attitude is one that includes a wish that the object not obtain, with key examples being disliking, feeling bad about, avoiding, and destroying. And a positive attitude is one that includes a wish that the object obtain, and is exemplified by wanting, appreciating, and pursuing. The fundamentality theory implies that the more judgement-sensitive attitudes that one positively orients towards fundamentality, the more meaningful one's life.

That said, I acknowledge that cognition and intentional action are the *central* attitudes, by which I mean that their absence would rarely be compensated by the presence of other attitudes, and their presence would usually compensate for the absence of other attitudes. Consider the meaningfulness of help. It would not be enough for real meaning, say, merely to want others to be helped and to be pleased upon seeing them helped; one must, of course, do some helping. And doing some helping would confer a decent share of meaning, even if one did not particularly want to help others and was not pleased that one had (10.2). Even though deliberation and decision (thought and action, theory and practice) play a major role in making a life meaningful, it would be incorrect to think that they exhaust the rational responses one can exhibit so as to obtain meaning.

(FT$_1$) can already be seen to satisfy Desiderata #2 and #8, and to have done so by appealing to an underlying, shared property. Desideratum #8 is the point that a theory should account for the judgement that rational reflection and volition come in degrees and that when they are, say, fairly informed, intricate, skilled, and immune to mental illness, they can confer substantial meaning on life, even when not directed towards an intuitively desirable end. (FT$_1$) so accounts, in that it strongly recommends, but does not require, that one positively orient one's rational nature towards fundamentality. And Desideratum #2 is the idea that a theory ought to accommodate the judgement that exhibiting propositional attitudes in response to certain conditions can enhance meaning in life, even though they are not necessary for it. The fundamentality theory accounts for both points by considering them to be about different aspects of one's rational self.

Consider how appealing this account of the meaningfulness of subjective conditions is, when compared to rivals in the literature. On the one hand, friends of subjectivity have often been vague in characterizing the relevant responses, speaking of 'being subjectively attracted to', 'being attached to', 'identifying with', or 'engaging with' worthwhile projects. On the other hand, when friends of subjectivity have been fairly specific about the relevant attitudes, taking care to specify that they are focusing on desires, goals, or emotions such as love and enjoyment, and spelling out their elements, they have nearly always failed to motivate why that particular attitude is relevant and others are not.[1]

In contrast, the fundamentality theory is specific about the meaning-conferring attitudes, namely, ones that can be rationally controlled, and it is attractive in two respects. First, it includes a wide variety of them. If wanting something and getting it can increase meaning in life, why not also obtaining the object of a goal one has set, or recognizing that a certain state of affairs would be meaningful, or appreciating the fact that one has realized it? Second, the fundamentality theory powerfully explains why all these various attitudes are relevant: *the more aspects of one's rational self that are positively directed towards a certain object, the more meaningful one's life*. In short, (FT_1) unifies much of the scattered discussion of subjective attraction in the literature, and does so on the basis of a plausible principle.

In earlier work (Metz 2003: 67-9), I proposed a different comprehensive view, according to which the relevant subjective factors are a function of one's self *in general*, and not merely one's *rational* self. Mintoff has recently suggested a version of the former view, according to which the relevant subjective conditions are forms of attachment. 'Attachment...is a diluted form of personal identity: X is attached to something else Y if....X has "enough" identity-relevant connections to Y' (2008: 70), such as X remembering, causing, and liking Y.

The difference between Mintoff's view and my current one is that, for him, non-rational aspects of one's identity can be conferring and detracting of meaning, something that I deny. To see the difference, consider a medical doctor who is an incurable workaholic, i.e., motivated to help others largely out of an ineradicable neurotic compulsion. For me, the respects in which the doctor's actions are a product of non-rational factors reduce the degree of meaningfulness they otherwise could have had, but this need not be so for Mintoff, since the doctor's identity as a person just is to be a self-aware individual moved by such factors.

In addition to accommodating Desiderata #2 and #8, (FT_1) can already been seen to account for Desideratum #6, the agent-relative thesis that a theory of meaning in life is better for entailing not only that some final goods confer the most meaning when realized in oneself as opposed to another, but also that producing them beyond oneself confers the most meaning when one does so in certain, active ways. (FT_1) captures the first point about self-regard by focusing on one's *own* rational nature, and not rational nature anywhere it may be. Helping others to exercise their rational nature can be a way to use one's rational nature in a meaning-conferring way (as will become clear below), but (FT_1) should be understood to imply that one's primary focus with regard to meaning should be the exercise of *one's* rational nature. Exhibiting virtue in oneself is *ceteris paribus* more meaningful than enabling someone else to do so.

---

[1] Notable exceptions include Starkey (2006) and Mintoff (2008).

And the idea that using one's reason comes in degrees, particularly with regard to deliberation and decision, accounts for the other point that some mechanisms by which one can affect the world are more meaning-conferring than others. To be sure, writing a cheque is not something that a rat or monkey can do,[2] but it is not a project that is particularly complex or sophisticated or that involves much strength of will or dedication. The non-consequentialist intuition that how one (permissibly) brings benefit to others makes a difference is well explained by the idea that doing so ought to be a function of dynamic rational behaviour.

In order to demonstrate that (FT_1) also accounts for Desiderata #1, #4, #7, and #9, I need to spell out its other major idea, the concept of the fundamental conditions of human existence. The basic idea of fundamentality, or depth, is that of conditions that are largely responsible for many other conditions in a given domain. In the first instance, I have a metaphysical relation in mind, such that an event or power is fundamental insofar as it principally is what brings about many other events. Metaphysically basic objects are substantially instrumental for and contributory to many other objects within a certain realm.

Sometimes I speak in epistemological terms, so that a judgement (or proposition) is fundamental insofar as it accounts for many other judgements in a given context. I expect that, at least most of the time, the metaphysical and epistemological relations co-vary, so that a fundamental event is one that is the proper object of an explanatorily basic judgement. However, it is conceivable that the two relations sometimes come apart, and, in that event, I take the metaphysical relationship between conditions to be key.

A fundamental condition should not be conflated with a necessary condition. A necessary condition of X is something that is required in order for X to obtain, whereas a fundamental condition of X is something that is responsible for the obtaining of X. Not every necessary condition is a fundamental condition. For instance, the fact that no asteroid has wiped out the human race is a necessary condition for a wide array of aspects of human existence, but it is not a fundamental condition of them, as it is not the main cause of a wide array of them (and nor does it explain them). Conversely, it seems that a condition could be fundamental for X and yet not be necessary for it. It is conceivable that a condition has caused X, but that it was not the only way X could have come about.

Now that I have spelled out the concept of fundamentality, I need to specify the relevant domains of it. (FT_1) prescribes positively orienting our rational nature towards conditions of human existence that are largely responsible for many of its other conditions. More specifically, the relevant conditions of human existence are threefold: those of *a living human person, human life as a collective* and *the environment in which humans live*. Let me illustrate by explaining how the positive orientation of one's reason towards such conditions figures into the good, the true, and the beautiful.

---

[2] But note that pressing a button on a result machine would be!

12.3.2 The good

I start with the good, the meaningful exemplar of which is outstanding moral achievement. Recall that, intuitively, great meaning was conferred on Mandela's life by virtue of having sacrificed so much to overcome apartheid and on Mother Teresa's life by virtue of having acted so compassionately with respect to large numbers of people in wretched conditions. In contrast, their lives would not have been notably important had they striven to ensure that everyone's toenails were regularly trimmed or that no one suffered from bad breath, even if these conditions were universally desired (or needed!). Why are the former plausible candidates for substantial significance, while the latter are not?

My proposal is that freeing people from discrimination and tyranny and providing them urgent medical assistance are forms of positively orienting one's rationality towards conditions of a characteristic human being's life that are responsible for much else about her life, of two sorts. First, they are positive responses towards others' rational nature. Rational nature, particularly as it is manifested in autonomous decision-making, is largely responsible for much about the course of a normal human's life. Promoting equality and liberty, and improving others' health, are substantial, constructive responses to people's reflecting and their choosing consequent to it, whereas trimming toenails and freshening breath have little, if any, bearing on that. The more intensely and widely one uses one's reason to foster and otherwise respect people's reason (along with exhibiting subjective factors such as being moved for the right reasons and taking pride in what one has done), the more meaning that will accrue to one's life with regard to the good. Such is a plausible explanation of the accomplishments of Mandela and Mother Teresa. Consider how much stronger the fundamentality explanation is compared to the most attractive rival considered at the end of the previous chapter, Gewirth's idea of satisfying a universal interest, which could well include toenail trimming.

Rational nature, prized by Kantian ethicists, is one large determinant of an individual's life, but what I am calling 'sharing' is another (cf. Metz 2010: 59-60). The basic idea is that, although a person is largely responsible for much about her life by virtue of her internal deliberations and decisions, her life is also a substantial function of her relationships with others. Those in the communitarian and feminist traditions are fond of pointing out how much our self-conceptions, choices, and lives in general are a product of various kinds of sharing, whether it be a common language, a culture in which one is embedded, a sense of identity with others (in which one thinks of oneself as part of a 'we'), a bond in which others sympathize with one and care for one's needs, or a variety of collective projects, practices, and rituals in which one takes part. To explain who a human person is requires not merely appealing to her deliberations and decisions, but also from where she has come from and with whom she has lived. Again, a good account of why toenail trimming would not confer much meaning on a life is, roughly, that it is not a positive response to people's heritage, to their continuity with others.

Now, another respect in which at least Mandela's life was meaningful in the realm of the good concerns the way his rational nature was positively directed towards shared conditions of life, as above. The colonial and apartheid systems that he fought against were substantially forms of splintering, ways of destroying, and otherwise degrading shared lives. For example, students were forced to learn Afrikaans and not permitted to use their native tongues when in school and when dealing with government officials. Traditional African cultures were not merely ignored, but denigrated as inferior to white ways of life. People were racially segregated not merely spatially, but also in terms of marriage, thereby dividing people into 'us' and 'them'. Families were split apart, with black households required to live in certain areas and breadwinners then forced to travel long distances to white areas in order to work. Finally, many organizations and other large-scale gatherings in civil society were banned or heavily regulated.

Beyond accounting for intuitively great instances of meaning in the realm of the good, the fundamentality theory also captures more everyday respects in which people obtain meaning from beneficence. Reasoning and relationships as objects to be promoted and more generally honoured explain much of the significance of romantic love, friendships, and child-rearing. To spell out the latter, consider how helping a child to become a healthy adult is both to nurture his capacity to make decisions for himself and to maintain a relationship that will largely determine who he is. Taking care to ensure that one's child avoids mental illness, supporting him so that he has a font of self-esteem to draw upon, educating him so that he knows how to read and write, teaching him how to navigate difficult situations, and helping him realize his goals are all ways to support a child's rational nature. And ways of upholding a child's shared life include being in the same house with him continuously throughout his childhood, spending time in which one attends to him and cares for his needs, enjoying rituals such as reading before bedtime or making pancakes every Saturday, and providing a context of values held by other members of the family.

A critic could object that moral achievement or beneficence confers meaning not merely insofar as it protects, develops, or otherwise respects people's reasoning and relationships, but also insofar as it enhances their quality of life. Relieving people's pain, say, could make one's life very important, and arguably explains part of Mother Teresa's importance.

In reply, often judgements about when someone's life is going poorly are a function of more basic judgements that her decision-making is frustrated in some way,[3] which better capture the significance involved. For one, when people lack food, health care and education, and when they are in pain, their voluntary choosing is thereby stunted.[4] For another, making competent adults well-off against their considered judgement, viz., in hard paternalist fashion, is intuitively not something capable of conferring meaning on a person's life, which is best explained by the idea that what really matters is helping

---

[3] Or that a relationship with her is not being prized.
[4] As is their sharing with others.

people achieve their highly ranked goals, one of which is often to be well off. In short, some standard Kantian resources are available to help the fundamentality theory here.

12.3.3 The true

Think now about the role of fundamentality in intellectual reflection. Cognition that confers great importance on a life ascertains, or does what is likely to ascertain, facts that are largely responsible for a wide array of other facts.[5] More specifically, there are two domains in which apprehending basic conditions is significant: roughly, human nature and the human environment. First, contemplation is particularly important (for non-consequentialist considerations) when and because it is about those facets of humanity that are largely responsible for much else about human life as such. Knowledge of ourselves is intuitively interesting or particularly worth pursuing insofar as it discloses those features that account for many other aspects of us as a species or kind,[6] for example, DNA, natural selection, rationality, sharing, socialization, personality, communication, time keeping, deviance, power, wealth, war, exchange value, neurosis, and other concepts central to the human and social sciences.

Second, contemplation is also particularly important (for reasons other than its results) when and because it reveals facets of our world that are largely responsible for much else about it, for example, space-time, energy, light, waves, gravity, atomic structure, and other concepts central to the natural sciences, as well as causation, necessity, properties, and other core notions of metaphysics. The more an individual's informed and skilled intellectual rationality is positively directed towards fundamental conditions of human life and human reality, and the more subjective factors such as wanting, enjoying, and appreciating are exhibited, the more meaningful her existence in the realm of the true.

This appeal to fundamentality well explains why great meaning would not come from discovering mere coincidences, despite their apparently 'universal' scope, and, conversely, why it would come from making discoveries of the sort that Darwin and Einstein did. Darwin revealed not merely the origin of our species, but an origin that accounts for much else about our existence. In short, human life is in large part a function of natural selection. Similarly, Einstein discovered basic facts about the spatio-temporal universe, ones that account for a large array of events in it. In addition, fundamentality seems to capture which ideas a university's researchers should be striving to discover, what academic departments there ought to be (psychology yes, tourism no), and how teachers ought to be instructing tertiary students.

Still more, the present account of which intellectual reflection is important avoids the problem recurrent in the previous chapter of overly prizing knowledge that is extensive

---

[5] Consider W. D. Ross's remark that 'our states of knowledge and opinion seem to derive some of their value from the nature of the fact apprehended, or believed to exist. The only rule I have to suggest here is that the more general the principle—the more facts it is capable of explaining—the better the knowledge' (1930: 147). See also Nozick (1981: 417–18, 625–6) and Hurka on dominance (1993: 115–18).

[6] Note that what is largely responsible for the development of a species need not be the same as what is largely responsible for the development of a characteristic member of it. For instance, war might not affect typical individuals, but its presence among enough of them could well affect the course of the species.

in size. By considering knowledge of human life and of human reality to be two distinct and comparable areas in which fundamentality should be sought, it does not categorically elevate physics over, say, the biological and natural sciences, as many of the other non-consequentialist theories do. Of course, human reality in its entirety is the physical universe, so that the most fundamental facts are those responsible for not merely earthly considerations, but the entire world in which we live. That does make theoretical or particle physics, *ceteris paribus*, more important than plate tectonics, which is not implausible. Note, however, that the latter does count insofar as it is about a large segment of the human environment. It is also important (partly for its own sake, and partly as a means) to learn fundamental facts not only about the local environment of the human species, but also about the lives of particular groups of human beings, viz., about their societies, making sociology and economics, for instance, fairly important disciplines, even when they do not in themselves shed light on human life as such.

One might object that, *contra* the fundamentality theory, discovering many piecemeal facts could be important. Consider, for instance, taxonomic work in botany, or the discovery of a new particle in physics. These cases are not captured by the appeal to propositions that explain lots of disparate data in a given domain.

Now, it is true that the fundamentality theory would not deem such work to be, apart from its consequences, a source of superlative significance. However, one may plausibly account for the value of this kind of knowledge as being instrumental for meaningful knowledge; that is, recording and organizing bits of initially unrelated information could be extremely useful for the realization of deep knowledge that is meaningful for its own sake. It is, therefore, well worth undertaking.

12.3.4 *The beautiful*

Consider now fundamentality in the context of aesthetic creation. It is a commonplace to say, with Gewirth, that great art, whether it is literature, painting, or music, addresses 'universal themes' or perspectives that are not merely particular or personal.[7] However, as I noted above, an artwork about excreting or dust on the face of it would address something that fits under this rubric. A compelling artwork cannot be about just any facet of human life that all of us (are disposed to) experience, but, instead, I submit, is about those facets of the human experience responsible for much else about the human experience. Such a qualification best enables us to distinguish between a novel about excreting or a painting depicting dust, on the one hand, and works about topics such as morality, war, death, love, family, and the like, on the other. To be sure, excreting is necessary for much else about what it is like to live a human life, but it is instrumental for and explains little of it; i.e., it is not fundamental. A central part of why Dostoyevsky's *Crime and Punishment* and Picasso's *Guernica* are great is that they

---

[7] For a theory that appeals to perennial theme as a way to distinguish literature from mere fiction, see Lamarque and Olsen (1994), and for one that appeals to generality to distinguish great music from the not so great, see Storr (1992: 72–3, 178–9).

convey an intimate awareness of deep themes. The appeal to universal theme is on the right track, but is too broad: fundamental topics are ones that transcend particularity, but not all topics that transcend particularity are fundamental.

One might suggest that my account of the beautiful applies only to contentful artworks and that meaning can come from non-representational pieces. Minimalist painting and music, for instance, appear to be excluded altogether by the fundamentality theory, for they do not seem to be about anything. In short, the present theory appears no better off than Nozick's in this respect, which, recall, cashes out aesthetic value in terms of organic unity (11.3.4).

This concern raises complex issues in aesthetic theory, and the easiest way to reply would be to narrow the theory down to something intended to account merely for contentful artworks. However, consider the following more robust replies, which I submit are much more promising than the appeal to organic unity. First, there are some who argue that 'non-representational' artworks are about themselves *qua* artworks or about the possibilities open to art (Kivy 1990: 202-18). Insofar as the aesthetic itself is responsible for much else about human experience, producing an artwork that is about art could be important, by the fundamentality theory.

Second, others point out that non-representational artworks are often intentional presentations of pattern and abstractions (White 1992), which conditions are largely responsible for a wide array of other facts of what it is like to live a human life. In drawing attention (even her own) to the artwork, an artist is in effect saying 'Look at this!', or 'Listen to this!', and when the 'this' picks out (perhaps via synecdoche) purposiveness without a purpose, it is about a condition that is basic to human experience in general. Similar remarks go for the experience of natural beauty, which is typically an apprehension of form.

Finally, and of most appeal to the friend of the fundamentality theory, there are those who suggest that ostensibly non-representational works can be understood 'as touching, in some fashion or other, on the most fundamental and pressing aspects of human existence, for example, death, fate, the inexorability of time, the space between aspiration and attainment' (Levinson 1992: 59). How artworks can do this without literal representation is, of course, a notoriously difficult question, but the point is that resources are available within the philosophy of art for taking the fundamentality theory seriously.

12.3.5 Additional desiderata

With this explication of (FT_1) and its illustration in the contexts of the good, the true, and the beautiful, I return to the desiderata for a theory of what constitutes meaning in life. Most obviously, (FT_1) satisfies Desideratum #9, the idea that the object of rationality plays a crucial role in its ability to confer meaning on life. In the ideal case, one wants an informed, skilled, and healthy rational process oriented towards fundamental conditions of human existence, whether those of our environment, us as a group of some kind, or a given one of us. In the previous sections, I have argued that fundamentality as an object of rationality is more defensible than is the transcendence of particularity, which is a broader idea.

(FT_1) also satisfies Desideratum #4, the idea that a theory ought to account for the idea that promoting desirable consequences in the long run can sometimes confer meaning on one's life. In the first instance, (FT_1) accounts for this notion by virtue of its account of the good; it entails that it is meaningful to positively orient one's behaviour towards people's reasoning and relationships, where successfully doing so is a greater such orientation than is merely trying to do so and failing.

And at this point I highlight a facet of (FT_1) I have not discussed as yet, the fact that it says that meaning comes from not merely exercising one's own reason, but also from getting others to do so. It does not limit the contouring of reason towards fundamentality to just the agent's rational nature, and instead speaks of an agent positively orienting 'rationality' towards fundamentality. For example, in the realm of the true, one is to discover or learn fundamental facts and then, in the ideal case, pass them on to others, viz., by publishing or teaching, thereby orienting others' rationality towards fundamentality. In regard to the beautiful, meaning would come from creating a work of art about a fundamental theme, and then still more meaning would accrue by exposing it to others, say, by finding a printer for one's novel, holding an exhibition of one's paintings, or engaging in a recital of one's music in which others take pleasure. Although there would be some meaning in the case of the true and the beautiful even if others did not enter in, meaning in both realms would be enhanced by prompting others to direct their rational selves towards fundamentality, a point that also enables (FT_1) to account for the kernel of truth in consequentialism.

It follows from these considerations that (FT_1) also takes care of Desideratum #7, the judgement that both self-regarding and other-regarding conditions are constitutive of meaning. Remember that it appears that meaning is located neither merely within oneself, say, as a function of one's own virtue, nor merely beyond oneself, for example, in the form of long-term results in the material world. Instead, it appears that both locations are relevant. The fundamentality theory captures that thought by entailing that rational processes of deliberation and decision and of the exhibition of other propositional attitudes enhance meaning in life on their own, that they particularly do so when directed towards fundamental conditions, and that they do so all the more when one gets others' rationality contoured to fundamentality. For instance, creative thinking is meaningful in itself, but all the more meaningful when it results in a separate art-object about a theme fundamental to human experience, and still more so when that art-object is presented to others and appreciated by them.

Finally, for now, (FT_1) explains Desideratum #1, that a naturalist should not ignore the potential relevance of supernatural conditions for meaning in life. If God existed, it would be incredibly important to know about Him and to make works of art about Him, as He would be largely responsible for nearly everything that goes on in the physical and spiritual worlds and in the human experience. Similar remarks apply to the possible world in which our identities are a function of the configurations of an immortal, spiritual substance. To know about human nature would then require knowing about the soul, and to love another individual would be a function of beholding hers. If spiritual persons do in fact exist, then I readily admit that an awful lot of meaning in our lives would depend on relating to them. That point, however, is consistent with naturalism, the view that a meaningful life does not require the existence of anything spiritual.

So far in this chapter, I have adumbrated nine desiderata for a theory of life's meaning that emerged from critical analysis of extant theories in the field, spelled out the central elements of the fundamentality theory, and argued that they entail and plausibly explain seven of the nine desiderata. In the next section, I refine the fundamentality theory a bit so that it retains these advantages, while also capturing the two desiderata still at large.

12.4 Refining the fundamentality theory

In this section, I first slightly adjust the fundamentality theory so that it can readily be seen to accommodate Desideratum #3, the relevance of negative conditions such as con-attitudes and anti-matter, and Desideratum #5, regarding moral constraints on the ways one can obtain meaning by pursuing fundamentality. Having completed my account of the part-life and *pro tanto* facets of meaning, I then provide a definitive, fuller statement of the view that addresses the whole-life and on balance facets of meaning.

The remaining desiderata are handled well by revising the fundamentality theory this way:

> (FT_2) A human person's life is more meaningful, the more that she, without violating certain moral constraints against degrading sacrifice, employs her reason and in ways that either positively orient rationality towards fundamental conditions of human existence, or negatively orient it towards what threatens them; in addition, the meaning in a human person's life is reduced, the more it is negatively oriented towards fundamental conditions of human existence.

I have explained what the positive orientation of one's rationality involves, and so the reader can readily grasp what 'negative orientation' means. With regard to intentional action, a negative orientation would be a destructive or otherwise inhibitive response, and in terms of propositional attitudes, it would include sentiments such as dislike and judgements of disvalue. (FT_2) expresses the idea that meaning can come from exhibiting con-attitudes and even harmful behaviour, when they serve the function of rebutting similarly negative or otherwise undermining orientations towards fundamentality.[8]

---

[8] By negating a negation, one might get a positive, as it were (for a similar move made in the context of developing a moral theory, see Metz 2002b). Hence, this point might already be implicit in (FT_1)'s notion of positive orientation towards fundamentality.

Hence, (FT$_2$) accounts for the judgement that con-attitudes can confer meaning on life, when directed towards the appropriate objects. Recall that the best illustration of this point was that of hating injustice, roughly, despising a condition in which people's reasoning or sharing is impaired, or otherwise treated disrespectfully. I capture the significance of that emotion by cashing it out as a facet of one's rational nature negatively responding to something negatively responding to a fundamental aspect of human nature.

However, note that (FT$_2$) does not, strictly speaking, mention being negatively oriented towards 'what is negatively oriented' towards fundamental conditions, but rather of being negatively oriented towards 'what threatens' them. The reason is that negative orientation is comprised of an agent's behaviour and attitudes, and sometimes forces of nature perform a similar function. Concretely, if meaning can come from hating injustice, then it probably can come from hating a tsunami or a geography that inhibits agriculture and hence induces poverty, and I intend (FT$_2$) to account for that.

(FT$_2$) therefore satisfies the part of Desideratum #3 about the way that con-attitudes can confer some meaning on life. It also satisfies the other part, the idea that a meaningful life is one that not merely has the good of meaning, but also avoids the bad of anti-matter. (FT$_2$)'s last clause, indicating how meaning in one's life could be reduced, is, so far as I know, the first explicit theoretical statement of what constitutes anti-matter. All other theories up to now have indicated merely what constitutes meaningfulness, but not of what anti-matter is composed. Supposing the desirable is well represented with a positive number and the undesirable with a negative one, in order to know when a life is meaningful on balance one simply must have a conception of anti-matter, conditions worthy of emotions such as shame, or ends besides one's own pain *qua* pain that are extremely worthy of avoidance.

According to (FT$_2$), anti-matter consists of an agent being negatively oriented towards fundamentality. This plausibly captures the respect in which the following actions and attitudes would subtract from the overall meaning of a life, beyond merely failing to enhance its meaning: blowing up the Sphinx; spreading nuclear waste; holding sexist and racist beliefs and emotions; hating others by, say, viewing them largely in terms of their weaknesses; torturing others for fun; burning science books; spreading disinformation about the nature of the world, for example, to the effect that one is a god.

Now, included in anti-matter would be actions that are particularly degrading; for example, prostituting oneself to feed a drug habit, or killing innocents and using their blood in one's paintings to make a statement about the value of human life. So, one could read (FT$_2$), for all that has been said, as implicitly satisfying Desideratum #5, that a theory must account for the idea that meaning is undercut when certain moral constraints are violated. However, I suspect the extra clause about moral constraints is necessary, for the point is not merely that severe degradation is a form of anti-matter, but also that, when one behaves in a severely degrading way, it prevents one from acquiring positive meaning that one might have otherwise had.

I submit that (FT$_2$) is the most defensible account of the *pro tanto* and part-life aspects of meaning in life. Unlike other principles in the literature, it entails and plausibly explains all nine desiderata, while also providing a fair characterization by virtue of what the good, the true, and the beautiful confer meaning on life. What remains to be done, in order to have a truly complete theory, is to add in the whole-life factors (set aside since Chapter 3), and to indicate what a meaningful life on balance would be (bracketed since Chapter 8).

Not only do theories of life's meaning by and large neglect to account for anti-matter, they also typically fail to address the whole-life facets of meaning, which, I have argued, are an independent dimension of meaningfulness beyond that of its parts. Now, from one perspective, my account of the whole-life elements is rich, but, from another, it is still impoverished. It seems rich when compared to rival theories that one finds in the literature in that I accept several respects in which a whole-life can exhibit meaning, facets canvassed earlier and ordered in developmental fashion (3.4). So, I accept the claims that a life is *pro tanto* more meaningful insofar as it avoids repetition in its parts, its bad parts cause its later, good parts, and they do so by virtue of personal growth or some other pattern that makes for a compelling life-story that is original.

However, the theory I offer is poor in absolute terms, as I do not yet know how to specify a number of these variables. For instance, I still lack a general and basic account of how to distinguish compelling life-stories from ones that are not so hot. In addition, if anti-matter can inhere in the parts of a life, it can also do so in the life as a whole, and I lack a satisfactorily detailed account of that, for now merely acknowledging the opposites of narrative value such as being random, being repetitive, ending on a low note, or intending to replicate another's whole-life, which I sum up as 'narrative disvalue'. At this point, I simply modify the fundamentality theory to indicate the most important whole-life aspects, recognizing that they still need development.

(FT$_3$) A human person's life is more meaningful, the more that she, without violating certain moral constraints against degrading sacrifice, employs her reason and in ways that either positively orient rationality towards fundamental conditions of human existence, or negatively orient it towards what threatens them, such that the worse parts of her life cause better parts towards its end by a process that makes for a compelling and ideally original life-story; in addition, the meaning in a human person's life is reduced, the more it is negatively oriented towards fundamental conditions of human existence or exhibits narrative disvalue.

(FT$_3$) is the most complete theoretical statement I have to offer about what makes a life *pro tanto* more meaningful. To all those who have asked me over the past ten years, 'So, what *is* the meaning of life (wise guy)?', there you have it! (FT$_3$) is the best answer I can muster, at this stage, about which ends, apart from one's own pleasure as such, are most worthy of pursuit, how to transcend one's animal nature, and which conditions are worthy of great pride or admiration.

To round out the account, I also include a statement of what constitutes a meaningful life on balance. Recall that I provided a principle about how much *pro tanto* meaning a life must have in order to be meaningful on balance, to have a life that, all things considered, counts as meaningful rather than meaningless. In particular, remember (IT_8), the view that a life counts as meaningful on balance insofar as it is not very far below the maximum amount of *pro tanto* meaning that a being that was born human is likely to achieve in the physical universe, given the laws of nature (8.6). Now I can say that the amount of *pro tanto* meaning in one's life is equal to: (A) the extent to which one has, without violating certain moral constraints against degrading sacrifice, employed one's reason and in ways that have either positively oriented rationality towards fundamental conditions of human existence or negatively oriented them towards what threatens them, such that the worse parts of her life have caused better parts towards its end by a process that makes for a compelling and ideally original life-story; minus (B) the extent to which one's rational self has been negatively oriented towards fundamental conditions of human existence and one's life has exhibited narrative disvalue.

12.5 Reconsidering the fundamentality theory

In the final section of this chapter, I consider some plausible objections to make to (FT_3). Sometimes I contend that the theory as it stands can handle them, while other times I explore the possibility of modifying it in various ways. Recall that my primary aim is to demonstrate that the fundamentality theory is an improvement over extant rivals; I do not mean to suggest that it is the last word on what matters.

12.5.1 Fundamentality not sufficient for meaning?

One criticism that naturally arises is to doubt whether positive orientation towards fundamental objects is sufficient for meaning, or at least notable meaning. After all, a critic might point out, great meaning would not come from any of the following: promoting people's agency merely by donating money that one can easily forgo; apprehending that 2+2=4; or reading trashy romantic fiction or crime thrillers. All these actions involve rationality, for lower animals cannot do these things, and in every case they seem to be positively oriented towards a fundamental condition.

In reply, note that, for superlative meaning, the fundamentality theory requires more than the bare fact of rationality directed towards a fundamental object; it also requires the *substantial* contouring of one's intelligence towards such an object. So, in the realm of moral achievement, it would probably require some kind of intense connection between one's rational self and the reasoning or sharing of others—for example, hard work, devotion, and sophisticated planning—which are all lacking when a person who has inherited massive wealth donates some of it to Oxfam.

Regarding particularly important intellectual reflection, the theory demands the rigorous exercise of one's theoretical capacities, which is missing in the case of 2+2=4.

Furthermore, substantially contouring one's theoretical reflection towards a fundamental condition involves not merely knowing a proposition that is fundamental or even that a proposition is fundamental, but also *why* it is fundamental; one ought to be in a position to understand how a piece of theoretical knowledge explains a diverse array of data, a key difference between a professional physicist and an amateur who has memorized '$E=MC^2$'.

And in respect of aesthetic creation, the content of one's artwork, or the thought that went into it, ought not to be merely something fundamental to human experience, but also revealing with regard to it. It is a banality in aesthetic theory to note that clichés, stereotypes, formulaic plots, and hackneyed techniques indicate a lack of sensitivity to the subject matter. Such essentially plagiaristic material means that the interpretation of the fundamental object is not illuminating.

Another way to avoid many of the above putative counterexamples would be to integrate another element into the fundamentality theory, one requiring some kind of advancement in order for great meaning to obtain. With regard to the realm of morality, I could require not merely outcomes that are likely to be supportive of reasoning and sharing, but those that actually are and in large-scale ways, for example, that have successfully brought about substantial changes in many people's decision-making and being-with-others in the form of newly democratic relationships or recovery from severe illness. With respect to enquiry, I could require discovering a new, well-supported theory about us or the cosmos, not merely coming to apprehend an existing one. And in terms of artistry, I could require a style or technique that breaks with the past, as opposed to one that is interesting but still fairly continuous with tradition.

12.5.2 *Fundamentality not necessary for meaning?*

Iddo Landau (2013) has responded to an earlier statement of the fundamentality theory (Metz 2011b), and done so principally by arguing that conforming to it is not necessary to obtain meaning in life, not even great meaning. Picking up from my suggestion about adding an 'advancement clause' to account for great meaning, Landau maintains that it would render the theory too narrow. He plausibly suggests that preventing a substantial decline could be just as meaning-conferring as producing a substantial increase. And he defends this view by appealing to the asteroid case I mentioned above (12.3). Specifically, Landau argues that great meaning would accrue to a scientist who works to prevent an asteroid from hitting the earth.

In reply, one might find advancement in the research the scientist undertook in order to prevent the destruction. Or it might be worth thinking about whether a project has made an advancement relative not to a state at the present, but instead in comparison to what likely would have happened in the absence of the project. Or, finally, perhaps the best suggestion would be to think of great meaning in terms of either a large increase or the prevention of a large decrease.

An additional reason Landau (2013) provides for thinking that positively directing one's rationality towards fundamentality is not required for meaning is that meaning can come from saving others' lives. Mother Teresa, he proposes, had great meaning in her life because she prevented much death, and the same would be true of someone who discovered a cure for cancer. But saving life, or, equivalently, preventing death, is a necessary condition for much else about human existence, not a fundamental one. So, Landau concludes that fundamentality is too narrow a ground for meaning and that positively directing one's agency towards at least some kinds of necessary conditions should be included.

I try to reply by remaining within the ambit of the fundamentality theory, as I have spelled it out so far. I suggest that it is not the *bare fact* of preventing death or saving life that is meaning-conferring. To see why not, suppose that those whose lives were saved then suffer fates worse than death, and that such an outcome was foreseeable (even if not foreseen) to the one who saved the lives. In that case, it is reasonable to maintain that no meaning would accrue to his life for having saved life, which suggests that it is not saving life as such that counts.[9]

Another objection to the fundamentality theory is that it fails to account for the intuition that retributive justice can be a source of meaning, and possibly its absence a source of anti-matter. Recall the quotations from *Ecclesiastes* (at the start of Chapter 4), where the author deems life to be 'vanity', not merely since the wicked and the righteous ultimately share the same fate, namely, death, but also since, while alive, the wicked often prosper and the righteous often languish. Retribution would be served if instead the wicked suffered or otherwise underwent harm, and if the righteous were benefited, both in proportion to their just deserts. However, it appears that such a distribution of unhappiness and happiness has nothing to do with fundamentality. Fundamentality can account for the meaningfulness of moral behaviour and virtue, roughly, as that which is positively oriented towards others' reasoning and sharing, but it appears unable to account for fitting punishment or reward in response to the exercise of agency.

It is tempting for me to reply that *giving* people what they deserve is nothing other than expressing respect for agency, and hence is ultimately a matter of responding positively to it in the case where the good are given recompense, and of responding negatively to negative responses to agency in the case where the bad are given a penalty (see, e.g., Cupit 1997; Metz 2002b, 2007b: 700-3). While that promises to go some distance, it apparently cannot go far enough; for the intuition that retribution would be meaningful persists even supposing that no agent intentionally brought it about. Imagine a quasi-Karmic, non-conscious force that allocated benefits and burdens according to the morality of one's life. Many would say that our existence would be more meaningful if such a force existed (at least for the good who have some compensation or reward coming). One's life apparently would mean more if the world were structured so that, without anyone intending it, the bad guys simply never got

---

[9] It does not follow, though, from this point that meaning is indeed a function of supporting conditions of human life largely responsible for much else about human life. See Landau (2013) for additional critical points that are well worth taking up in future work.

away with it. But that is beyond the purview of the fundamentality theory, which bases meaning on the exercise of intelligence. In reply, perhaps the unconscious, Karmic-like allocation of desert throughout the universe counts as a candidate for a cosmic account of meaning, and not an individual one (1.2).

A final difficult case for the fundamentality theory, for now, is the case of animal pain. Some readers will not be happy with my account of the relevance of relieving human pain (12.3.2), but at least I have an account to give. In the present case, where it appears as though one's life would be more meaningful for reducing animal suffering, à la Peter Singer or Temple Grandin, the fundamentality theory is empty-handed. This could be a respect in which the fundamentality theory does not well capture Desideratum #4 regarding good consequences.

Of course, such people would be exercising their rationality, and presumably doing so in sophisticated, strategic ways. However, that point fails to capture the respect in which the objects of their activities, as well as the consequences of them, would confer meaning on their lives. They are not, or need not be, positively oriented towards fundamental conditions of human existence, and yet their lives would be quite significant all the same, or at least more so than if they had employed their reason merely to count blades of grass.

The only way to resolve this problem would be to revise (FT_3) in some way, but I am unsure how. One might consider fundamentality with respect to human and animal existence, or with respect to the life of organisms, or perhaps with regard to organic unities, where there are hierarchies of complexity among beings in the natural world that would ground differential amounts of meaning upon positive orientation towards their fundamental conditions.[10]

In this chapter, I have developed a new theory of what makes a life meaningful, roughly according to which meaning in life is a matter of positively orienting one's and others' rational nature towards fundamental objects, conditions of human existence that are largely responsible for many other of its conditions. I have argued that this theory plausibly captures the meaningfulness of morality, enquiry, and creativity, and that it not only avoids the major counterexamples to all other theories of life's meaning, but also incorporates their salient kernels of truth.

I do not suppose that the fundamentality theory is so complete and air-tight as to require belief at this point. It is still vague in some respects, and, being a new principle, it has yet to survive a volley of counterexamples and other objections. However, almost nothing is everything. I submit that there is enough evidence reasonably to think that the fundamentality theory is more justified than its closest rivals and that the theory warrants systematic attempts to make it less vague, more clearly defensible, and more wide-ranging in its application.

---

[10] A mix of ideas from Nozick's (NCT_4) and Smith's (NCT_7) in 11.3.

13

Conclusion: The Fine Game of Nil[1]

'A man who considers that his life is of very wonderful importance is awfully close to a padded cell. Let anybody study the ordinary, everyday details of life; see how closely he is bound and fettered; see how little it all amounts to.'

Clarence Darrow ('Facing Life Fearlessly')

'The best part of every mind is not that which he knows, but that which hovers in gleams, suggestions, tantalizing, unpossessed, before him....But this dancing chorus of thoughts and hopes is the quarry of his future, is his possibility, and teaches him that his man's life is of a ridiculous brevity and meanness.'

Ralph Waldo Emerson (On Man and God)

13.1 Remaining concerns

In parts two and three I critically explored theories of all those conditions that, strictly speaking, *would* make a life meaningful, which theories do not imply that anyone's life is actually meaningful. Of course, a large majority of those who believe in those theories believe that the relevant conditions for meaning in fact obtain, and have posited those conditions often *because* they best account for the meaning that we judge actually to exist in our lives. Recall how often I have appealed to intuitions, that I presumed were shared by most readers, about what does and does not confer meaning on life, not merely what would confer it (e.g., 5.5, 8.4-8.5). And if the fundamentality theory, or something close to it, is correct, then many lives have meaning in them.

Such an approach to meaning in life leaves unaddressed three major issues in which readers might have an interest. Having given an account of meaning in life as an *end*, it is natural to want to know which *means* would be particularly *efficient* to realize it. For example, would having some faith that one's projects will be successful help to promote

---

[1] A well-known anagram of 'the meaning of life'.

meaning in one's life (as per Cottingham 2003; cf. Crisp 2008), or would a failure to apportion one's beliefs to the evidence subtract from meaning (à la Kekes 2010)? At a political level, would a state that sought to promote meaning in its residents' lives necessarily be counterproductive, since the use of coercion would invariably undercut the kind of voluntary decision-making or personal endorsement essential to meaning (a view advanced by Dworkin 2000: 267-74 and Smith 2002-2003)? Which kinds of laws and policies could a state, or other large-scale institutions such as major corporations or the United Nations, adopt that would foster meaning?

A second, unexplored issue concerns the *permissible* means to use in order to achieve meaningful goals. Supposing that a state could maximize meaning in its residents' lives by using coercion, should it do so? Or would such a non-liberal approach instead be objectionably disrespectful of people's autonomy (cf. Metz 2001b)?

Third, the supposition made in this book that some people's lives have in fact been meaningful does not address those who deny that claim. In this, the last chapter of the book, I step back and briefly consider whether we truly have sufficient reason to believe that anyone's life is meaningful. There are several arguments in the literature for the conclusion that no one's life is meaningful, which I label 'nihilism' or 'pessimism'. I explore two of the most influential rationales, ones about which I believe I have something fresh to contribute.[2]

One argument, grounded on the writings of Tolstoy and Camus, appeals to supernaturalism (13.2). Roughly, if God is necessary for meaning, and if God does not exist, then life is meaningless. Although I spent much time arguing against supernaturalism, I explain below that those arguments by and large presumed that some lives have meaning in them, something that I cannot presume when objecting to an argument for the view that no lives have meaning in them. Given a nihilist dialectic, supernaturalism is still a live option as an account of meaning in life.

The other pessimistic argument, found in the work of Thomas Nagel and David Benatar, appraises human life in light of an extremely objective or encompassing point of view, that of 'the universe'. From the point of view of the universe, our lives appear meaningless, so the argument goes (13.3).

I argue that both of these nihilist rationales evince a kind of incoherence, from which it follows that neither provides enough reason for someone who does not already believe nihilism to change her mind and adopt that view. That is not quite to say, though, that I provide enough reason for someone who already believes nihilism, like Clarence Darrow, to change his mind. And in the final section, I confess what continues to nag at me about the nihilist perspective (13.4).

---

[2] I do not address the Schopenhauerian arguments (1851a, 1851b) that one's life is meaningful only if one's strongest desires are satisfied, which no one's are, or that life is senseless since either one has not achieved one's goals and is merely striving or one has, in which case one is bored (see also Gordon 1980; Martin 1993; cf. Benatar 2006: 76–7). For additional arguments for nihilism, see Murphy (1982: 9–17); Smith (2003); Benatar (2006: 18–59).

13.2 Supernaturalism redux

Two of the philosophers most concerned about suicide in the face of meaninglessness are naturally read as being crushed by the apparent absence of a spiritual realm. Tolstoy (1884) is well known for having nearly committed suicide upon coming to believe that neither God nor a soul exists, and Camus famously claims, 'There is but one truly serious philosophical problem, and that is suicide' (1942: 94), a problem that arises in the face of an apparently chaotic, irrational, and mortal existence not governed by a benevolent creator.

Supernaturalism is a straightforward and common motivation for nihilism in the modern era. If one believes that a certain kind of relationship with God or a soul is necessary for meaning in life, and if one cannot find evidence that they exist, then one is driven to believe that life is meaningless. Supernaturalism is often mistakenly thought to be essentially tied to theism, the view that we know that God (and perhaps a soul) exists, but that is not true; it is logically consistent to hold the combination of supernaturalism and atheism and, hence, nihilism, as Tolstoy and Camus appear to have done.[3]

The natural way to question this argument for nihilism is to cast doubt on supernaturalism. However, that turns out to be difficult to do, as the objections to it that I have found the most powerful beg the question against the nihilist. Recall that one objection maintained that certain conditions would intuitively be meaningful in the absence of God or a soul (8.4), but the nihilist argument expresses the opposite judgement, that nothing would be meaningful in their absence. And remember that another objection maintained that we know that certain aspects of life are meaningful but do not know whether God or a soul exists (8.5), but, again, the present argument denies the discrepancy.[4]

I did have another objection to supernaturalism that would, in principle, be more telling at this stage, namely, the claim that life would be meaningless if a soul existed (8.3). I do find some force in the suggestion that murder would never be (all that) wrong and that 'saving life' would never be (all that) meaningful, if we would survive the deaths of our bodies. However, I find myself unable to execute this kind of strategy with regard to God, and so I at least need an additional argument against a God-centred motivation for nihilism at this point.

Another option would be to grant supernaturalism but contend that there is knowledge of God, but I find myself unable to defend that perspective. Even supposing that God would not necessarily have created the most desirable physical world, and even

---

[3] Tolstoy in the end elects to believe in God and a soul *on faith* (1884: 17–20), and hence is able to *judge* his life to be meaningful. However, his reasons for thinking that the *existence* of God and a soul are necessary for meaning are simply sidestepped by adopting a fideist orientation towards life.

[4] This does not mean that my earlier objections to supernaturalism are weak; they apply to the large majority of readers who do not yet believe nihilism, but they do not readily compel the rare person already convinced of that view on supernaturalist grounds.

supposing that considerations of free will and the goods of compassion and justice would justify the existence of substantial evil, I continue to doubt that if God existed, He would let things get as bad as they have in this world. No perfectly knowledgeable, powerful, and moral being would have allowed the German Holocaust or the Rwandan massacre. Nor, to continue with stock examples, would one have created a world in which sentient animals must kill and eat one another in order to survive, or in which they burn alive in forest fires touched off by lightning strikes.

In any event, I will not get bogged down in metaphysical debates, instead directing the reader to them when they would, given a lifetime to reflect on all relevant matters, shed light on issues of meaning in life. Since I accept the premise that the evidence favours the view that neither God nor a soul exists, and since the argument for nihilism is valid, and since my previous objections to supernaturalism lack dialectical force against the nihilist, I must come up with fresh reasons to question supernaturalism if I am to block the inference to nihilism.

What I try to do is to show that *there is little or no reason to adopt supernaturalism, if atheism is true*. If the atheist premise that is needed to entail a nihilist conclusion undermines the supernaturalist premise, then there would be little or no reason to accept supernaturalism in the context of this argument. My strategy against the supernaturalist argument for nihilism points out that the strongest argument for supernaturalism works only if it is assumed that atheism is false. Recall that I have contended that the best motivation for supernaturalism is the perfection thesis, the claim that engagement with some kind of maximally conceivable value is necessary for meaning in life (6.5, 7.4, 7.6). Specifically, I pointed out that one has particular reason to accept a God-centred theory only if meaning turns on the existence of a person who is utterly simple or infinite, something possible only in a spiritual realm. And soul-centred theory, I contended, makes sense to adopt only if meaning depends on communing with God. In the following, I argue that one has good reason to believe the perfection thesis only if one has good reason to believe that perfection actually exists, and hence only if one rejects the atheist premise of the present argument for nihilism.

Why believe that the perfection thesis about what is required for meaning in life is implausible if it is assumed that atheism is true? The answer is a meta-ethical rationale about the likely source of our value judgements in a purely physical world. If neither God nor a soul exists, then human beings and their central characteristics are (all the more) probably a product of natural selection; that is, the reason we have the fundamental properties we do is that they were instrumental for our ancestors' genes to flourish. Now, we are creatures who are essentially, among other things, valuers. A salient and ineradicable feature of our species is that we make evaluative and normative judgements, and, absent God, it is very likely that we do so now because having done so helped our distant relatives to pass on their genes. Indeed, it is common for contemporary moral theorists to maintain that cooperating made it much more likely for us to succeed as a species, and that cooperating was made much more likely by having acquired, biologically and socially, the disposition to make certain, emotionally coloured judgements of what is good and bad, right and wrong, innocent and blameworthy.

Now, if this naturalist story about the origin of our value judgements is broadly correct, then it is unlikely that these judgements would be informed by the perfection thesis. Cooperation, or any other action of the sort that would have enabled our ancestors' genes to be passed on, would not have done so had it been predicated on facts about a maximally conceivable ideal that could obtain only in a spiritual realm, which the friend of the present argument for nihilism asserts does not exist. Early members of *Homo sapiens* would not have judged their own or others' behaviour in light of standards that, *ex hypothesi*, could never be fulfilled. They would not have judged their lives to be worthy of great esteem in light of a state of perfection that is non-existent. Instead, they would have appealed to imperfect standards that could be satisfied by earthly lives. What would have facilitated survival and flourishing are judging behaviour to be worthy of great esteem insofar as it exhibited, roughly, respect for reasoning and sharing and judging behaviour to be worthy of great shame to the extent that it has been degrading of the fundamental conditions of human life.

If there is no God, as per the present argument for nihilism, then the source of our value judgements is such that their content is unlikely to be informed by the perfection thesis, viz., by anything about God. The kinds of value judgements that would have enabled purely physical creatures to be naturally selected are ones appealing to imperfect standards that could be fulfilled on earth. Hence, the atheist premise of the supernaturalist argument for nihilism undercuts the best reason for believing its supernaturalist premise. The present argument for nihilism is incoherent, in that one premise provides good reason to reject the other.

13.3 The point of view of the universe

Nagel famously contends that our lives are absurd because of two perspectives inherent to human nature that give opposing answers to the question of whether our lives are meaningful (1986: 208-32). A life is absurd, for Nagel, when its meaninglessness from a very objective or external standpoint is juxtaposed to the importance ascribed to it from another standpoint (roughly, the individual's). However, from this reasoning, Nagel does not try to conclude that life is in fact meaningless, unlike Benatar (2006: 82-6).[5] I therefore focus on the latter's argumentation.

Arguing against objective naturalist conceptions of well-being and of meaning, Benatar points out that they are invariably constructed from a human point of view. When determining what is objectively good and to what degree, theorists invariably take the relevant mind-independent conditions to be good from a human perspective, not the *sub specie aeternitatis*, the point of view of the universe (or eternity). The

---

[5] Cf. Hanfling (1987: 22-4).

CONCLUSION: THE FINE GAME OF NIL   245

latter perspective is an all-encompassing stance with respect to one's life, a viewpoint in which one steps back and views oneself as one member of an enormous class, say, as one sentient inhabitant of the spatio-temporal world among countless others.

When viewing one's life as occupying one of innumerable spots in the purview of the Hubble Telescope, nothing of one's short-lived existence and its puny effects appears to matter. What one does in a certain community on the third rock from the sun over an approximately 70 year span just does not amount to much, for Benatar, when considering the likely gazillions of beings who (logically) could live forever in a universe that is probably infinite.

Now, why appraise life from the point of view of the universe? Benatar presents several interesting arguments for doubting value judgements that are informed by a merely human perspective (2006: 82-4). First, he points out that they are likely to be informed by the limits of what human beings can expect or what is within our control, but he reasonably asks what reason there is to think that impossible standards should be excluded *a priori* when ascertaining the quality of human life.

Second, Benatar notes that most of those writing on the meaning of life reject the subjectivist view that an individual's life is meaningful for obtaining the objects of her contingent propositional attitudes, e.g., desiring something and getting it. Most maintain that the individual's viewpoint is too arbitrary to ground judgements of meaning, as it counterintuitively entails that, say, counting blades of grass could be a very meaningful project. By analogy, Benatar suggests, most should maintain that humanity's viewpoint is too arbitrary, and should rather opt for a more encompassing standpoint.

Third, most of us believe that our lives are much better than those of non-human primates, let alone those of beings such as cats or mosquitos. If so, Benatar asks, why should not we not reasonably judge other possible lives to be much better than ours, so that our quality of life is seen to be poor on the whole?

These arguments merit careful and thoughtful responses, but rather than rebut them directly here,[6] I provide more food for thought by offering two new reasons to doubt that the point of view of the universe is the relevant perspective to invoke when appraising the quality of human of life. The first reason to reject the point of view of the universe invokes part of the argument I just made against supernaturalism (13.2). If our value judgements are a product of natural selection, so that we became disposed to make them because of their usefulness for survival, then a resolutely human perspective no doubt informs them; only that standpoint, and not the point of view of the

---

[6] I have, in effect, responded to these arguments in 8.6, where I argued that intuitions about when a life is on balance meaningful are best systematized so that, although judgements of when a life counts as meaningful on balance are not a function of what a given human being can obtain, they probably are a function of what is maximally possible, given the laws of nature, for beings that were born human but that could morph into something non-human. While I reject the view of those such as Martha Nussbaum (1989) and Leon Kass (2001), that value judgements must be grounded on human nature, I accept the view, *contra* Benatar, that they must be grounded in what is *available to* human beings.

universe, would have enabled members of our species to evolve by means of judging the behaviour of oneself and others.

A second argument is dialectical, in the sense of appealing to premises that, perhaps unlike the previous rationale, Benatar and most readers already accept. For one, notice that Benatar, along with others such as Nagel, who invoke the point of view of the universe, make plenty of judgements about what we *ought* to think, which views are more *justified* than others, and which theories are *better* than others. For instance, Benatar maintains that we have sufficient reason to believe pessimism, that it is more worthy of belief than optimism. Benatar is comfortable making judgements of shoulds and goods in the context of *belief* without appealing to any standpoint of eternity that would include all rational enquirers, which evinces a tension when he suggests that we must appeal to such a standpoint when making judgements of shoulds and goods in the context of *action*.

In addition, consider that Benatar and also Nagel routinely make judgements of what is immoral and what is harmful, again, without appealing to the point of view of the universe. In particular, Benatar (2006) claims to know that murder is immoral, that procreation is immoral, that abortion is morally required in the early stages of pregnancy, that it is wrong to treat people merely as a means, that pleasure is good, that pain is bad, etc. If he can know *which* conditions are good or bad without appealing to a non-human standpoint, then why can one not make judgements of *how* good or bad something is without appealing to such a standpoint?

Summing up, to defend an argument for nihilism that appeals to the point of view of the universe, theorists do, and—at least in the context of belief—*must* appeal to goods and shoulds that are not a function of it, which makes it inconsistent to set a different, and much higher, standard at other points.[7]

If one rejects the point of view of the universe, and accepts an objective naturalist account of meaning in life, then it is far from obvious that the best human lives are all that badly off or meaningless. Many people are not utterly neurotic, are not entirely controlled from without, are not completely ignorant, are not systematically shunned by other human beings, are not devastatingly sick, and so on. And, given that, most are able to be reflective, beneficent, creative, and the like. Naturally, I would not like to have been airdropped into the Congo in the 1990s and suffered with everyone else there from war, crimes against humanity, malnutrition, bugs and parasites, humidity and heat, and the lack of quality education, anaesthetic, health care, infrastructure, and transport. And I certainly do not look forward to my life petering out, as I indicate below. But I presume most readers are like me in deeming themselves to have acquired

---

[7] For other objections to adopting the standpoint of the universe, see Solomon (1993: 12–52); Schmidtz (2001: 174–9); Kekes (2010: 232–45); and Pritchard (2010). There are also those who purport to see something different from Benatar and Nagel when they take up the objective standpoint; some see us as children of a caring God (Quinn 2000: 65–6), as beings whose sufferings all count equally (Singer 1993: 333–4, 1995: 222–33), or as people capable of realizing goods that others can appreciate (Wolf 1997b: 19–21).

enough objective goods to judge their lives to be, if not meaningful on balance, then at least not very bad and full of anti-matter, from a human standpoint.

13.4 Concerns about remaining

I have addressed two interesting and influential arguments for nihilism, and contended that both evince different kinds of incoherence. The first argument, according to which nihilism follows from the combination of supernaturalism and atheism, is incoherent, in that holding the atheistic view that value judgements have a non-theistic source commits one to rejecting a supernaturalist conception of their content. The second argument, that nihilism follows once one takes up the standpoint of the universe, is vulnerable to the point that essential elements of making that very claim appeal to value judgements not grounded in such a standpoint. I conclude that, if one is not already a nihilist, one's mind should not be changed by at least these rationales.

I do not accept nihilism, but there is a line of argument characteristic of nihilism that bothers me. In general, nihilists tend to get a hook into their interlocutor by appealing to considerations of magnitude or impact. There are theorists who deny that a life's consequences, particularly posthumously, affect its meaning (e.g., Schmidtz 2001; Trisel 2004; cf. Smilansky 2012), but I find their position implausible. Even if one need not make a difference to the world that would be appreciated from the point of view of the universe, one's life would matter more, the greater and longer its earthly impact. As I mentioned earlier, it seems to most of us better that 5,000 people benefit from and recognize one's accomplishments now and another 5,000 also do so in the next generation than that 10,000 do so now but none does so posthumously. We tend to want to make ripples that extend far into the future, or, in Nozick's characteristically elegant words, to leave longstanding 'traces' behind upon our deaths (7.4.3). And, of course, many of us would prefer not to die in the first place.

Now, if meaning is possible in a purely natural world, and indeed in the context of what is more or less available to human beings, as I have argued, then why should I still crave immortality, or at least to have some significant influence on the long-term future? I presume I am not speaking merely for myself; even those who have written essays rejecting the relevance of traces surely want their essays to continue to be read after they have died.

I suspect the lingering longing for something greater than what is available to us human beings, as we are known by scientific means, is a function of our tendency to idealize, which Ralph Waldo Emerson characterizes with grace above. Our capacity to conceive of ever better states of affairs reaches to perfection, and, furthermore, our capacity to wish is insatiable. Those of us who are resolute deontologists about morality still wish that the consequences had turned out well, after all. There are some who argue that wishing is rational, in the sense that it ought to cohere tightly with our judgements (Adams 2006). If that were true, then persistent wishing to make a major difference to the world or to live forever would need to have some kind of bearing on one's views of meaning in life. I deny that, however; a wish is just a wish (Metz 2009).

Few are so lucky as to make noticeable ripples into the future, at least beyond one's children, but most are nonetheless able to develop their rational nature and to contour it towards fundamental objects while alive. One can know that this is what one should do so as to confer meaning on one's life, despite not being able to convince sceptics otherwise, and one can make one's life on balance meaningful by doing so, regardless of whether doing so engages with perfection or satisfies the point of view of the universe. So, yes, dear reader, you may reasonably conclude that your life is, or at least could be, meaningful (while continuing to hanker to make an enormous splash).

Epilogue

This page culminates more than a decade's worth of reflection on the meaning of life. At a psychological level, I pretty much *have to* think that the search for life's meaning has itself been a source of meaning in my life, and a substantial one at that. However, it would be all the better for me if there were a philosophical justification for my judgement, indeed, one grounded on the very theory of meaning in life that I have argued is the most justified relative to existing rivals in the literature. I conclude by briefly pointing out how the fundamentality theory entails that the search for the most justified theory of meaning in life[1] is itself a source of meaning.

According to the fundamentality theory, the search for what confers meaning on life itself confers meaning on one's life insofar as the pursuit of meaning is fundamental to human nature, that is, largely accounts for a wide array of actions and attitudes typical of human life as such. Now, the pursuit of meaning is indeed fundamental to our lives in this way.[2] For one, it underwrites much of what we most prize and are willing to make sacrifices for, e.g., marriage, children, religion, beneficence, justice, art, beauty, education, knowledge, and ripples. For another, the pursuit of meaning also explains much of human life that we abhor, as Arthur Koestler explained with insight a while back (1967); the most destructive projects undertaken in the twentieth century (ranging from the Holocaust in Europe to the Great Purge in the Soviet Union to the Cultural Revolution in China) have been mainly done not out of a search for happiness, but rather a misdirected attempt to participate in something greater than oneself in the form of *Volk*, *Vaterland*, God, political party, utopian society, or the like. It would be impossible to grasp human nature to any comprehensive degree without invoking people's drive to seek out ends beyond their own pleasure and to make their lives worthy of great esteem and admiration, so far as they can tell in what these consist.

Is the fundamentality theory the holy grail of Western normative philosophy, *the* respect in which the good, the true, and the beautiful genuinely constitute a unity, *the* principle that captures all and only the myriad factors that make up Meaning in Life? It would be presumptuous to say that the search for an adequate theory of what makes a life meaningful is over, given how few philosophers have undertaken the enquiry in earnest. However, the fundamentality theory is now the one to beat.

---

[1] Which includes having read this book up to this point!

[2] For thorough discussion from a psychological and anthropological perspective, see Becker (1971).

Bibliography

Adams, E. M. (2002) 'The meaning of life', *International Journal for Philosophy of Religion*, 51: 71–81.
Adams, Marilyn McCord (1975) 'Hell and the God of justice', *Religious Studies*, 11: 433–47.
Adams, Robert (1999) *Finite and Infinite Goods*, Oxford: Oxford University Press.
—— (2006) 'Love and the problem of evil', repr. in *The Positive Function of Evil*, Tabensky, Pedro (ed.), New York: Palgrave Macmillan, 2009, pp. 1–13.
Affolter, Jacob (2007) 'Human nature as God's purpose', *Religious Studies*, 43: 443–55.
Arendt, Hannah (1961) *Between Past and Future*, New York: Viking Press.
Audi, Robert (2005) 'Intrinsic value and meaningful life', *Philosophical Papers*, 34: 331–55.
Ayer, A. J. (1947) 'The claims of philosophy', repr. in *The Meaning of Life*, 2nd edn, Klemke, E. D. (ed.), New York: Oxford University Press, 2000, pp. 219–33.
—— (1990) *The Meaning of Life*, New York: Charles Scribner's Sons.
Baggini, Julian (2004) *What's It All About?: Philosophy and the Meaning of Life*, London: Granta Books.
Baier, Kurt (1957) 'The meaning of life', repr. in *The Meaning of Life*, 2nd edn, Klemke, E. D. (ed.), New York: Oxford University Press, 2000, pp. 101–32.
—— (1997) *Problems of Life and Death: A Humanist Perspective*, Amherst, NY: Prometheus Books.
Barnes, Hazel (1967) *An Existentialist Ethics*, New York: Alfred A. Knopf.
Baumeister, Roy (1991) *Meanings of Life*, New York: The Guilford Press.
Becker, Ernest (1971) *The Birth and Death of Meaning: An Interdisciplinary Perspective on the Problem of Man*, 2nd edn, New York: The Free Press.
Becker, Lawrence (1992) 'Good lives: prolegomena', in *The Good Life and the Human Good*, Paul, Ellen Frankel, Miller, Fred, and Paul, Jeffrey (eds), New York: Cambridge University Press, pp. 15–37.
Belliotti, Raymond (2001) *What is the Meaning of Human Life?*, Amsterdam: Rodopi.
Belshaw, Christopher (2005) *10 Good Questions About Life and Death*, Malden, MA: Blackwell.
Benatar, David (2004) 'Introduction', in *Life, Death & Meaning: Key Philosophical Readings on the Big Questions*, Benatar, David (ed.), Lanham, MD: Rowman & Littlefield Publishers, Inc., pp. 1–17.
—— (2006) *Better Never to Have Been: The Harm of Coming into Existence*, New York: Oxford University Press.
Bennett, James (1984) '"The meaning of life": a qualitative perspective', *Canadian Journal of Philosophy*, 14: 581–92.
Blackburn, Simon (1984) *Spreading the Word*, Oxford: Clarendon Press.
Blumenfeld, David (2009) 'Living life over again', *Philosophy and Phenomenological Research*, 79: 357–86.
Bond, E. J. (1983) *Reason and Value*, Cambridge: Cambridge University Press.
Bortolotti, Lisa (2010) 'Agency, life extension, and the meaning of life', *The Monist*, 92: 38–56.
Bortolotti, Lisa and Nagasawa, Yujin (2009) 'Immortality without boredom', *Ratio*, 22: 261–77.
Boyd, Richard (1988) 'How to be a moral realist', in *Essays on Moral Realism*, Sayre-McCord, Geoffrey (ed.), Ithaca, NY: Cornell University Press, pp. 181–228.

Boylan, Michael (2008) *The Good, the True and the Beautiful: A Quest for Meaning*, New York: Continuum.

Brännmark, Johan (2001) 'Good lives: parts and wholes', *American Philosophical Quarterly*, 38: 221–31.

—— (2003) 'Leading lives', *Philosophical Papers*, 32: 321–43.

Brink, David (1989) *Moral Realism and the Foundations of Ethics*, New York: Cambridge University Press.

Britton, Karl (1969) *Philosophy and the Meaning of Life*, Cambridge: Cambridge University Press.

Brogaard, Berit and Smith, Barry (2005) 'On luck, responsibility, and the meaning of life', *Philosophical Papers*, 34: 443–58.

Brown, Delwin (1971) 'Process philosophy and the question of life's meaning', *Religious Studies*, 7: 13–29.

Buber, Martin (1923) *I and Thou*, Kaufmann, Walter (tr.), repr. New York: Simon & Schuster Inc., 1970.

Camus, Albert (1942) *The Myth of Sisyphus*, O'Brian, Justin (tr.), repr. London: H. Hamilton, 1955.

Chappell, Timothy (2007) 'Infinity goes up on trial: must immortality be meaningless?', *European Journal of Philosophy*, 17: 30–44.

Clark, C. H. D. (1958) *Christianity and Bertrand Russell*, London: Lutterworth Press.

Cooper, David (2003) *Meaning*, Durham: Acumen Publishing.

—— (2005) 'Life and meaning', *Ratio*, 18: 125–37.

Copan, Paul (2004) 'Morality and meaning without God: another failed attempt', *Philosophia Christi*, 6: 295–304.

Cottingham, John (2003) *On the Meaning of Life*, London: Routledge.

—— (2005) *The Spiritual Dimension: Religion, Philosophy and Human Value*, Cambridge: Cambridge University Press.

—— (2008) 'The self, the good life and the transcendent', in *The Moral Life: Essays in Honour of John Cottingham*, Athanassoulis, Nafsika and Vice, Samantha (eds), New York: Palgrave Macmillan, pp. 231–74.

Craig, William Lane (1994) 'The absurdity of life without God', repr. in *The Meaning of Life*, 2nd edn, Klemke, E. D. (ed.), New York: Oxford University Press, 2000, pp. 40–56.

Crisp, Roger (2008) 'Meaning, morality, and religion', in *The Moral Life: Essays in Honour of John Cottingham*, Athanassoulis, Nafsika and Vice, Samantha (eds), New York: Palgrave Macmillan, pp. 167–83.

Cupit, Geoffrey (1997) *Justice as Fittingness*, Oxford: Clarendon Press.

Dahl, Norman (1987) 'Morality and the meaning of life', *Canadian Journal of Philosophy*, 17: 1–22.

Darwall, Stehpen (1983) *Impartial Reason*, Ithaca, NY: Cornell University Press.

Davis, William (1987) 'The meaning of life', *Metaphilosophy*, 18: 288–305.

Debord, Guy (1967) *The Society of the Spectacle*, Knabb, Ken (tr.), repr. Canberra: Hobgoblin Press, 2002.

Durant, Will (1932) *On the Meaning of Life*, repr. Carrollton, TX: Promethean Press, 2005.

Dworkin, Ronald (2000) *Sovereign Virtue: The Theory and Practice of Equality*, Cambridge, MA: Harvard University Press.

Eagleton, Terry (2007) *The Meaning of Life: A Very Short Introduction*, Oxford: Oxford University Press.

Edwards, Paul (1972) 'Why', in *The Encyclopedia of Philosophy, Volumes 7–8*, Edwards, Paul (ed.), New York: Macmillan Publishing Company, pp. 296–302.

Ellin, Joseph (1995) *Morality and the Meaning of Life*, Ft Worth, TX: Harcourt Brace College Publishers.

Fabry, Joseph (1968) *The Pursuit of Meaning*, Boston, MA: Beacon Press.

Fackenheim, Emil (1965) 'Judaism and the meaning of life', *Commentary*, 39: 49–55.

Feinberg, Joel (1980) 'Absurd self-fulfillment', repr. in *Freedom and Fulfillment: Philosophical Essays*, Feinberg, Joel (ed.), Princeton: Princeton University Press, 1994, pp. 297–330.

Feldman, Fred (1997) *Utilitarianism, Hedonism and Desert: Essays in Moral Philosophy*, Cambridge: Cambridge University Press.

Fischer, John Martin (1994) 'Why immortality is not so bad', *International Journal of Philosophical Studies*, 2: 257–70.

—— (2005) 'Free will, death, and immortality: the role of narrative', *Philosophical Papers*, 34: 379–403.

Flew, Anthony (1966) *God and Philosophy*, New York: Harcourt, Brace, & World Inc.

Ford, Dennis (2007) *The Search for Meaning: A Short History*, Berkeley, CA: University of California Press.

Frankfurt, Harry (1982) 'The importance of what we care about', repr. in *The Importance of What We Care About*, Frankfurt, Harry (ed.), New York: Cambridge University Press, 1988, pp. 80–94.

—— (2002) 'Reply to Susan Wolf', in *The Contours of Agency: Essays on Themes from Harry Frankfurt*, Buss, Sarah and Overton, Lee (eds), Cambridge, MA: The MIT Press, pp. 245–52.

Frankl, Viktor (1984) *Man's Search for Meaning*, rev. edn, New York: Simon & Schuster Inc.

Gale, Richard (1991) *On the Nature and Existence of God*, New York: Cambridge University Press.

Gewirth, Alan (1998) *Self-Fulfillment*, Princeton: Princeton University Press.

Gibbard, Allan (1990) *Wise Choices, Apt Feelings*, Cambridge, MA: Harvard University Press.

Gilbert, Alan (1990) *Democratic Individuality*, Cambridge: Cambridge University Press.

Godwin, William (1793) *Enquiry Concerning Political Justice*, repr. New York: Penguin Classics, 1976.

Gordon, Jeffrey (1980) 'In defense of despair: preliminary reflections on the meaning of life', *Southwest Philosophical Studies*, 5: 36–43.

—— (1983) 'Is the existence of God relevant to the meaning of life?', *The Modern Schoolman*, 60: 227–46.

Gordon, Robert (1987) *The Structure of Emotions*, Cambridge: Cambridge University Press.

Graham, Gordon (1990) *Living the Good Life*, New York: Paragon House.

Griffin, James (1981) 'On life's being valuable', *Dialectics and Humanism*, 8: 51–62.

Haber, Joram Graf (1997) 'Contingency and the meaning of life', *Philosophical Writings*, 5: 32–44.

Habermas, Jürgen (1983) *Moral Consciousness and Communicative Action*, Lenhardt, Christian and Nicholsen, Shierry Weber (trs), Cambridge, MA: The MIT Press, 1990.

Hanfling, Oswald (1987) *The Quest for Meaning*, New York: Blackwell.

Harman, Gilbert (1976) 'Practical reasoning', *The Review of Metaphysics*, 29: 431–63.

Harris, John (2002) 'Intimations of immortality—the ethics and justice of life-extending therapies', in *Current Legal Problems, Volume 55*, Freeman, Michael (ed.), New York: Oxford University Press, pp. 65–95.

Hartshorne, Charles (1984) 'God and the meaning of life', in *Boston University Studies in Philosophy and Religion, Volume 6: On Nature*, Rouner, Leroy (ed.), Notre Dame: University of Notre Dame Press, pp. 154–68.

―― (1996) 'The meaning of life', *Process Studies*, 25: 10–18.

Hepburn, R. W. (1966) 'Questions about the meaning of life', repr. in *The Meaning of Life*, 2nd edn, Klemke, E. D. (ed.), New York: Oxford University Press, 2000, pp. 261–76.

Hocking, William Ernest (1957) *The Meaning of Immortality in Human Experience*, New York: Harper & Brothers Publishers.

Hooker, Brad (2008) 'The meaning of life: subjectivism, objectivism, and divine support', in *The Moral Life: Essays in Honour of John Cottingham*, Athanassoulis, Nafsika and Vice, Samantha (eds), New York: Palgrave Macmillan, pp. 184–200.

Horgan, Terry and Timmons, Mark (1990-1991) 'New wave moral realism meets moral twin earth', *Journal of Philosophical Research*, 16: 447–65.

Horrobin, Steven (2005) 'The ethics of aging intervention and life-extension', in *Aging Interventions and Therapies*, Rattan, Suresh (ed.), Hackensack, NJ: World Scientific Publishers, pp. 1–27.

Hurka, Thomas (1993) *Perfectionism*, New York: Oxford University Press.

Huxley, Aldous (1932) *Brave New World*, London: HarperCollins.

Jacquette, Dale (2001) *Six Philosophical Appetizers*, Boston, MA: McGraw-Hill.

James, Laurence (2005) 'Achievement and the meaningfulness of life', *Philosophical Papers*, 34: 429–42.

―― (2009) 'Shape and the meaningfulness of life', in *Philosophy and Happiness*, Bortolotti, Lisa (ed.), New York: Palgrave Macmillan, pp. 54–67.

―― (2010) 'Activity and the meaningfulness of life', *The Monist*, 93: 57–75.

James, William (1899) 'What makes a life significant?', in *On Some of Life's Ideals*, James, William (ed.), New York: Henry Holt and Company, pp. 49–94.

Joske, W. D. (1974) 'Philosophy and the meaning of life', repr. in *The Meaning of Life*, 2nd edn, Klemke, E. D. (ed.), Oxford: Oxford University Press, 2000, pp. 283–94.

Kagan, Shelly (1998) 'Rethinking intrinsic value', *The Journal of Ethics*, 2: 277–97.

Kamm, Frances (2003) 'Rescuing Ivan Ilych: how we live and how we die', *Ethics*, 113: 202–33.

Kant, Immanuel (1787) *Critique of Pure Reason*, 2nd edn, Wood, Allen and Guyer, Paul (trs), repr. New York: Cambridge University Press, 1999.

―― (1788) *Critique of Practical Reason*, Beck, 3rd edn, Lewis White (tr.), repr. New York: Macmillan Publishing Company, 1993.

―― (1790) *Critique of Judgment*, Pluhar, Werner (tr.), repr. Indianapolis: Hackett Publishing Company, 1987.

―― (1793) *Religion Within the Limits of Reason Alone*, Greene, Theodore and Hudson, Hoyt (trs), repr. New York: Harper & Row Publishers, 1960.

Kass, Leon (2001) '*L'Chaim* and its limits: why not immortality?', *First Things*, 113: 17–24.

Kauppinen, Antti (2012) 'Meaningfulness and time', *Philosophy and Phenomenological Research*, 82: 345–77.

Kekes, John (1986) 'The informed will and the meaning of life', *Philosophy and Phenomenological Research*, 47: 75–90.

―― (2000) 'The meaning of life', in *Midwest Studies in Philosophy, Volume 24: Life and Death*, French, Peter and Wettstein, Howard (eds), Malden, MA: Blackwell, pp. 17–34.

―― (2010) *The Human Condition*, New York: Oxford University Press.

Kershnar, Stephen (2005) 'The injustice of hell', *International Journal for Philosophy of Religion*, 58: 103–23.

Kierkegaard, Søren (1850) 'Training in Christianity', Lowrie, Walter (tr.), repr. in *A Kierkegaard Anthology*, Bretall, Robert (ed.), Princeton: Princeton University Press, 1946, pp. 372–417.

Kivy, Peter (1990) *Music Alone*, Ithaca, NY: Cornell University Press.

Klemke, E. D. (2000) 'Living without appeal', rev. edn, in *The Meaning of Life*, 2nd edn, Klemke, E. D. (ed.), New York: Oxford University Press, 2000, pp. 186–97.

Koestler, Arthur (1967) *The Ghost in the Machine*, New York: Macmillan.

Korsgaard, Christine (1983) 'Two distinctions in goodness', *The Philosophical Review*, 92: 169–95.

Kretzmann, Norman and Stump, Eleonore (1985) 'Absolute simplicity', *Faith and Philosophy*, 2: 353–82.

Kurtz, Paul (1974) *The Fullness of Life*, New York: Horizon Press.

―― (2000) *Embracing the Power of Humanism*, Lanham, MD: Rowman & Littlefield Publishers.

Kushner, Harold (1987) *When All You've Ever Wanted Isn't Enough: The Search for a Life that Matters*, New York: Pocket Books.

Lamarque, Peter and Olsen, Stein Haugom (1994) *Truth, Fiction, and Literature*, Oxford: Oxford University Press.

Landau, Iddo (1997) 'Why has the question of the meaning of life arisen in the last two and a half centuries?', *Philosophy Today*, 41: 263–70.

―― (2013) 'Conceptualizing great meaning in life: Metz on the good, the true and the beautiful', *Religious Studies*, 49: 505–14.

LaPorte, Joseph (1997) 'Essential membership', *Philosophy of Science*, 64: 96–112.

Lenman, James (1995) 'Immortality: a letter', *Cogito*, 9: 164–9.

―― (2000) 'Consequentialism and cluelessness', *Philosophy and Public Affairs*, 29: 342–70.

Levine, Michael (1987) 'What does death have to do with the meaning of life?', *Religious Studies*, 23: 457–65.

Levinson, Jerrold (1992) 'Musical profundity misplaced', *Journal of Aesthetics and Art Criticism*, 50: 58–60

―― (2004) 'Intrinsic value and the notion of a life', *Journal of Aesthetics and Art Criticism*, 62: 319–29.

Levy, Neil (2005) 'Downshifting and meaning in life', *Ratio*, 18: 176–89.

Lewis, C. I. (1946) *An Analysis of Knowledge and Valuation*, LaSalle, IL: Open Court.

Lodzinski, Don (1998) 'The eternal act', *Religious Studies*, 34: 325–52.

MacDonald, Scott (ed.) (1991) *Being and Goodness*, Ithaca, NY: Cornell University Press.

Marcuse, Herbert (1964) *One-Dimensional Man: Studies in the Ideology of Advanced Industrial Society*, Boston, MA: Beacon Press.

Margolis, Joseph (1990) 'Moral realism and the meaning of life', *The Philosophical Forum*, 22: 19–48.

Markus, Arjan (2003) 'Assessing views of life: a subjective affair?', *Religious Studies*, 39: 125–43.

Martin, Michael (2002) *Atheism, Morality, and Meaning*, Amherst, NY: Prometheus Books.

Martin, Raymond (1993) 'A fast car and a good woman', repr. in *The Experience of Philosophy*, 2nd edn, Kolak, Daniel and Martin, Raymond (eds), Belmont, CA: Wadsworth Publishing Company, 1999, pp. 490–7.

Marx, Karl (1844) 'Economic and philosophical manuscripts', Bottomore, Tom (tr.), repr. in *Karl Marx: Early Writings*, Bottomore, Tom (ed.), New York: McGraw-Hill, 1964, pp. 61–194.

Mawson, Timothy (2010) 'Sources of dissatisfaction with answers to the question of the meaning of life', *European Journal for Philosophy of Religion*, 2: 19–41.

Metz, Thaddeus (2000) 'Could God's purpose be the source of life's meaning?', *Religious Studies*, 36: 293–313.

―――― (2001a) 'The concept of a meaningful life', *American Philosophical Quarterly*, 38: 137–53.

―――― (2001b) 'Respect for persons and perfectionist politics', *Philosophy and Public Affairs*, 30: 417–42.

―――― (2002a) 'Recent work on the meaning of life', *Ethics*, 112: 781–814.

―――― (2002b) 'The reasonable and the moral', *Social Theory and Practice*, 28: 277–301.

―――― (2003) 'Utilitarianism and the meaning of life', *Utilitas*, 15: 50–70.

―――― (2005a) 'Critical notice: Baier and Cottingham on the meaning of life', *Disputatio*, 19: 215–28.

―――― (2005b) 'Introduction', *Philosophical Papers*, 34: 311–29.

―――― (2006) 'Judging because understanding: a defence of retributive censure', in *Judging and Understanding: Essays on Free Will, Narrative, Meaning and the Ethical Limits of Condemnation*, Tabensky, Pedro (ed.), Aldershot: Ashgate Publishing Ltd, pp. 221–40.

―――― (2007a) 'The meaning of life', in *Stanford Encyclopedia of Philosophy*, Zalta, Edward (ed.), <http://plato.stanford.edu/entries/life-meaning/>

―――― (2007b) 'How to reconcile liberal politics with retributive punishment', *Oxford Journal of Legal Studies*, 27: 683–705.

―――― (2007c) 'God's purpose as irrelevant to life's meaning: reply to Affolter', *Religious Studies*, 43: 457–64.

―――― (2008) 'God, morality and the meaning of life', in *The Moral Life: Essays in Honour of John Cottingham*, Athanassoulis, Nafsika and Vice, Samantha (eds), New York: Palgrave Macmillan, pp. 201–27.

―――― (2009) 'Love and emotional reactions to necessary evil', in *The Positive Function of Evil*, Tabensky, Pedro (ed.), New York: Palgrave Macmillan, pp. 28–44.

―――― (2010) 'For the sake of the friendship: relationality and relationship as grounds of beneficence', *Theoria*, 57: 54–76.

―――― (2011a) 'Are lives worth creating?', *Philosophical Papers*, 40: 233–55.

―――― (2011b) 'The good, the true and the beautiful: toward a unified account of great meaning in life', *Religious Studies*, 47: 389–409.

―――― (2012) 'The meaningful and the worthwhile: clarifying the relationships', *The Philosophical Forum*, 43: 435–48.

Miller, Richard (1992) *Moral Differences*, Princeton: Princeton University Press.

Mintoff, Joseph (2008) 'Transcending absurdity', *Ratio*, 21: 64–84.

Moore, G. E. (1903) *Principia Ethica*, Cambridge: Cambridge University Press.

Moorhead, Hugh (ed.) (1988) *The Meaning of Life*, Chicago, IL: Chicago Review Press.

Moreland, J. P. (1987) *Scaling the Secular City: A Defense of Christianity*, Grand Rapids, MI: Baker Book House.

Morris, Thomas (1992) *Making Sense of it All: Pascal and the Meaning of Life*, Grand Rapids, MI: William B. Eerdmans Publishing Company.

Munitz, Milton (1986) *Cosmic Understanding*, Princeton: Princeton University Press.

—— (1993) *Does Life Have a Meaning?*, Buffalo, NY: Prometheus Books.
Murdoch, Iris (1970) *The Sovereignty of Good*, London: ARK.
Murphy, Jeffrie (1982) *Evolution, Morality, and the Meaning of Life*, Totowa, NJ: Rowman and Littlefield.
Nagasawa, Yujin (2008) 'A new defence of Anselmian theism', *The Philosophical Quarterly*, 58: 577–96.
Nagel, Thomas (1970) 'Death', repr. in *Mortal Questions*, Nagel, Thomas (ed.), New York: Cambridge University Press, 1979, pp. 1–10.
—— (1971) 'The absurd', *The Journal of Philosophy*, 68: 716–27.
—— (1986) *The View from Nowhere*, New York: Oxford University Press.
Nielsen, Kai (1981) 'Linguistic philosophy and "the meaning of life"', rev. edn, repr. in *The Meaning of Life*, 2nd edn, Klemke, E. D. (ed.), New York: Oxford University Press, 2000, pp. 233–56.
Nietzsche, Friedrich (1886) *Beyond Good and Evil*, Kaufmann, Walter (tr.), repr. New York: Rand House Inc., 1966.
Nozick, Robert (1974) *Anarchy, State and Utopia*, New York: Basic Books.
—— (1981) *Philosophical Explanations*, Cambridge, MA: Harvard University Press.
—— (1989) *The Examined Life*, New York: Simon and Schuster.
—— (1993) *The Nature of Rationality*, Princeton: Princeton University Press.
—— (2001) *Invariances: The Structure of the Objective World*, Cambridge, MA: Harvard University Press.
Nussbaum, Martha (1989) 'Mortal immortals: Lucretius on death and the voice of nature', *Philosophy and Phenomenological Research*, 50: 303–51.
Ostwald, Wilhelm (1906) *Individuality and Immortality*, Boston, MA: Houghton Mifflin Company.
Parfit, Derek (1984) *Reasons and Persons*, New York: Oxford University Press.
Partridge, Ernest (1981) 'Why care about the future?', in *Responsibilities to Future Generations: Environmental Ethics*, Partridge, Ernest (ed.), Buffalo, NY: Prometheus Books, pp. 203–19.
Pascal, Blaise (1669) *Pensées*, Trotter, W. F. (tr.), repr. Stilwell, KS: Digireads.com, 2005.
Perrett, Roy (1985) 'Tolstoy, death and the meaning of life', *Philosophy*, 60: 231–45.
—— (1986) 'Regarding immortality', *Religious Studies*, 22: 219–33.
Pogge, Thomas (1997) 'Kant on ends and the meaning of life', in *Reclaiming the History of Ethics: Essays for John Rawls*, Reath, Andrew, Herman, Barbara, and Korsgaard, Christine (eds), New York: Cambridge University Press, pp. 361–87.
Post, John (1987) *The Faces of Existence*, Ithaca, NY: Cornell University Press.
Pritchard, Duncan (2010) 'Absurdity, angst, and the meaning of life', *The Monist*, 93: 3–16.
Putnam, Hilary (1988) *Representation and Reality*, Cambridge, MA: The MIT Press.
Quigley, Muireann and Harris, John (2009) 'Immortal happiness', in *Philosophy and Happiness*, Bortolotti, Lisa (ed.), New York: Palgrave Macmillan, pp. 68–81.
Quinn, Philip (2000) 'How Christianity secures life's meanings', in *The Meaning of Life in the World Religions*, Runzo, Joseph and Martin, Nancy (eds), Oxford: Oneworld Publications, pp. 53–68.
Rabinowicz, Wlodek and Rønnow-Rasmussen, Toni (2000) 'A distinction in value: intrinsic and for its own sake', *Proceedings of the Aristotelian Society*, 100: 33–51.

Railton, Peter (1984) 'Alienation, consequentialism, and the demands of morality', *Philosophy and Public Affairs*, 13: 134–71.
Rao, P. Nagaraja (1965) 'Hinduism', in *The Meaning of Life in Five Great Religions*, Chalmers, R. C. and Irving, John (eds), Philadelphia: The Westminster Press, pp. 23–36.
Raz, Joseph (2001) *Value, Respect, and Attachment*, Cambridge: Cambridge University Press.
Rees, Marc (2011) 'The great man', MA thesis, University of the Witwatersrand, <http://wiredspace.wits.ac.za/bitstream/handle/10539/10097/The%20Great%20Man.pdf?sequence=2>.
Reid, Julie (2009) 'Superhuman meaningful pastimes for the everyday cyborg', MA thesis, University of the Witwatersrand, <http://wiredspace.wits.ac.za/handle/10539/6903>.
Rescher, Nicholas (1990) *Human Interests: Reflections on Philosophical Anthropology*, Stanford: Stanford University Press.
Rogers, Katherine (1996) 'The traditional doctrine of divine simplicity', *Religious Studies*, 32: 165–86.
―――― (2000) *Perfect Being Theology*, Edinburgh: Edinburgh University Press.
Ross, W. D. (1930) *The Right and the Good*, repr. Indianapolis: Hackett Publishing Company, 1988.
Runzo, Joseph (2000) 'Eros and meaning in life and religion', in *The Meaning of Life in the World Religions*, Runzo, Joseph and Martin, Nancy (eds), Oxford: Oneworld Publications, pp. 187–201.
Russell, Bertrand (1912) *The Problems of Philosophy*, repr. New York: Oxford University Press, 1959.
Sampson, Ronald (1969) 'The vanity of humanism', repr. in *Sources*, Roszak, Theodore (ed.), New York: Harper & Row Publishers, 1972, pp. 483–501.
Sartre, Jean-Paul (1946) 'Existentialism is a humanism', Kaufmann, Walter (tr.), repr. in *Existentialism from Dostoyevsky to Sartre*, Kaufmann, Walter (ed.), New York: The World Publishing Company, 1956, pp. 287–311.
Scanlon, T.M. (1998) *What We Owe to Each Other*, Cambridge, MA: Harvard University Press.
Schlick, Moritz (1927) 'On the meaning of life', Heath, Peter (tr.), repr. in *Life and Meaning: A Philosophical Reader*, Hanfling, Oswald (ed.), Cambridge: Basic Blackwell Inc., 1987, pp. 60–73.
Schmidtz, David (2001) 'The meanings of life', in *Boston University Studies in Philosophy and Religion, Volume 22: If I Should Die*, Rouner, Leroy (ed.), Notre Dame: University of Notre Dame Press, pp. 170–88.
Scholem, Gershom (1991) *'Devekut* or Communion with God', in *Essential Papers on Hasidism*, Hundert, Gershon David (ed.), New York: New York University Press, pp. 275–98.
Schopenhauer, Arthur (1851a) 'Additional remarks on the doctrine of the vanity of existence', Payne, E. F. J. (tr.), repr. in *Arthur Schopenhauer: Philosophical Writings*, Schirmacher, Wolfgang (ed.), New York: Continuum, 1996, pp. 19–26.
―――― (1851b) 'Additional remarks on the doctrine of the suffering of the world', Payne, E. F. J. (tr.), repr. in *Arthur Schopenhauer: Philosophical Writings*, Schirmacher, Wolfgang (ed.), New York: Continuum, 1996, pp. 27–41.
Seachris, Joshua (2009) 'The meaning of life as narrative: a new proposal for interpreting philosophy's "primary" question', *Philo*, 12: 5–23.
Sharpe, Bob (1999) 'In praise of the meaningless life', *Philosophy Now*, 24: 15.
Singer, Irving (1996) *Meaning of Life, Volume 1: The Creation of Value*, Baltimore: John Hopkins University Press.

Singer, Peter (1993) *Practical Ethics*, 2nd edn, New York: Cambridge University Press.
―― (1995) *How Are We to Live?*, Amherst, NY: Prometheus Books.
Slote, Michael (1983) *Goods and Virtues*, Oxford: Oxford University Press.
Smart, J. J. C. (1999) 'Meaning and purpose', *Philosophy Now*, 24: 16.
Smilansky, Saul (2012) 'On the common lament, that a person cannot make much difference in this world', *Philosophy*, 87: 109–22.
Smith, Adam (1759) *The Theory of Moral Sentiments*, Raphael, D. D. and Macfie, A. L. (eds), repr. Indianapolis: Liberty Classics, 1976.
Smith, Barry (2002-2003) 'The measure of civilizations', *Academic Questions*, 16: 16–22.
Smith, Huston (2000) 'The meaning of life in the world's religions', in *The Meaning of Life in the World Religions*, Runzo, Joseph and Martin, Nancy (eds), Oxford: Oneworld Publications, pp. 255–68.
Smith, Quentin (1997) *Ethical and Religious Thought in Analytic Philosophy of Language*, New Haven: Yale University Press.
―― (2003) 'Moral realism and infinite spacetime imply moral nihilism', in *Time and Ethics: Essays at the Intersection*, Dyke, Heather (ed.), Dordrecht: Kluwer Academic Publishers, pp. 43–54.
Solomon, Robert (1993) *The Passions: Emotions and the Meaning of Life*, Indianapolis: Hackett Publishing Company.
―― (1986) *The Big Questions*, New York: Harcourt Brace Jovanovich.
Starkey, Charles (2006) 'Meaning and affect', *The Pluralist*, 1: 88–103.
Storr, Anthony (1988) *Solitude: A Return to the Self*, New York: Free Press.
―― (1992) *Music and the Mind*, New York: Ballantine Books.
Sturgeon, Nicholas (1988) 'Moral explanations', in *Essays on Moral Realism*, Sayre-McCord, Geoffrey (ed.), Ithaca, NY: Cornell University Press, pp. 229–55.
Suckiel, Ellen (2003) 'William James on the cognitivity of feelings, religious pessimism, and the meaning of life', *Journal of Speculative Philosophy*, 17: 30–9.
Swenson, David (1949) *Kierkegaardian Philosophy in the Faith of a Scholar*, Philadelphia: The Westminster Press.
Swinburne, Richard (1991) *The Existence of God*, Oxford: Oxford University Press.
―― (1993) *The Coherence of Theism*, rev. edn, Oxford: Oxford University Press.
Tabensky, Pedro (2003) 'Parallels between living and painting', *The Journal of Value Inquiry*, 37: 59–68.
Tännsjö, Torbjörn (1988) 'The moral significance of moral realism', *The Southern Journal of Philosophy*, 26: 247–61.
Taylor, Charles (1989) *Sources of the Self: The Making of the Modern Identity*, Cambridge, MA: Harvard University Press.
―― (1992) *The Ethics of Authenticity*, Cambridge, MA: Harvard University Press.
Taylor, Richard (1970) 'The meaning of life', in *Good and Evil*, Taylor, Richard (ed.), repr. Amherst, NY: Prometheus Books, 2000, pp. 319–34.
―― (1981) 'The meaning of human existence', in *Values in Conflict: Life, Liberty, and the Rule of Law*, Leiser, Burton (ed.), New York: Macmillan Publishing, Co., pp. 3–26.
―― (1985) *Ethics, Faith, and Reason*, Englewood Cliffs, NJ: Prentice-Hall Inc.
―― (1987) 'Time and life's meaning', *The Review of Metaphysics*, 40: 675–86.
―― (1999) 'The meaning of life', *Philosophy Now*, 24: 13–14.
Teichman, Jenny (1993) 'Humanism and the meaning of life', *Ratio*, 6: 155–64.

Theron, Stephen (1985) 'Happiness and transcendent happiness', *Religious Studies*, 21: 349–67.
Thomas, Laurence (2005) 'Morality and a meaningful life', *Philosophical Papers*, 34: 405–27.
Thomson, Garrett (2003) *On the Meaning of Life*, South Melbourne: Wadsworth.
Tolstoy, Leo (1884) *My Confession*, Wiener, Leo (tr.), repr. in *The Meaning of Life*, 2nd edn, Klemke, E. D. (ed.), New York: Oxford University Press, 2000, pp. 11–20.
Trisel, Brooke Alan (2002) 'Futility and the meaning of life debate', *Sorites*, 14: 70–84.
──── (2004) 'Human extinction and the value of our efforts', *The Philosophical Forum*, 35: 371–91.
──── (2012) 'Intended and unintended life', *The Philosophical Forum*, 43: 395–403.
Van Schaik, Louis (1977) 'The concept of care', in *Meaning in Life*, Macnamara, Michael (ed.), Cape Town: A.D. Donker, pp. 139–50.
Velleman, David (1991) 'Well-being and time', *Pacific Philosophical Quarterly*, 72: 48–77.
──── (1999) 'Love as a moral emotion', *Ethics*, 109: 338–74.
──── (2009) *How We Get Along*, New York: Cambridge University Press.
Vice, Samantha (2003) 'Literature and the narrative self', *Philosophy*, 78: 93–108.
Walker, Lois Hope (1989) 'Religion and the meaning of life and death', in *Philosophy: The Quest for Truth*, Pojman, Louis (ed.), Belmont, CA: Wadsworth Publishing Co., pp. 167–71.
White, David (1992) 'Toward a theory of profundity in music', *Journal of Aesthetics and Art Criticism*, 50: 23–34.
Wielenberg, Erik (2005) *Value and Virtue in a Godless Universe*, Cambridge: Cambridge University Press.
Wiggins, David (1988) 'Truth, invention, and the meaning of life', rev. edn, in *Essays on Moral Realism*, Sayre-McCord, Geoffrey (ed.), Ithaca, NY: Cornell University Press, pp. 127–65.
Williams, Bernard (1973) 'The Makropulos case: reflections on the tedium of immortality', in *Problems of the Self*, Williams, Bernard (ed.), Cambridge: Cambridge University Press, pp. 82–100.
──── (1976) 'Persons, character and morality', in *The Identities of Persons*, Rorty, Amélie Oksenberg (ed.), Berkeley: University of California Press, pp. 197–216.
──── (1985) *Ethics and the Limits of Philosophy*, Cambridge, MA: Harvard University Press.
──── (2007) 'Life as narrative', *European Journal of Philosophy*, 17: 305–14.
Williams, Garrath (1999) 'Kant and the question of meaning', *The Philosophical Forum*, 30: 115–31.
Wisdom, John (1965) 'The meanings of the questions of life', repr. in *The Meaning of Life*, 2nd edn, Klemke, E. D. (ed.), New York: Oxford University Press, 2000, pp. 257–60.
Wisnewski, J. Jeremy (2005) 'Is the immortal life worth living?', *International Journal of Philosophy of Religion*, 58: 27–36.
Wittgenstein, Ludwig (1914-1916) *Notebooks 1914-16*, 2nd edn, Anscombe, G. E. M. (tr.), repr. Malden, MA: Blackwell, 1998.
──── (1921) *Tractatus Logico-Philosophicus*, Pears, David and McGuinness, Brian (trs), repr. London: Routledge & Kegan Paul, 1961.
──── (1965) 'A lecture on ethics', *The Philosophical Review*, 74: 3–12.
Wohlgennant, Rudolf (1981) 'Has the queston about the meaning of life any meaning?', Hanfling, Oswald (tr.), repr. in *Life and Meaning: A Reader*, Hanfling, Oswald (ed.), Cambridge: Basic Blackwell Inc., 1987, pp. 34–8.
Wolf, Susan (1997a) 'Happiness and meaning: two aspects of the good life', *Social Philosophy and Policy*, 14: 207–25.

—— (1997b) 'Meaningful lives in a meaningless world', *Quaestiones Infinitae*, Volume 19, Utrecht: Utrecht University, pp. 1–22.
—— (1997c) 'Meaning and morality', *Proceedings of the Aristotelian Society*, 97: 299–315.
—— (2002) 'The true, the good, and the lovable: Frankfurt's avoidance of objectivity', in *The Contours of Agency, Essays on Themes from Harry Frankfurt*, Buss, Sarah and Overton, Lee (eds), Cambridge, MA, The MIT Press, pp. 227–44.
—— (2010) *Meaning in Life and Why it Matters*, Princeton: Princeton University Press.
Wollheim, Richard (1984) *The Thread of Life*, Cambridge, MA: Harvard University Press.
Wong, Wai-hung (2008) 'Meaningfulness and identities', *Ethical Theory and Moral Practice*, 11: 123–48.

Index

absolute morality *see* divine command theory/
  theorists; invariant morality
absurdity 6, 98, 100, 104, 105, 106, 142, 188, 244
Adams, E. M. 147, 148, 149
Adams, Robert 19, 47, 101
admiration *see* 'esteem/admiration'
aesthetics *see* 'beauty'; 'creativity'
Affolter, Jacob 115–17
afterlife 33, 80, 83, 108, 123, 124, 125, 132, 143
  see also 'death'; 'immortality'; 'soul-centred
    theory/theories'
agent-relativity 12, 181, 197–8, 199, 221, 225
Albee, Edward 1
Alexis, Alexander 175, 179
anti-matter 53, 64, 69, 71, 72, 73, 134, 146, 147,
  151, 174, 201, 219, 220, 233, 234, 235, 238,
  247
  definition of 64
  existence of 64, 220
  negative dimension of 63–4, 222, 234
  see also 'disvalue (of meaningfulness, of pain)';
    'immorality'
Aquinas, Thomas 79, 87, 107, 111, 115
Arnason, Kathleen 37
atheism/atheists 80, 88, 89, 242, 243, 247
attitudes *see* 'subjective conditions'
attraction *see* 'subjective conditions';
  'subjectivism/subjectivists'
attractiveness *see* 'objectivism/objectivists'
authenticity 29, 65, 69, 173–4, 179
autonomy 4, 29–30, 69, 99–104, 127, 174, 227,
  241

Baier, Kurt 24, 102–3, 141, 148, 150, 151, 152, 153,
  154, 155, 187
  see also 'imperfection thesis'
bearer of meaning 3, 4, 37, 38, 60, 65–8, 141,
  167
  definition of 38, 66
  intentional action as primary 65–6, 223–4
  mixed view 10, 39–40, 51–8, 77, 146
  narrative structure 37–9, 41, 52–7
  part-life meaning 4, 10, 37, 38, 39, 41, 42, 44,
    49, 50, 51, 55, 57, 78, 146, 147, 163, 198,
    219, 220, 222, 233, 235
  pure part-life view 10, 38, 39, 49–51, 57–8
  pure whole-life view 38–49, 57
  whole-life meaning 4, 9, 10, 37, 38, 39, 45–9,
    50, 51, 52, 54, 55, 56, 57, 58, 63, 70, 78,
    146, 147, 198, 222, 233, 235

beauty 5, 87, 89, 95, 131, 156, 203, 204, 208, 210,
  214, 230–1, 249
  see also 'creativity'; 'Dostoyevsky, Fyodor';
    'Picasso, Pablo'
Becker, Lawrence 55, 249
Benatar, David 158, 241, 244, 245, 246
Bentham, Jeremy 186
bias toward the future 10, 70–3
Blake, William 139
boredom 20, 62, 69, 134–5, 174, 182, 183, 196,
  241
Brännmark, Johan 38, 45, 55, 56, 57
Brogaard, Berit and Barry Smith 67, 176,
  177
Buber, Martin 19, 107

Camus, Albert 6, 70, 241, 242
care *see* 'love'
coherentism 8
conation/conative 30, 166, 167, 182, 198,
  223
  see also 'subjective conditions'; 'subjectivism/
    subjectivists'
concept of meaning 3–6, 17–36, 77
  apt emotions 31–4
  contrasted with conceptions 18–21
  pluralist analysis of 34–5
  purposiveness 24–8
  transcendence 28–31
conceptions of meaning 11, 18–21, 23, 24, 26,
  32, 77–8, 119, 133, 147, 150, 186
  contrasted with concept 18–21
  see also 'meaningfulness, theories of'
consequentialism/consequentialists 4, 12, 184–5,
  192, 193, 194, 195, 196, 197, 198, 200, 221,
  232
  arguments against 194–8
  arguments for 184–5, 186–7, 194, 232
  definition of 181, 184–5
  perfectionist 193–4
  side-constrained 190, 194
  utilitarian 185–7
  see also 'objectivism/objectivists'
contingency 83–4, 85–6, 94, 116
Cottingham, John 19, 80, 85–97, 107, 125, 146,
  175, 191
  see also 'divine command theory/theorists';
    'invariant morality'; 'purpose theory/
    theorists'
Craig, William Lane 83–4, 145

264 INDEX

creativity 6, 51, 63, 65–8, 71, 87, 115, 131, 174, 181, 194, 205, 211–12, 214, 215, 221, 222, 223, 232, 246
  see also 'beauty'; 'Dostoyevsky, Fyodor'; 'Picasso, Pablo'
Crisp, Roger 97

Darrow, Clarence 240, 241
Darwall, Stephen 168, 178, 179
Darwin, Charles 2, 5, 137, 144, 204, 205, 207, 212, 213, 214, 229
death 48–9, 54, 70, 111, 119, 125–6, 128, 130–1, 136–7, 142, 148, 150, 153–4, 177, 238, 247–8
  see also 'afterlife'; 'immortality'; 'soul-centred theory/theories'
Dennett, Daniel 110
deontology 26, 57, 181, 198
  see also 'agent-relativity'; 'non-consequentialism/non-consequentialists'; 'side-constraint'
depth see fundamental conditions
desiderata for a theory of meaning 220–2, 224, 225, 226, 232, 233, 235
  agent-relativity 221
  deliberation and decision 222
  good consequences 221
  internal and external 221
  moral constraints 221
  negative conditions 220
  object of rationality 222
  spiritual realm 220
  subjective conditions 220
dignity 4, 22, 26, 63, 99, 116, 117, 189, 215
disvalue (of meaningfulness, of pain) 10, 48, 63–4, 72, 153, 233, 235, 236
divine command theory/theorists 10, 85, 87, 96, 97, 99, 103, 139, 145, 158, 159, 170, 194
  see also 'Cottingham, John'; 'invariant morality'; 'meta-ethics'; 'purpose theory/theorists'
Dostoyevsky, Fyodor 2, 5, 8, 45, 137, 144, 146, 204, 211, 212, 214, 230
Dworkin, Ronald 187, 241

Eagleton, Terry 201, 203, 215
Ecclesiastes 59, 124, 126, 132, 238
Einstein, Albert 2, 4, 63, 96, 137, 144, 150, 194, 204, 212, 213, 229
Emerson, Ralph Waldo 240, 247
enquiry see 'truth'
esteem/admiration 9, 18, 31, 32, 33, 34, 35, 37, 53, 64, 69, 72, 73, 77, 79, 137, 139, 169, 185, 186, 235, 244, 249
Euthyphro objections 10, 87–88, 159, 170
evil 80, 89, 124, 243
excellence see 'virtue'

existence see 'God's existence'; 'human existence/life'
existentialism 8, 19, 80, 143, 169
extension 87, 111, 115, 117, 133, 205, 214–15, 217, 229–30

Fackenheim, Emil 19
Fischer, John Martin 41, 51, 134–5
Flew, Anthony 128
foundationalism 8
Frankfurt, Harry 20, 170, 171
Frankl, Viktor 65, 188
fundamental conditions 226
  contrasted with necessary conditions 226, 237–8
  contrasted with universal conditions 227, 229, 230–1
fundamentality theory 13, 222–39, 241, 249
  aesthetic creation 230–1
  arguments against 219, 236–9
  arguments for 13, 220–35, 249
  basic definition of 13, 146, 222–6
  complete definition of 235
  intellectual reflection 229–30
  moral achievement 227–9
  see also 'non-consequentialism/non-consequentialists'; 'objectivism/objectivists'
futility 6, 140, 141

Gauguin, Paul 190, 191
Gewirth, Alan 210, 211, 215, 216, 217, 218, 227, 230
God 63, 79–80, 81, 110–15
  definition of 79–80, 81, 113–14
  eternality 105–6
  in relation to universe 104, 105, 108, 109, 113, 114, 115
  omnipotence 104–5
  perfect nature 87, 88, 110–15
  qualitative properties 110–12, 114–15
God-centred theory/theories 10, 11, 79–82, 85, 99, 106–12, 114, 118, 119, 120–2, 123, 124, 128, 130, 132, 133, 138, 243
  see also 'divine command theory/theorists'; 'purpose theory/theorists'; 'supernaturalism/supernaturalists'
God's existence 81, 83, 88, 89, 90, 91, 97, 107, 108, 109, 110, 145, 146, 170, 242–3
  see also 'atheism/atheists'; 'spiritual realm'; 'theism/theists'
God's purpose see 'divine command theory/theorists'; 'purpose theory/theorists'; 'purposiveness'
goodness see 'morality'
Gorecki, Henri 46
Griffin, James 168

happiness 2, 5, 6, 10, 17, 19, 21, 27, 30, 33, 34, 35, 39, 60, 73–4, 78, 83, 107, 126, 142, 192, 193, 197, 238, 249
  definition of 33, 60, 73–4
  different from virtue 192
  different from meaning 5–6, 27–9, 60–2, 65–74, 249
  partly constitutive of meaning 33, 62, 126, 197, 238
  similar to meaning 60–5, 142
  see also 'pleasure'; 'well-being'
Hartshorne, Charles 107, 109
Heller, Joseph 70
Hepburn, R.W. 180, 181, 184, 197
Hitler, Adolf 5, 26, 29, 52, 101
Hocking, William Ernest 124, 129
Hooker, Brad 175, 216
human existence/life 3, 4, 6, 21, 24, 25, 31, 37, 38–9, 41, 45, 53, 65, 83–4, 94, 96, 102–3, 104, 109, 111, 115, 120, 128, 137, 140, 144–5, 150, 153, 154, 168, 175, 179, 185, 187, 189, 223, 229, 231, 239, 242, 243, 245
  fundamental conditions of 13, 222, 226, 233, 235, 236–8, 249
  significant 5, 18, 19, 20, 21, 26, 27, 30, 33, 40, 49, 54, 69, 70, 78, 79, 80, 96, 97, 106, 112, 123, 142, 144, 150, 164, 197, 210, 216, 221
  see also 'meaningfulness'
Humeanism 86, 93, 169
Hurka, Thomas 44, 71, 192, 193, 212, 214, 223, 229

immorality 5, 85–6, 99–104, 190–1, 227–9
  disrespect 99–104, 190–1, 227–9
  in relation to meaning 5, 46, 49, 52, 85–6, 190–1, 212, 221, 224, 234
  see also 'anti-matter'; 'justice'; 'Kantian ethics'; 'morality'
immortality 11, 120, 123, 124, 126, 127, 128, 129, 130, 131, 132, 134, 140, 156, 247
  see also 'afterlife'; 'death'; 'soul-centred theory/theories'
immortality requirement see 'soul-centred theory/theories'
impact 26, 185, 187, 240, 247
  see also 'permanence'
imperfection thesis 11, 12, 140, 146–60, 163
  definition of 11, 140
  favoured conception of 12, 154–8, 236
  implications of 158–60
  individual conceptions 147–50
  social conceptions 150–8
  see also 'naturalism/naturalists'
incoherence arguments 88–91, 94, 96–7, 139, 145–6, 158–9, 163, 220
  against consequentialism 194
  against divine command theory 88–91, 94, 96–7, 158
  against nihilism 241–6
  against non-naturalism 158–9
  against supernaturalism 139, 145–6, 158
integrity 5, 29, 30, 61, 112, 191, 192, 193, 210
intersubjectivity see 'subjectivism/subjectivists, intersubjectivity'
invariant morality 10, 78, 85, 86, 87–97, 114
  see also 'divine command theory/theorists'; 'meta-ethics'; 'moral realism/realists'; 'objectivity'

James, William 19, 167
justice 6, 33, 59, 124–7, 132, 133, 137, 166, 183, 238
  compensatory 125, 132
  distributive 137, 207, 216, 234
  retributive 33, 59, 83, 100–1, 114, 124–7, 132, 133, 197, 238
  see also 'Karma'; 'morality'
justification/justified theory 1, 2, 8, 12, 13, 140, 177, 184, 239, 249

Kamm, Frances 55–6, 57
Kant, Immanuel 26, 42, 85, 88, 102, 126, 127, 132, 159
Kantian ethics 92, 100
  as justified by moral realism 91–5
  respect for rational nature 26, 40, 42, 88, 165, 167, 176, 192, 193, 210, 215, 221, 223, 224, 225, 226, 227, 228, 232, 234, 239, 248
Kantian perspective 4, 42, 68, 99, 100, 126, 229
Karma 83, 109, 125, 238–9
  see also 'justice, retributive'
Kass, Leon 136, 245
Kauppinen, Antti 22, 31, 38
Kierkegaard, Søren 19, 98, 99, 107
knowledge see 'justification/justified theory'; 'truth'
Koestler, Arthur 249
Kripke, Saul 91, 210, 215

Landau, Iddo 237, 238
laws of nature 12, 29, 91–2, 95, 131, 155, 163, 202, 215, 245
Led Zeppelin 119
Levinson, Jerrold 48, 49
Levy, Neil 29, 30, 32, 205–7, 215
libertarian freedom 29, 104, 159
love 1, 5, 25, 27, 31, 61, 77, 81, 121, 122, 128, 131, 136, 137, 143, 144, 164, 170, 171, 172, 174, 183, 201–2, 203–4, 206, 207, 214, 216, 223, 224, 228, 230, 233
  definition of 121, 201–2
  God's 26, 109, 122, 139
  ground of theories of meaning 170–1, 201–2
luck 68, 73

266 INDEX

Mandela, Nelson 2, 4, 137, 144, 207, 210, 217, 227, 228
Markus, Arjan 34, 167
Marx, Karl 20, 205
meaningfulness
 bearer of *see* 'bearer of meaning'
 cognitivism about 21–2
 concept of *see* 'concept of meaning'
 conceptions of *see* 'conceptions of meaning'
 cosmic (holist) 3, 37, 239
 definition of *see* 'concept of meaning'
 degrees of 4, 6, 22, 38–9, 63, 66–8, 84, 123, 140, 141–2, 146, 186, 222
 final (non-instrumental) value of 4, 6, 34, 62, 68, 107, 129, 133, 142
 individualist 3, 37, 38
 on balance 39, 146–58, 236
 pro tanto 39, 78, 163, 220, 235
 scales of 63–4, 234
 search for 1–3, 7, 13, 36, 184, 249
 source of 66–8
 synonyms of 3, 18, 21–2
 theories of 6–9, 10, 18–21, 77–9, 106–8, 164–5, 180–1, 200, 220–2, 249
 see also 'human existence/life, significant'
meaninglessness 4, 5, 6, 13, 32, 41, 46–7, 59, 64, 66, 68, 95, 151, 152–3
 see also 'anti-matter'; 'disvalue (of meaningfulness, of pain)'; 'nihilism/nihilists'
meta-ethics 22, 88, 90, 91, 93, 95, 169–73, 179, 243–4
 see also 'divine command theory/theorists'; 'invariant morality'; 'moral realism/realists'
metaphysics 7, 8, 72, 78, 111, 114, 120, 125, 134, 154, 159, 163, 170, 209, 210, 214, 220, 226, 229, 243
 free will 29, 41, 82, 84, 103, 104, 105, 113, 143, 159
 personal identity 125, 126, 134, 157, 225
 see also 'contingency'; 'God'; 'God's existence'; 'immortality'; 'naturalism/naturalists'; 'non-naturalism'; 'qualitative properties'; 'soul'; 'spiritual realm'; 'supernaturalism/supernaturalists'
Mill, John Stuart 62, 186
Mintoff, Joseph 202–5, 206, 209, 210, 214, 215, 225
monotheism/monotheists 77, 79, 81, 83, 87, 97, 99, 107, 116, 150
 see also 'God'; 'God's existence'; 'theism/theists'
Moore, G. E. 25, 26, 28, 38
moral realism/realists 7, 91–6, 171–2, 179
morality v, 5, 6, 17, 19, 21, 35, 56, 68, 94, 96, 125, 186, 191, 213, 224, 227–9, 237–9, 244

 see also 'immorality'; 'justice'; 'Kantian ethics'; 'Karma'; 'love'; 'Mandela, Nelson'; 'Mother Teresa'; 'virtue'
Morris, Thomas 110
Mother Teresa 2, 4, 49, 96, 135, 137, 144, 175, 183, 196, 197, 198, 212, 220, 227, 228, 237
Munitz, Milton 166
Murray, Bill 50, 52, 56

Nagel, Thomas 6, 153, 241, 244, 246
narrative structure 37, 38, 39, 41, 52–8
naturalism/naturalists 12, 19, 23, 78, 79, 84, 86, 91, 92, 93, 94, 99, 108, 114, 123, 139–58, 163, 164–5, 171–2, 220, 233
 arguments against 82–6, 93–4, 124–33, 137–8
 arguments for 140–6, 243–4
 as a kind of metaphysics 92, 93
 as a kind of moral realism 7, 91–6, 171–2, 243–4
 conceptions of 19–20, 164–5, 200
 definition of 12, 19, 78–9, 93, 164, 220
 see also 'imperfection thesis'; 'objectivism/objectivists'; 'subjectivism/subjectivists'
Nielsen, Kai 24, 26, 28
Nietzsche, Friedrich 20, 156
nihilism/nihilists 13, 80, 124, 146, 241, 242, 243, 244, 246, 247
 arguments for 13, 242–8
 definition of 13, 241
non-consequentialism/non-consequentialists 13, 199–218, 221, 230
 arguments for 190–1, 194–8
 creativity 211–12
 culture 202–5
 definition of 13, 199–200
 fundamentality *see* 'fundamentality theory'
 love 201–2
 one's rational capacities 210–11
 open-ended activities 205–7
 organic unity 207–10
 rationality toward universality 215–18
 theoretical and practical rationality 212–15
 see also 'agent-relativity'; 'deontology'; 'objectivism/objectivists'; 'side-constraint'
non-naturalism 20, 78, 79, 91, 140, 158–9, 194
 argument against 91, 158–9, 220
 arguments for 159
 definition of 78, 79, 158–9
normativity 86, 87, 92–5
Nozick, Robert 20, 27, 28, 29, 50, 94, 109, 110, 112, 113, 130, 131, 132, 133, 194, 207–10, 214, 215, 221, 229, 231, 247
 see also 'transcendence'; 'unity, organic'
Nussbaum, Martha 137, 156, 157, 245

objective morality *see* 'invariant morality'
objective standpoint *see* 'point of view of the universe'
objectivism/objectivists 12, 13, 19, 20, 24, 26, 29, 30, 33, 146, 164–5, 169–74, 175–82, 199–200, 221, 244–7
  arguments against 169–74, 244–6
  arguments for 175–9
  conceptions of 12–13, 181, 197, 199, 221
  definition of 12, 20, 165, 182
  pure 182, 186
  see also 'consequentialism/consequentialists'; 'fundamentality theory'; 'naturalism/naturalists'; 'non-consequentialism/non-consequentialists'; 'subjective attraction to objective attractiveness'
objectivity 85, 86, 87, 91–6, 147, 171, 202–3, 210
  see also 'invariant morality'
organic unity *see* 'unity, organic'

pain 60, 62, 63, 64, 69, 70–2, 137, 144, 146, 153, 183, 193, 202, 203, 212, 228, 239
Parfit, Derek 10, 70–3, 172
part-life meaning *see* 'bearer of meaning'
Pascal, Blaise 88, 107
perfection thesis 11, 12, 120, 137–8, 139, 141, 142, 143, 144, 145, 146, 147, 158, 163, 243, 244
  arguments against 11, 140–6, 243–4
  definition of 11, 120, 138
  fundamental motivation for supernaturalism 11, 132–3, 137–8
  see also 'supernaturalism/supernaturalists'
perfectionism/perfectionists 192–4, 212–13
  arguments against 194–8
  arguments for 192–4, 212–13
  definition of 192
  see also 'consequentialism/consequentialists'
permanence 110, 128–30, 132, 144, 203, 204, 205, 247–8
Perrett, Roy 134
pessimism *see* 'nihilism/nihilists'
Picasso, Pablo 2, 5, 96, 137, 144, 204, 207, 214, 230
Plato 85, 127
pleasure 5, 10, 27, 28, 29, 33, 34, 35, 36, 39, 48–9, 59–74, 131, 134–5, 164, 223
  different from meaning 5–6, 9, 10, 27–9, 30, 60–2, 65–74, 129, 188, 249
  in relation to happiness 33, 60, 73–4, 78, 186
  partly constitutive of meaning 33, 62, 126, 197, 223, 238
  similar to meaning 60–5, 142
  see also 'happiness'; 'well-being'
point of view of the universe 185, 188, 196, 198, 241, 244–8
posthumous meaning 23, 50, 54, 70, 177, 200, 247
  see also 'ripples'; 'traces'

pride 31, 32, 33, 34, 69, 77, 79, 137, 169, 197, 210, 216, 227, 235
purpose theory/theorists 10–11, 77–97, 98–118
  arguments against 10, 11, 80–2, 98–118
  arguments for 10, 78, 80, 82–6, 108–10
  definition of 10, 78, 80–2
  God-centred theories in addition to 117–22
  incompatible with eternality 105–6
  incompatible with morality 99–104
  incompatible with omnipotence 104–5
  incompatible with qualitative properties 106–14, 115–17
  see also 'Cottingham, John'; 'divine command theory/theorists'; 'God-centred theory/theories'; 'metaphysics, free will'; 'purposiveness'
purposiveness 11, 18, 24–8, 34, 35, 36, 60, 77, 99, 106, 112–17
  aesthetic 231
  as the concept of meaning 24–8, 34, 35, 77
  God's purpose and the qualitative properties 11, 99, 106, 112–14, 115–17
  see also 'purpose theory/theorists'
Putnam, Hilary 91, 210, 215

qualitative properties 11, 108–22
  atemporality 63, 81, 87, 99, 106, 108, 110–17, 122, 124, 133, 134
  definition of 11, 110
  in relation to God's purpose 112–14, 115–17
  in relation to meaning 118–22
  in relation to perfection thesis 118–20, 122
  infinity 109, 111–12
  simplicity 111, 113

rational nature 26, 40, 42, 88, 165, 167, 176, 192, 193, 210, 215, 221, 223, 224, 225, 226, 227, 228, 232, 234, 239, 248
  see also 'fundamentality theory'; 'Kantian ethics'; 'rationality'
rationality 30, 42, 43, 44, 45, 83, 93, 105, 192, 193, 210, 211–15, 218, 222, 223, 224, 227, 229, 231, 232, 233, 235, 236, 237, 239
  as unity 40–5
  creative 211–12, 215
  fundamentality as object of 226–32, 233–4, 236–9, 249
  in relation to emotions 121, 223–5
  in relation to love 121, 201–2, 213–14, 216, 223–4
  in relation to others 195–6, 198, 210–11, 221, 225, 232
  in relation to virtue 30, 210–11, 218, 221, 225
  Kant on 42
  object of 203, 204, 205, 207–18, 222, 226–32, 233–4, 236–9

rationality (*Cont.*)
    process of 194–5, 206, 210, 211–12, 214, 221–2, 224, 226, 236–7
    theoretical and practical 212–15
    universality as object of 215–18, 222, 227, 229, 230–1
    see also 'rational nature'; 'truth'
reasons *see* 'normativity'
relativism/relativists 7, 8, 84–5, 93, 94, 116, 176–8
ripples 50, 54, 247–8, 249
    see also 'posthumous meaning'; 'traces'
Ross, W. D. 229
Russell, Bertrand 23, 212, 214, 216, 217

Sartre, Jean-Paul 4, 19, 82, 99, 117, 143, 169
Scanlon, T. M. 62, 223
self-esteem 32, 148, 228
self-respect 29, 189
side-constraint 57, 190, 194, 195, 212
    see also 'deontology'
Sidgwick, Henry 185, 188
Singer, Irving 103, 105, 106, 112, 186, 189
Singer, Peter 20, 186, 187, 188, 193, 246
Sisyphus 20, 30, 31, 61, 174, 175, 176, 178, 201
Small, David 1
Smith, Quentin 192, 193, 194, 212, 213, 214, 215, 223, 241
soul 79, 122–4
soul-centred theory/theories 11, 79–80, 107, 119, 122–4, 125, 126, 127, 128, 130, 131, 132, 133, 135, 136, 137, 138, 142, 143, 243
    arguments against 11, 133–7, 140–6
    arguments for 11, 124–33, 137–8
    definition of 11, 79, 122–4
    immortality requirement 123, 133–4
    see also 'afterlife'; 'death'; 'immortality'; 'supernaturalism/supernaturalists'
spiritual realm 10, 19, 33, 78, 79, 97, 121, 123, 127, 134, 140, 143, 144, 145, 159, 200, 220, 229, 233, 242, 243, 244
    see also 'God's existence'; 'supernaturalism/supernaturalists'; 'theism/theists'
Starkey, Charles 166, 225
Steiner, Rudolf 199, 200
subjective attraction to objective attractiveness 12, 173, 182–4, 196, 197, 221
    see also 'objectivism/objectivists'; 'Wolf, Susan'
subjective conditions 166–7, 220, 224–5
    negative 166, 220, 233–4
    partly constitutive of meaning 182–3, 196–7, 198, 200, 220, 224–5, 233–4
    positive 166, 196–7, 224–5
    unnecessary for meaning 183–4, 197, 220

subjectivism/subjectivists 12, 19, 20, 24, 26, 30, 33, 99, 147, 164–5, 167, 169, 170, 172, 173, 174, 175, 176, 178, 180, 200, 220
    argument against 175–9
    arguments for 169–74
    definition of 19–20, 164–5, 220
    dominant form of 165, 169, 173
    intersubjectivity 12, 20, 167, 168, 173, 176, 178–9
    types of 165–9
supernaturalism/supernaturalists 10, 11, 19, 20, 23, 24, 26, 29, 30, 33, 78, 79–82, 91, 97, 107, 108, 117, 118, 119–38, 141, 142, 143, 144, 145, 146, 158, 159–60, 163, 169, 180, 194, 220, 241, 242–4, 245, 247
    arguments against 11, 140–6, 243–4
    arguments for 11, 82–6, 93–4, 124–33, 137–8
    definition of 19, 77–9
    implications of imperfection thesis for 158–60, 220, 232–3
    in relation to atheism 80, 88–90, 242, 243–4, 247
    in relation to nihilism 80, 124, 146, 242–4
    non-purposive 119–32
    purpose theory *see* 'purpose theory/theorists'
    see also 'God-centred theory/theories'; 'perfection thesis'; 'soul-centred theory/theories'

Tabensky, Pedro 42–4
Taylor, Charles 31, 32, 46, 47, 173
Taylor, Richard 19, 20, 30, 31, 61, 174, 211–12, 215
theism/theists 19, 23, 80, 81, 84, 87, 88, 89, 90, 111, 144, 145, 146, 158, 242
    see also 'God's existence'; 'spiritual realm'
theory of meaning *see* 'conceptions of meaning'; 'meaningfulness, theories of'
Thomson, Garrett 34, 47, 110
Tolstoy, Leo 19, 31, 107, 128–9, 130, 132, 144, 145, 241, 242
traces 130–1, 247
    see also 'posthumous meaning'; 'ripples'
transcendence 9, 28–9, 34, 35, 36, 60, 77, 131–2, 203, 204
    as argument for soul-centred theory 131–2
    as concept of meaning 28–32
    as conception of meaning 131–2, 207–10
    of one's animal nature 9, 30, 34, 35, 36, 53, 60, 61, 79, 133, 144, 185, 210, 211, 235
    of one's limits 28, 124, 130–3, 208
    of oneself 18, 34, 35, 215
transcendent ends 202–5, 210
Trisel, Brooke Alan 84, 140–1, 148, 203
truth 1, 5, 13, 22, 23, 24, 45, 48, 86, 87, 94–6, 131, 132, 180, 181, 191, 193, 199, 201–3, 204, 205, 208, 209, 210, 214, 216, 217, 221, 229–30, 232, 236–7, 239

biology 146, 209, 217, 229, 230
  definition of 5, 181, 191, 229
  metaphysics 209–10, 214, 229
  natural sciences 209, 210, 217, 229, 230
  philosophy 202, 209, 210, 215
  physics 146, 204–6, 209, 214, 229, 230
  social sciences 205, 209, 217, 229
  see also 'Darwin, Charles'; 'Einstein, Albert'; 'rationality, object of'; 'rationality, theoretical and practical'

unity 6, 7, 111–12, 249
  narrative 46, 54, 55
  organic 207–10, 231, 239
  rational 40–5
  theoretical 6, 7, 10, 124, 184, 249
  value of 111–12
universality 81, 85, 86, 91, 94, 95, 116, 205, 215–18, 222, 227, 229, 230–1
Updike, John 17, 36
utilitarianism 184–98
  arguments for 184–5, 186–7, 193
  arguments against 189–98
  as a moral theory 100, 186–7, 191
  as a theory of meaning 12, 181, 184–92
  see also 'consequentialism/consequentialists'

value–theoretic conditions 62–73
  appropriate attitude 69
  bearer 65–6

bias toward the future 70–3
  degrees 63
  luck 68–9
  personal 62–3
  posthumous benefit 70
  scales 63–4
  source 66–8
van Gogh, Vincent 23, 62, 70
Velleman, David 121, 122, 201, 216
virtue 29, 68, 125, 127, 136–7, 191, 193–4, 195–6, 198, 210–11, 215, 216, 221, 225, 232, 238

well-being v, 12, 40, 66, 74, 108, 181, 186, 187, 188, 189, 192, 197, 198, 228–9, 244
  definition of 74
  promotion of 12, 181, 188, 190, 193, 198
  relation to happiness 74
  see also 'happiness'; 'pleasure'
whole-life meaning *see* 'bearer of meaning'
Wiggins, David 163, 164
Williams, Bernard v, 134, 135
Wittgenstein, Ludwig 23, 133
Wolf, Susan v, 20, 170, 171, 181, 182, 183, 184, 196, 197
  see also 'subjective attraction to objective attractiveness'
Wong, Wai-hung 3, 41, 52, 177

Zenk, Gabriel 219

Printed and bound by CPI Group (UK) Ltd, Croydon, CR0 4YY